T0202040

Complete
Pure Mathematics

2/3 for Cambridge International AS & A Level

Second Edition

Jean Linsky
James Nicholson
Brian Western

Oxford excellence for Cambridge AS & A Level

Great Clarendon Street, Oxford, OX2 6DP, United Kingdom

Oxford University Press is a department of the University of Oxford.
It furthers the University's objective of excellence in research,
scholarship, and education by publishing worldwide. Oxford is a registered
trade mark of Oxford University Press in the UK and in certain other countries

British Library Cataloguing in Publication Data
Data available

978-0-19-842513-7

12

Paper used in the production of this book is a natural, recyclable product
made from wood grown in sustainable forests.
The manufacturing process conforms to the environmental regulations
of the country of origin.

Printed in India by Multivista Global Pvt. Ltd.

Acknowledgements

The publisher would like to thank the following for permission to reproduce photographs:

p2: NASA/ESA/STSCI/Science Photo Library; **p19t:** Andrey Burmakin/Shutterstock; **p19b:**
James King-Holmes/Science Photo Library; **p26:** Sipa Press/Rex; **p40:** 1_Weblogiq/Shutterstock;
p38: Sebastian Kaulitzki; **p70:** Stephen Rees/Shutterstock; **p71c:** Joe Mercier/Shutterstock;
p71b: Sipa Press/Rex; **p72:** 3D Sculptor/Shutterstock; **p94:** Junne/Shutterstock; **p120:** Jacques
Descloitres/MODIS Land Rapid Response Team/GSFC/NASA; **p146t:** Tim Gainey/Alamy; **p146c:**
Irafael/Shutterstock; **p146b:** NASA; **p147t:** Yury Dmitrienko/Shutterstock; **p148:** ntdanai/
Shutterstock p147b: Gabriel Georgescu/Shutterstock; **p166:** Dinodia Photos/Alamy; **p194:** Nicolas
Boyer; **p226:** Mark Evans/Getty Images; **p252:** Alfred Pasieka/Science Photo Library; **p294tr:**
sfam_photo/Shutterstock; **p294cr:** Zurijeta/Shutterstock; **p294l:** Nerthuz/Shutterstock; **p294cl:**
Peter Dazeley/Getty; **p294br:** Norman Chan/Shutterstock; **p294b:** mamahoohooba/Shutterstock;
p295: KPG_Payless;

This Student Book refers to the Cambridge International AS & A Level Mathematics (9709) Syllabus
published by Cambridge Assessment International Education.

This work has been developed independently from and is not endorsed by or otherwise connected
with Cambridge Assessment International Education.

Contents

Introduction

About this book

This book has been written to cover the **Cambridge AS & A level International Mathematics (9709)** course, and is fully aligned to the syllabus. The first six chapters of the book cover material applicable to both Pure 2 and Pure 3, and the final five chapters cover Pure 3 material only.

In addition to the main curriculum content, you will find:

- 'Maths in real-life', showing how principles learned in this course are used in the real world.
- Chapter openers, which outline how each topic in the Cambridge 9709 syllabus is used in real-life.
- 'Did you know?' boxes (as shown below), which give interesting side-notes beyond the scope of the syllabus.

The book contains the following features:

Note	Did you know?
Advice on calculator use	Pure 2
Pure 3	

Throughout the book, you will encounter worked examples and a host of rigorous exercises. The examples show you the important techniques required to tackle questions. The exercises are carefully graded, starting from a basic level and going up to exam standard, allowing you plenty of opportunities to practise your skills. Together, the examples and exercises put maths in a real-world context, with a truly international focus.

At the start of each chapter, you will see a list of objectives that are covered in the chapter. These objectives are drawn from the Cambridge AS and A level syllabus. Each chapter begins with a *Before you start* section and finishes with a *Summary exercise* and *Chapter summary*, ensuring that you fully understand each topic.

Each chapter contains key mathematical terms to improve understanding, highlighted in colour, with full definitions provided in the Glossary of terms at the end of the book.

The answers given at the back of the book are concise. However, when answering exam-style questions, you should show as many steps in your working as possible. All exam-style questions, as well as *Exam-style papers 2A, 2B, 3A* and *3B*, have been written by the authors.

About the authors

Brian Western has over 40 years of experience in teaching mathematics up to A Level and beyond, and is also a highly experienced examiner. He taught mathematics and further mathematics, and was an Assistant Headteacher in a large state school. Brian has written and consulted on a number of mathematics textbooks.

James Nicholson is an experienced teacher of mathematics at secondary level, having taught for 12 years at Harrow School and spent 13 years as Head of Mathematics in a large Belfast grammar school. He is the author of several A Level texts, and editor of the *Concise Oxford Dictionary of Mathematics*. He has also contributed to a number of other sets of curriculum and assessment materials, is an experienced examiner and has acted as a consultant for UK government agencies on accreditation of new specifications.

Jean Linsky has been a mathematics teacher for over 30 years, as well as Head of Mathematics in Watford, Herts, and is also an experienced examiner. Jean has authored and consulted on numerous mathematics textbooks.

A note from the authors

The aim of this book is to help students prepare for the Pure 2 and Pure 3 units of the Cambridge International AS and A Level mathematics syllabus, although it may also be useful in providing support material for other AS and A Level courses. The book contains a large number of practice questions, many of which are exam-style.

In writing the book we have drawn on our experiences of teaching mathematics and Further mathematics to A Level over many years as well as on our experiences as examiners, and our discussions with mathematics educators from many countries at international conferences.

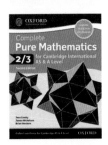

Student Book: *Complete Pure Mathematics 2 & 3 for Cambridge International AS & A Level*

Syllabus: Cambridge International AS & A Level Mathematics: Pure Mathematics 2 & 3 (9709)

PURE MATHEMATICS 2 & 3	Student Book

Syllabus overview

Unit P2: Pure Mathematics 2 (Paper 2)

Knowledge of the content of unit P1 is assumed and candidates may be required to demonstrate such knowledge in answering questions.

1. Algebra

• understand the meaning of $|x|$, sketch the graph of $y = |ax + b|$ and use relations such as $|a| = |b| \Leftrightarrow a^2 = b^2$ and $|x - a| < b \Leftrightarrow a - b < x < a + b$ when solving equations and inequalities;	Pages 2–17
• divide a polynomial, of degree not exceeding 4, by a linear or quadratic polynomial, and identify the quotient and remainder (which may be zero);	Pages 2–17
• use the factor theorem and the remainder theorem, e.g. to find factors, solve polynomial equations or evaluate unknown coefficients.	Pages 2–17

2. Logarithmic and exponential functions

• understand the relationship between logarithms and indices, and use the laws of logarithms (excluding change of base);	Pages 18–39
• understand the definition and properties of e^x and $\ln x$, including their relationship as inverse functions and their graphs;	Pages 18–39
• use logarithms to solve equations and inequalities in which the unknown appears in indices;	Pages 18–39
• use logarithms to transform a given relationship to linear form, and hence determine unknown constants by considering the gradient and/or intercept.	Pages 18–39

3. Trigonometry

4. Differentiation

5. Integration

6. Numerical solution of equations

OXFORD
UNIVERSITY PRESS

Student book & Cambridge syllabus
matching grid

ASPIRE
SUCCEED
PROGRESS

Unit P3: Pure Mathematics (Paper 3)

Knowledge of the content of unit P1 is assumed and candidates may be required to demonstrate such knowledge in answering questions.	

1. Algebra

• understand the meaning of $|x|$, sketch the graph of $y = |ax + b|$ and use relations such as $|a| = |b| \Leftrightarrow a^2 = b^2$ and $|x - a| < b \Leftrightarrow a - b < x < a + b$ when solving equations and inequalities;	Pages 2–17
• divide a polynomial, of degree not exceeding 4, by a linear or quadratic polynomial, and identify the quotient and remainder (which may be zero);	Pages 2–17
• use the factor theorem and the remainder theorem, e.g. to find factors, solve polynomial equations or evaluate unknown coefficients;	Pages 2–17
• recall an appropriate form for expressing rational functions in partial fractions, and carry out the decomposition, in cases where the denominator is no more complicated than: - $(ax + b)(cx + d)(ex + f)$, - $(ax + b)(cx + d)^2$, - $(ax + b)(x^2 + c^2)$, and where the degree of the numerator does not exceed that of the denominator;	Pages 136–153
• use the expansion of $(1 + x)^n$, where n is a rational number and $|x| < 1$ (finding a general term is not included, but adapting the standard series to expand e.g. $(2 - \frac{1}{2}x)^{-1}$ is included).	Pages 152–169

2. Logarithmic and exponential functions

• understand the relationship between logarithms and indices, and use the laws of logarithms (excluding change of base);	Pages 18–39
• understand the definition and properties of e^x and $\ln x$, including their relationship as inverse functions and their graphs;	Pages 18–39
• use logarithms to solve equations of the form $a^x = b$, and similar inequalities;	Pages 18–39
• use logarithms to transform a given relationship to linear form, and hence determine unknown constants by considering the gradient and/or intercept.	Pages 18–39

OXFORD
UNIVERSITY PRESS

Student book & Cambridge syllabus
matching grid

ASPIRE
SUCCEED
PROGRESS

3. Trigonometry

• understand the relationship of the secant, cosecant and cotangent functions to cosine, sine and tangent, and use properties and graphs of all six trigonometric functions for angles of any magnitude;	Pages 40–65
• use trigonometrical identities for the simplification and exact evaluation of expressions and in the course of solving equations, and select an identity or identities appropriate to the context, showing familiarity in particular with the use of: - $\sec^2 \theta \equiv 1 + \tan^2 \theta$ and $\operatorname{cosec}^2 \theta \equiv 1 + \cot^2 \theta$, - the expansions of $\sin(A \pm B)$, $\cos(A \pm B)$ and $\tan(A \pm B)$, - the formulae for $\sin 2A$, $\cos 2A$ and $\tan 2A$, - the expressions of $a \sin \theta + b \cos \theta$ in the forms $R \sin(\theta \pm a)$ and $R \cos(\theta \pm a)$.	Pages 40–65

4. Differentiation

• use the derivatives of e^x, $\ln x$, $\sin x$, $\cos x$, $\tan x$, $\tan^{-1} x$, together with constant multiples, sums, differences and composites;	Pages 68–90
• differentiate products and quotients;	Pages 68–90
• find and use the first derivative of a function which is defined parametrically or implicitly.	Pages 68–90

5. Integration

• extend the idea of 'reverse differentiation' to include the integration of $e^{ax + b}$, $\dfrac{1}{ax + b}$, $\sin(ax + b)$, $\cos(ax + b)$, $\sec^2(ax + b)$ and $\dfrac{1}{x^2 + a^2}$;	Pages 91–116 and Pages 170–197
• use trigonometrical relationships (such as double-angle formulae) to facilitate the integration of functions such as $\cos^2 x$;	Pages 97–122
• integrate rational functions by means of decomposition into partial fractions (restricted to the types of partial fractions specified in paragraph 1 above);	Pages 154–181
• recognise an integrand of the form $\dfrac{kf'(x)}{f'(x)}$, and integrate, for example, $\dfrac{x}{x^2 + 1}$ or $\tan x$;	Pages 170–197
• recognise when an integrand can usefully be regarded as a product, and use integration by parts to integrate, for example, $x \sin 2x$, $x^2 e^x$ or $\ln x$;	Pages 170–197
• use a given substitution to simplify and evaluate either a definite or an indefinite integral;	Pages 170–197
• use the trapezium rule to estimate the value of a definite integral, and use sketch graphs in simple cases to determine whether the trapezium rule gives an over-estimate or an under-estimate.	Pages 97–122

OXFORD
UNIVERSITY PRESS

Student book & Cambridge syllabus
matching grid

ASPIRE
SUCCEED
PROGRESS

6. Numerical solution of equations

• locate approximately a root of an equation, by means of graphical considerations and/or searching for a sign change;	Pages 117–133
• understand the idea of, and use the notation for, a sequence of approximations which converges to a root of an equation;	Pages 117–133
• understand how a given simple iterative formula of the form $x_{n+1} = F(x_n)$ relates to the equation being solved, and use a given iteration, or an iteration based on a given rearrangement of an equation, to determine a root to a prescribed degree of accuracy (knowledge of the condition for convergence is not included, but candidates should understand that an iteration may fail to converge).	Pages 117–133

7. Vectors

• use standard notations for vectors, i.e. $\begin{pmatrix} x \\ y \end{pmatrix}$, $x\mathbf{i} + y\mathbf{j}$, $\begin{pmatrix} x \\ y \\ z \end{pmatrix}$, $x\mathbf{i} + y\mathbf{j} + z\mathbf{k}$, $\overrightarrow{AB}$, $\mathbf{a}$;	Pages 182-214
• carry out addition and subtraction of vectors, and multiplication of a vector by a scalar, and interpret these operations in geometrical terms;	Pages 182-214
• calculate the magnitude of a vector, and use unit vectors, displacement vectors and position vectors;	Pages 182-214
• understand the significance of all the symbols used when the equation of a straight line is expressed in the form $\mathbf{r} = \mathbf{a} + t\mathbf{b}$, and find the equation of a line, given sufficient information;	Pages 182-214
• determine whether two lines are parallel, intersect or are skew, and find the point of intersection of two lines when it exists;	Pages 182-214
• use formulae to calculate the scalar product of two vectors, and use scalar products in problems involving lines and points.	Pages 182-214

8. Differential equations

• formulate a simple statement involving a rate of change as a differential equation, including the introduction if necessary of a constant of proportionality;	Pages 215–240
• find by integration a general form of solution for a first order differential equation in which the variables are separable;	Pages 215–240
• use an initial condition to find a particular solution;	Pages 215–240
• interpret the solution of a differential equation in the context of a problem being modelled by the equation.	Pages 215–240

9. Complex numbers

• understand the idea of a complex number, recall the meaning of the terms real part, imaginary part, modulus, argument, conjugate, and use the fact that two complex numbers are equal if and only if both real and imaginary parts are equal;	Pages 241–275
• carry out operations of addition, subtraction, multiplication and division of two complex numbers expressed in cartesian form $x + iy$;	Pages 241–275
• use the result that, for a polynomial equation with real coefficients, any non-real roots occur in conjugate pairs;	Pages 241–275
• represent complex numbers geometrically by means of an Argand diagram;	Pages 241–275
• carry out operations of multiplication and division of two complex numbers expressed in polar form $r(\cos \theta + i \sin \theta) \equiv r\,e^{i\theta}$;	Pages 241–275
• find the two square roots of a complex number;	Pages 241–275
• understand in simple terms the geometrical effects of conjugating a complex number and of adding, subtracting, multiplying and dividing two complex numbers;	Pages 241–275
• illustrate simple equations and inequalities involving complex numbers by means of loci in an Argand diagram, e.g. $\lvert z - a \rvert < k$, $\lvert z - a \rvert = \lvert z - b \rvert$, $\arg(z - a) = \alpha$.	Pages 241–275

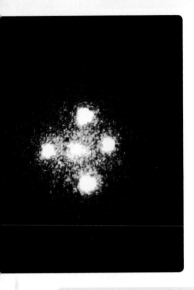

Algebra is used extensively in mathematics, chemistry, physics, economics and social sciences. For example, the study of polynomials in astrophysics has led to our understanding of gravitational lensing.

Gravitational lensing occurs when light from a distant source bends around a massive object (such as a galaxy) between a source and an observer. Multiple images of the same object may be seen. Here, the 'Einstein Cross', four images of a very distant supernova, is seen in a photograph taken by the Hubble telescope. The supernova is at a distance of approximately 8 billion light years, and is 20 times further away than the galaxy, which is at a distance of 400 million light years. The light from the supernova is bent in its path by the gravitational field of the galaxy. This bending produces the four bright outer images. The bright central region of the galaxy is seen as the central object. This phenomena was predicted by Einstein's general theory of relativity published in 1915, but was not observed until 1979.

Objectives

- Understand the meaning of $|x|$, sketch the graph of $y = |ax + b|$ and use relations such as $|a| = |b| \Leftrightarrow a^2 = b^2$ and $|x - a| < b \Leftrightarrow a - b < x < a + b$ when solving equations and inequalities.
- Divide a polynomial, of degree not exceeding 4, by a linear or quadratic polynomial, and identify the quotient and remainder (which may be zero).
- Use the factor theorem and the remainder theorem, e.g. to find factors, solve polynomial equations or evaluate unknown coefficients.

Before you start

You should know how to:

1. Do long division,

    e.g. $357 \div 21$

$$
\begin{array}{r}
17 \\
21\overline{)357} \\
\underline{21} \\
147 \\
\underline{147} \\
0
\end{array}
$$
    Therefore $\dfrac{357}{21} = 17$

2. Find the remainder,

    e.g. $461 \div 37$

$$
\begin{array}{r}
12 \\
37\overline{)461} \\
\underline{37} \\
91 \\
\underline{74} \\
17
\end{array}
$$
    Remainder $= 17$

Skills check:

1. Find the following using long division.

    a) $608 \div 19$

    b) $2774 \div 38$

    c) $1081 \div 23$

    d) $1392 \div 24$

2. Find the remainder of the following after doing long division.

    a) $923 \div 21$

    b) $742 \div 32$

    c) $1527 \div 43$

    d) $4258 \div 26$

1.1 The modulus function

> The **modulus** of a real number is the magnitude of that number.

If we have a real number x, then the modulus of x is written as $|x|$. We say this as 'mod x'.

Thus $|2| = 2$ and $|-2| = 2$, and if we write $|x| < 2$ this means that $-2 < x < 2$.

> The **modulus function** $f(x) = |x|$ is defined as
> $$|x| = x \quad \text{for} \quad x \geq 0$$
> $$|x| = -x \quad \text{for} \quad x < 0$$

Consider the impact that the modulus function has by looking at the graphs of $y = x - 1$ and $y = |x - 1|$.

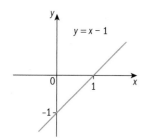
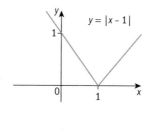

Note: For $f(x) < 0$, $|f(x)| = -f(x)$.

When graphing $y = |f(x)|$, we reflect the graph of $y = f(x)$ in the x-axis whenever $f(x) < 0$.

Example 1

Solve the equation $|x + 2| = |3x|$.

· ·

Method 1: Using a graph

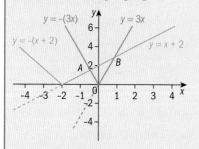

Sketch the graphs and find where they intersect.

The lines cannot be drawn below the x-axis.

For $x < 0$, $|3x| = -(3x)$

For $x < -2$, $|x + 2| = -(x + 2)$

Graphs intersect at A and B.

At A, $x + 2 = -3x$ ←   At A, the line $y = x + 2$ intersects the line $y = -(3x)$.

$\qquad 4x = -2 \Rightarrow x = -\dfrac{1}{2}$

At B, $x + 2 = 3x$ ←   At B, the line $y = x + 2$ intersects the line $y = 3x$.

$\qquad 2x = 2 \Rightarrow x = 1$

$x = -\dfrac{1}{2}$ or $x = 1$

▶ Continued on the next page

Method 2: Squaring both sides of the equation

$$(x + 2)^2 = (3x)^2$$

$$x^2 + 4x + 4 = 9x^2$$

$$8x^2 - 4x - 4 = 0$$

$$2x^2 - x - 1 = 0$$

$$(2x + 1)(x - 1) = 0$$

$$x = -\frac{1}{2} \text{ or } x = 1$$

> We do this to ensure both sides of the equation are positive.

> **Note:** You can only use this method if the variable (e.g. x) is inside the modulus expression.

Method 3: Removing modulus signs by equating the left-hand side with both 'plus and minus' the right-hand side

$$x + 2 = 3x \quad \text{or} \quad x + 2 = -3x$$

$$2x = 2 \quad \text{or} \quad 4x = -2$$

$$x = 1 \quad \text{or} \quad x = -\frac{1}{2}$$

> We get the same result if we say $3x = \pm(x + 2)$.

Example 2

Solve the inequality $|4x + 3| > |2x - 1|$.

Method 1: Using a graph

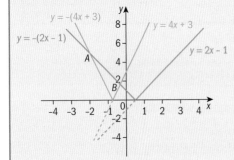

Sketch the graphs and find where they intersect.

The lines cannot be drawn below the x-axis.

For $x < -\frac{3}{4}$, $|4x + 3| = -(4x + 3)$

For $x < \frac{1}{2}$, $|2x - 1| = -(2x - 1)$

Graphs meet at A and B.

At A, $-(4x + 3) = -(2x - 1)$

$$2x = -4, \quad x = -2$$

> At A, the line $y = -(4x + 3)$ intersects the line $y = -(2x - 1)$.

At B, $4x + 3 = -(2x - 1)$

$$6x = -2, \quad x = -\frac{1}{3}$$

> At B, the line $y = 4x + 3$ intersects the line $y = -(2x - 1)$.

We want the region where $|4x + 3| > |2x - 1|$.

This is where the 'blue' lines are above the 'green' lines.

$$x < -2 \text{ and } x > -\frac{1}{3}$$

▶ Continued on the next page

The modulus function

Method 2: Squaring both sides of the inequality

$$(4x + 3)^2 > (2x - 1)^2$$

$$16x^2 + 24x + 9 > 4x^2 - 4x + 1$$

$$12x^2 + 28x + 8 > 0$$

$$3x^2 + 7x + 2 > 0$$

$$(3x + 1)(x + 2) > 0$$

We can do this because we know both $|4x + 3|$ and $|2x - 1|$ are positive so squaring them will not change the inequality.

Draw a rough sketch to see where $y > 0$. There are two separate blue regions that satisfy $y > 0$.

$$x < -2 \text{ or } x > -\frac{1}{3}$$

There are two separate regions that satisfy $y > 0$.

Method 3: Removing modulus signs by equating the left-hand side with both 'plus and minus' the right-hand side

First find where the graphs of $y = |4x + 3|$ and $y = |2x - 1|$ intersect (the values of x at these points are called the critical values).

$$4x + 3 = 2x - 1 \quad \text{or} \quad 4x + 3 = -(2x - 1)$$

$$2x = -4 \quad \text{or} \quad 6x = -2$$

$$x = -2 \quad \text{or} \quad x = -\frac{1}{3}$$

This gives us the two critical values.

To find the correct inequalities you need to take values of x on either side of the critical values.

Take values on either side of $x = -2$ and $x = -\frac{1}{3}$.

e.g.

$x = -3$, $|4x + 3| = 9$ and $|2x - 1| = 7$

Thus $|4x + 3| > |2x - 1|$, so $x < -2$

One of the solution regions is less than -2.

is one region where $|4x + 3| > |2x - 1|$

When $x = -1$, $|4x + 3| = 1$ and $|2x - 1| = 3$

$|4x + 3| > |2x - 1|$ is false so there is no solution between -2 and $-\frac{1}{3}$, so we do **not** require this region.

When $x = 0$, $|4x + 3| = 3$ and $|2x - 1| = 1$

Thus $|4x + 3| > |2x - 1|$ so $x > -\frac{1}{3}$

One of the solution regions is greater than $-\frac{1}{3}$.

$$x < -2 \quad \text{or} \quad x > -\frac{1}{3}$$

Exercise 1.1

1. Solve each of these equations algebraically.

   a) $|1 - 2x| = 3$

   b) $|x - 3| = |x + 1|$

   c) $|5x - 2| = |2x|$

   d) $|5 - 4x| = 4$

   e) $|2x - 1| = |x + 2|$

   f) $|x| = |4 - 2x|$

   g) $|3x + 1| = |4 - 2x|$

   h) $|2x - 6| = |3x + 1|$

   i) $|x + 4| = |3x + 1|$

   j) $|1 - 3x| = |5x - 3|$

   k) $3|x - 4| = |x + 2|$

   l) $5|2x - 3| = 4|x - 5|$

2. Solve each of these inequalities algebraically.

   a) $|2x - 3| < |x|$

   b) $|x - 1| \geq 4$

   c) $|x + 3| \geq |2x + 2|$

   d) $|2x + 3| > x + 6$

3. Solve each of these inequalities graphically.

   a) $|x + 6| \leq 3|x - 2|$

   b) $|3x - 2| < |x + 4|$

   c) $|2x| < |1 - x|$

   d) $5 \leq |2x - 1|$

   e) $2|x - 1| < |x + 3|$

   f) $|2x + 1| \geq |1 - 4x|$

   g) $|x + 2| < 2|x + 1|$

   h) $|3x - 1| \leq |x + 3|$

1.2 Sketching linear graphs of the form $y = a|x| + b$

We have sketched graphs of the form $y = |ax + b|$ in section 1.1. Using the information given in P1 Chapter 3 on transformations, we can sketch graphs of the form $y = a|x| + b$.

Example 3

Sketch the following graphs.

a) $y = |2x - 4|$

b) $y = 2|x| - 4$

a)

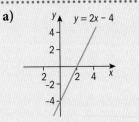

First sketch the graph without the modulus.

▶ Continued on the next page

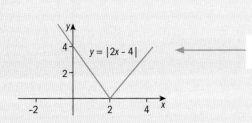

Then reflect the negative y values in the x-axis.

b)

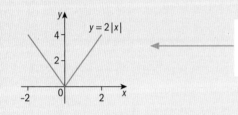

First sketch the graph of $|x|$.

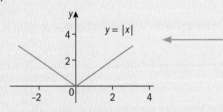

Then stretch the graph by a stretch factor of 2 in the direction of the y-axis.

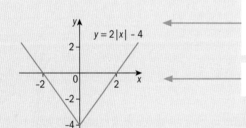

Then translate the graph by the vector $\begin{pmatrix} 0 \\ -4 \end{pmatrix}$.

When $y = 0$, $x = 2$ or -2.

Exercise 1.2

1. Sketch the following.

   a) i) $y = |x + 1|$ **ii)** $y = |x| + 1$

   b) i) $y = |3x + 2|$ **ii)** $y = 3|x| + 2$

   c) i) $y = |2x - 2|$ **ii)** $y = 2|x - 2|$

   d) i) $y = \left| \dfrac{1}{2}x + 3 \right|$ **ii)** $y = \dfrac{1}{2}|x| + 3$

   e) i) $y = |-x|$ **ii)** $y = -|x|$

   f) i) $y = |3 - x|$ **ii)** $y = 3 - |x|$

1.3 Division of polynomials

We can use long division to divide a polynomial by another polynomial.

Example 4

Divide $x^3 - 5x^2 + x + 10$ by $(x - 2)$.

$$
\begin{array}{r}
x^2 - 3x - 5 \\
x - 2 \overline{)\, x^3 - 5x^2 + x + 10} \\
x^3 - 2x^2 \\
\hline
-3x^2 + x + 10 \\
-3x^2 + 6x \\
\hline
-5x + 10 \\
-5x + 10 \\
\hline
\end{array}
$$

$x^3 \div x = x^2$, so multiply $(x - 2)$ by x^2.

Subtract $(x^3 - 2x^2)$ from $(x^3 - 5x^2)$ and bring down $(+ x + 10)$.

$-3x^2 \div x = 3x$, so multiply $(x - 2)$ by $-3x$.

Subtract $(3x^2 + 6x)$ from $(3x^2 + x)$ and bring down $+ 10$.

We cannot continue the process because $(-5x + 10) - (-5x + 10) = 0$.

Thus $(x^3 - 5x^2 + x + 10) \div (x - 2) = x^2 - 3x - 5$

The expression $(x^3 - 5x^2 + x + 10)$ is called the dividend, $(x - 2)$ is called the divisor, and $(x^2 - 3x - 5)$ is called the quotient. When we subtract $(-5x + 10)$ from $(-5x + 10)$ we are left with nothing, so we say there is no remainder. Because there is no remainder, we can say that $(x - 2)$ is a factor of $x^3 - 5x^2 + x + 10$.

Example 5

Find the remainder when $4x^3 - 7x - 1$ is divided by $(2x + 1)$.

$$
\begin{array}{r}
2x^2 - x - 3 \\
2x + 1 \overline{)\, 4x^3 + 0x^2 - 7x - 1} \\
4x^3 + 2x^2 \\
\hline
-2x^2 - 7x - 1 \\
-2x^2 - x \\
\hline
-6x - 1 \\
-6x - 3 \\
\hline
+ 2
\end{array}
$$

As there is no term in x^2 in the dividend, it is useful to write $0x^2$ as part of the dividend.

$(4x^3 \div 2x) = 2x^2$, so multiply $(2x + 1)$ by $2x^2$.

Subtract $(4x^3 + 2x^2)$ from $(4x^3 + 0x^2)$ and bring down $-7x - 1$.

$(-2x^2 \div 2x) = -x$, so multiply $(2x + 1)$ by $-x$.

Subtract $(-2x^2 - x)$ from $(-2x^2 - 7x)$ and bring down -1.

Subtract $(-6x - 3)$ from $(-6x - 1)$ to get a remainder of $+2$.

We cannot continue the process as 2 cannot be divided by $(2x + 1)$.

Thus, when $(4x^3 - 7x - 1)$ is divided by $(2x + 1)$, the remainder is 2.

Looking at Example 5 we can write
$(4x^3 - 7x - 1) = (2x^2 - x - 3)(2x + 1) + 2$

> In general:
> $$f(x) = \text{quotient} \times \text{divisor} + \text{remainder}$$

Exercise 1.3

1. Divide
   a) $x^3 + 3x^2 + 3x + 2$ by $(x + 2)$
   b) $x^3 - 2x^2 + 6x + 9$ by $(x + 1)$
   c) $x^3 - 3x^2 + 6x - 8$ by $(x - 2)$
   d) $x^3 + x^2 - 3x - 2$ by $(x + 2)$
   e) $2x^3 - 6x^2 + 7x - 21$ by $(x - 3)$
   f) $3x^3 - 20x^2 + 10x + 12$ by $(x - 6)$
   g) $6x^4 + 5x^3 + 5x^2 + 10x + 7$ by $(3x^2 - 2x + 4)$.

Hint: In part (g), use the same method as when dividing by a linear expression. State any remainder.

2. Find the remainder when
   a) $6x^3 + 28x^2 - 7x + 10$ is divided by $(x + 5)$
   b) $2x^3 + x^2 + 5x - 4$ is divided by $(x - 1)$
   c) $x^3 + 2x^2 - 17x - 2$ is divided by $(x - 3)$
   d) $2x^3 + 3x^2 - 4x + 5$ is divided by $(x^2 + 2)$
   e) $4x^3 - 5x + 4$ is divided by $(2x - 1)$
   f) $3x^3 - x^2 + 1$ is divided by $(x + 2)$.

3. Show that $(2x + 1)$ is a factor of $2x^3 - 3x^2 + 2x + 2$.

4. a) Show that $(x - 1)$ is a factor of $x^3 - 6x^2 + 11x - 6$.
   b) Hence factorise $x^3 - 6x^2 + 11x - 6$.

5. Show that when $4x^3 - 6x^2 + 5$ is divided by $(2x - 1)$ the remainder is 4.

6. Divide $x^3 + 1$ by $(x + 1)$.

7. Find the quotient and the remainder when $x^4 + 2x^3 + 3x^2 + 7$ is divided by $(x^2 + x + 1)$.

8. Find the quotient and the remainder when $2x^3 + 3x^2 - 4x + 5$ is divided by $(x + 2)$.

9. **a)** Show that $(2x - 1)$ is a factor of $12x^3 + 16x^2 - 5x - 3$.

   b) Hence factorise $12x^3 + 16x^2 - 5x - 3$.

10. The expression $2x^3 - 5x^2 - 16x + k$ has a remainder of -6 when divided by $(x - 4)$.

   Find the value of k.

11. Find the quotient and the remainder when $2x^4 - 8x^3 - 3x^2 + 7x - 7$ is divided by $(x^2 - 3x - 5)$.

12. The polynomial $x^4 + x^3 - 5x^2 + ax - 4$ is denoted by p(x). It is given that p(x) is divisible by $(x^2 + 2x - 4)$.

   Find the value of a.

1.4 The remainder theorem

You can find the remainder when a polynomial is divided by $(ax - b)$ by using the **remainder theorem**.

We know that if f(x) is divided by $(x - a)$ then f(x) = quotient $\times$ $(x - a)$ + remainder.

When $x = a$, f(a) = quotient $\times$ $(a - a)$ + remainder = remainder.

Thus f(a) = remainder.

> When a polynomial f(x) is divided by $(x - a)$, the remainder is f(a).
>
> When a polynomial f(x) is divided by $(ax - b)$, the remainder is $f\left(\dfrac{b}{a}\right)$.

Example 6

Find the remainder when $4x^3 + x^2 - 3x + 7$ is divided by $(x + 2)$.

. . . .

$f(x) = 4x^3 + x^2 - 3x + 7$ ←   Write the polynomial as a function.

$f(-2) = 4(-2)^3 + (-2)^2 - 3(-2) + 7$ ← $(x + 2) = 0 \Rightarrow x = -2$, so calculate f(−2).

       $= -32 + 4 + 6 + 7 = -15$

The remainder is -15.

Example 7

When $16x^4 - ax^3 + 8x^2 - 4x - 1$ is divided by $(2x - 1)$, the remainder is 3.

Find the value of a.

$f(x) = 16x^4 - ax^3 + 8x^2 - 4x - 1$ ← Write the polynomial as a function.

$f\left(\dfrac{1}{2}\right) = 16\left(\dfrac{1}{2}\right)^4 - a\left(\dfrac{1}{2}\right)^3 + 8\left(\dfrac{1}{2}\right)^2 - 4\left(\dfrac{1}{2}\right) - 1$ ← $(2x - 1) = 0 \Rightarrow x = \dfrac{1}{2}$, so calculate $f\left(\dfrac{1}{2}\right)$.

$= 1 - \dfrac{1}{8}a + 2 - 2 - 1 = 3$ ← Equate this to 3, since the remainder is 3.

$-\dfrac{1}{8}a = 3 \Rightarrow a = -24$

Exercise 1.4

1. Find the remainder when

   a) $2x^3 + 8x^2 - x + 4$ is divided by $(x - 3)$

   b) $5x^4 - 3x^3 - 2x^2 + x - 1$ is divided by $(x + 1)$

   c) $x^3 + 4x^2 + 8x - 3$ is divided by $(2x + 1)$

   d) $3x^3 - 2x^2 - 5x - 7$ is divided by $(2 - x)$

   e) $9x^3 - 8x + 3$ is divided by $(1 - x)$

   f) $243x^4 - 27x^3 + 6x + 4$ is divided by $(3x - 2)$.

2. When $ax^3 + 16x^2 - 5x - 5$ is divided by $(2x - 1)$ the remainder is -2.
   Find the value of a.

3. The polynomial $4x^3 - 4x^2 + ax + 1$, where a is a constant, is denoted
   by $p(x)$. When $p(x)$ is divided by $(2x - 3)$ the remainder is 13.
   Find the value of a.

4. The polynomial $x^3 + ax^2 + bx + 1$, where a and b are constants, is denoted
   by $p(x)$. When $p(x)$ is divided by $(x - 2)$ the remainder is 9 and when
   $p(x)$ is divided by $(x + 3)$ the remainder is 19. Find the value of a and the
   value of b.

5. When $5x^3 + ax + b$ is divided by $(x - 2)$, the remainder is equal to the
   remainder obtained when the same expression is divided by $(x + 2)$.

a) Explain why b can take any value.

b) Find the value of a.

6. The polynomial $2x^4 + 3x^2 - x + 2$ is denoted by p(x). Show that the remainder when p(x) is divided by $(x + 2)$ is 8 times the remainder when p(x) is divided by $(x - 1)$.

7. The polynomial $x^3 + ax + b$, where a and b are constants, is denoted by p(x). When p(x) is divided by $(x - 1)$ the remainder is 14 and when p(x) is divided by $(x - 4)$ the remainder is 56. Find the values of a and b.

8. The polynomial $x^3 + ax^2 + 2$, where a is a constant, is denoted by p(x). When p(x) is divided by $(x + 1)$ the remainder is one more than when p(x) is divided by $(x + 2)$. Find the value of a.

9. When $6x^2 + x + 7$ is divided by $(x - a)$, the remainder is equal to the remainder obtained when the same expression is divided by $(x + 2a)$, where $a \neq 0$. Find the value of a.

1.5 The factor theorem

We can deduce the factor theorem directly from the remainder theorem (section 1.4).

> For any polynomial f(x), if f(a) = 0 then the remainder when f(x) is divided by $(x - a)$ is zero. Thus $(x - a)$ is a factor of f(x).
>
> For any polynomial f(x), if $f\left(\dfrac{b}{a}\right) = 0$, then $(ax - b)$ is a factor of f(x).

Example 8

The polynomial $x^3 - ax^2 + 2x + 8$, where a is a constant, is denoted by p(x).

It is given that $(x - 2)$ is a factor of p(x).

a) Evaluate a.

b) When a has this value, factorise p(x) completely.

a) p(2) = 8 − 4a + 4 + 8 = 0

$4a = 20 \Rightarrow a = 5$

b) We can factorise $x^3 - 5x^2 + 2x + 8$ using either (**i**) long division or (**ii**) testing other factors using the factor theorem.

▶ Continued on the next page

i)

$$\begin{array}{r} x^2 - 3x - 4 \\ x - 2 \overline{)x^3 - 5x^2 + 2x + 8} \\ x^3 - 2x^2 \\ \hline -3x^2 + 2x + 8 \\ -3x^2 + 6x \\ \hline -4x + 8 \\ -4x + 8 \end{array}$$

Put $a = 5$.

You would expect there to be no remainder since $x - 2$ is a factor.

and $x^2 - 3x - 4 = (x - 4)(x + 1)$

Factorise the quotient.

So $p(x) = (x - 2)(x - 4)(x + 1)$

ii) $f(+1) = 1 - 5 + 2 + 8 \neq 0$ $(x - 1)$ is not a factor

Try a value of x.

$f(-1) = -1 - 5 - 2 + 8 = 0$ $(x + 1)$ is a factor

$f(4) = 64 - 80 + 8 + 8 = 0$ $(x - 4)$ is a factor

So $p(x) = (x - 2)(x - 4)(x + 1)$

Note: Instead of performing this last trial we could work out that the final factor is $(x - 4)$ as we know $x \times x \times x = x^3$ and $(-2) \times (+1) \times (-4) = +8$.

Example 9

Solve $x^3 - 3x^2 - 4x + 12 = 0$.

Let $f(x) = x^3 - 3x^2 - 4x + 12$.

$f(1) = 1 - 3 - 4 + 12 \neq 0$

To solve, we must first factorise $x^3 - 3x^2 - 4x + 12$.

so $(x - 1)$ is not a factor.

Trial any value of x that is a factor of 12.

$f(2) = 8 - 12 - 8 + 12 = 0$

so $(x - 2)$ is a factor.

$f(-2) = -8 - 12 + 8 + 12 = 0$

Alternatively, at this stage you could also do a long division to find the other two factors since you already know one factor.

so $(x + 2)$ is a factor.

We can deduce that the third factor is $(x - 3)$.

$12 \div 2 \div -2 = -3$ (and the coefficient of x^3 is 1).

$f(3) = 27 - 27 - 12 + 12$

so $(x - 3)$ is a factor.

Thus $(x - 2)(x + 2)(x - 3) = 0$

$x = 2$ or $x = -2$ or $x = 3$

Example 10

The polynomial $ax^3 + x^2 + bx + 6$, where a and b are constants, is denoted by p(x). It is given that $(2x - 1)$ is a factor of p(x) and that when p(x) is divided by $(x - 1)$ the remainder is -4. Find the values of a and b.

$p(x) = ax^3 + x^2 + bx + 6$

$p\left(\dfrac{1}{2}\right) = \dfrac{a}{8} + \dfrac{1}{4} + \dfrac{b}{2} + 6 = 0$ ⟵ $p\left(\dfrac{1}{2}\right) = 0$ as $(2x - 1)$ is a factor.

$a + 2 + 4b + 48 = 0$ ⟵ Multiply each term by 8.

$a + 4b = -50 \qquad (1)$

$p(1) = a + 1 + b + 6 = -4$ ⟵ $p(1) = -4$ as the remainder is -4.

$a + b = -11 \qquad (2)$

$(1) - (2) \Rightarrow 3b = -39$ ⟵ Solve the two equations simultaneously.

$\qquad\qquad b = -13$

$(2) \Rightarrow \quad a - 13 = -11$

$\qquad\qquad a = 2$

$a = 2$ and $b = -13$

Exercise 1.5

1. Factorise the following as a product of three linear factors.
    In each case, one of the factors has been given.

    a) $2x^3 - 5x^2 - 4x + 3$ One factor is $(x - 3)$.

    b) $x^3 - 6x^2 + 11x - 6$ One factor is $(x - 2)$.

    c) $5x^3 + 14x^2 + 7x - 2$ One factor is $(5x - 1)$.

    d) $2x^3 + 3x^2 - 18x + 8$ One factor is $(x + 4)$.

    e) $x^3 + x^2 - 4x - 4$ One factor is $(x + 2)$.

    f) $6x^3 + 13x^2 - 4$ One factor is $(3x + 2)$.

2. Solve the following equations.

    a) $2x^3 + 7x^2 - 7x - 12 = 0$ b) $2x^3 - 5x^2 - 14x + 8 = 0$

    c) $x^3 - 6x^2 + 3x + 10 = 0$ d) $x^3 + 3x^2 - 6x - 8 = 0$

    e) $2x^3 - 15x^2 + 13x + 60 = 0$ f) $3x^3 - 2x^2 - 7x - 2 = 0$

3. Show that $(x - 3)$ is a factor of $x^5 - 3x^4 + x^3 - 4x - 15$.

4. Factorise $x^4 + x^3 - 7x^2 - x + 6$ as a product of four linear factors.

5. $(x - 2)$ is a factor of $x^3 - 3x^2 + ax - 10$. Evaluate the coefficient a.

6. **a)** Show that $(2x - 5)$ is a factor of $4x^3 - 20x^2 + 19x + 15$.

   b) Hence factorise $4x^3 - 20x^2 + 19x + 15$ as a product of three linear factors.

7. The polynomial $ax^3 - 3x^2 - 5ax - 9$ is denoted by p(x) where a is a real number. It is given that $(x - a)$ is a factor of p(x). Find the possible values of a.

8. The polynomial $3x^3 + 2x^2 - bx + a$, where a and b are constants, is denoted by p(x). It is given that $(x - 1)$ is a factor of p(x) and that when p(x) is divided by $(x + 1)$ the remainder is 10. Find the values of a and b.

9. The polynomial $ax^3 + bx^2 - 5x + 3$, where a and b are constants, is denoted by p(x). It is given that $(2x - 1)$ is a factor of p(x) and that when p(x) is divided by $(x - 1)$ the remainder is -3. Find the remainder when p(x) is divided by $(x + 3)$.

10. Factorise $2x^4 + 5x^3 - 5x - 2$ as a product of four linear factors.

Summary exercise 1

1. Solve algebraically the equation $|5 - 2x| = 7$.

2. Solve algebraically the equation $|3x - 4| = |5 - 2x|$.

3. Sketch the following graphs:
   a) $y = 2|x| + 5$
   b) $y = 2 - |x|$.

4. Solve graphically the inequality $2|x - 2| < |x|$.

5. Solve graphically the inequality $|2x - 1| < |3x - 4|$.

EXAM-STYLE QUESTION

6. Solve the inequality $|x + 3| \geq 2|x - 3|$.

7. Solve the inequality $|x - 2| \leq 3|x + 1|$.

EXAM-STYLE QUESTION

8. Solve the inequality $2|x - a| > |2x + a|$ where a is a constant and $a > 0$.

9. Divide $2x^4 - 9x^3 + 13x^2 - 15x + 9$ by $(x - 3)$.

10. Find the quotient and the remainder when $x^3 - 3x^2 + 6x + 1$ is divided by $(x - 2)$.

EXAM-STYLE QUESTIONS

11. **a)** Show that $(x - 4)$ is a factor of $x^3 - 3x^2 - 10x + 24$.
    b) Hence factorise $x^3 - 3x^2 - 10x + 24$.

12. The expression $x^3 + 3x^2 + 6x + k$ has a remainder of -3 when divided by $(x + 1)$. Find the value of k.

13. The polynomial $ax^4 + bx^3 - 8x^2 + 6$ is denoted by p(x). When p(x) is divided by $(x^2 - 1)$ the remainder is $2x + 1$. Find the value of a and the value of b.

14. The polynomial $x^4 + ax^3 + bx^2 - 16x - 12$ is denoted by p(x). $(x + 1)$ and $(x - 2)$ are factors of p(x).
    a) Evaluate the coefficients a and b.
    b) Hence factorise p(x) fully.

Algebra   15

15. The polynomial $x^4 + x^3 - 22x^2 - 16x + 96$ is denoted by p(x).
    a) Find the quotient when p(x) is divided by $x^2 + x - 6$.
    b) Hence solve the equation p(x) = 0.

16. The polynomial $6x^3 - 23x^2 + ax + b$ is denoted by p(x). When p(x) is divided by ($x + 1$) the remainder is -21. When p(x) is divided by ($x - 3$) the remainder is 11.
    a) Find the value of a and the value of b.
    b) Hence factorise p(x) fully.

17. A polynomial is defined by
    p(x) = $x^3 + Ax^2 + 49x - 36$, where A is a constant. ($x - 9$) is a factor of p(x).
    a) Find the value of A.
    b) i)   Find all the roots of the equation p(x) = 0.
        ii) Find all the roots of the equation p(x^2) = 0.

18. The polynomial $x^3 - 15x^2 + Ax + B$, where A and B are constants, is denoted by p(x). ($x - 16$) is a factor of p(x). When p(x) is divided by ($x - 2$) the remainder is -56.
    a) Find the value of A and the value of B.
    b) i)   Find all 3 roots of the equation p(x) = 0.
        ii) Find the 4 real roots of the equation p(x^4) = 0.

19. i)   Find the quotient and remainder when
        $x^4 + 2x^3 + x^2 + 20x - 25$
        is divided by ($x^2 + 2x - 5$).

    ii) It is given that, when
        $x^4 + 2x^3 + x^2 + px + q$
        is divided by ($x^2 + 2x - 5$), there is no remainder.
        Find the values of the constants p and q.

    iii) When p and q have these values, show that there are exactly two real values of x satisfying the equation
        $x^4 + 2x^3 + x^2 + px + q = 0$
        and state what these values are. Give your answer in the form a $\pm \sqrt{b}$.

Chapter summary

The modulus function

- The modulus of a real number is the magnitude of that number.
- The modulus function $f(x) = |x|$ is defined as

$$|x| = x \quad \text{for} \quad x \geq 0$$
$$|x| = -x \quad \text{for} \quad x < 0$$

Sketching graphs of the modulus function

- When sketching the graph of $y = |f(x)|$ we reflect the section of the graph where $y < 0$ in the x-axis.
- When sketching the graph of $y = f(|x|)$ we sketch the section of the graph where $x > 0$ and then reflect this in the y-axis.

Division of polynomials

- When dividing algebraic expressions, for example $(4x^3 - 7x - 3) \div (2x + 1) = (2x^2 - x - 3)$, you need to know the following terms:
  - $(4x^3 - 7x - 3)$ is called the dividend.
  - $(2x + 1)$ is called the divisor.
  - $(2x^2 - x - 3)$ is called the quotient, and there is no remainder.
  - $(2x + 1)$ is a factor of $(4x^3 - 7x - 3)$.
- $f(x) = $ quotient $\times$ divisor $+$ remainder

The remainder theorem

- When a polynomial $f(x)$ is divided by $(x - a)$, the remainder is $f(a)$.
- When a polynomial $f(x)$ is divided by $(ax - b)$, the remainder is $f\left(\dfrac{b}{a}\right)$.

The factor theorem

- For any polynomial $f(x)$, if $f(a) = 0$ then the remainder when $f(x)$ is divided by $(x - a)$ is zero. Thus $(x - a)$ is a factor of $f(x)$.
- For any polynomial $f(x)$, if $f\left(\dfrac{b}{a}\right) = 0$, then $(ax - b)$ is a factor of $f(x)$.

The shell of a nautilus grows in a manner where each new chamber is an approximate copy of the last one, but is enlarged by a constant factor. This gives rise to the aesthetically pleasing shape known as a *logarithmic spiral*.

Many things in nature grow in the same way; that is, its rate of growth is proportional to its current size. Birth rates, for example, behave in this way: as a population increases in size, its birth rate increases proportionally.

As such, understanding the mathematics which describes the behaviour of entities like birth rates is crucially important if we are to predict how they will behave in future. A lot of important physical quantities, such as the energy released in an earthquake, have very large ranges and thus logarithmic scales (the Richter scale, in the case of the earthquake) are helpful in providing a way of making the numerical values of these measurements manageable.

Objectives

- Understand the relationship between logarithms and indices, and use the laws of logarithms (excluding change of base).
- Understand the definition and properties of e^x and $\ln x$, including their relationship as inverse functions and their graphs.
- Use logarithms to solve equations of the form $a^x = b$, and similar inequalities.
- Use logarithms to transform a given relationship to linear form, and hence determine unknown constants by considering the gradient and/or intercept.

Before you start

You should know how to:

1. Apply the laws of indices.

   e.g. **a)** Simplify $(x^3)^4$.

   $(x^3)^4 = x^{3 \times 4} = x^{12}$

   e.g. **b)** Simplify $x^5 \times x^4$.

   $x^5 \times x^4 = x^{5+4} = x^9$

2. Know the properties of powers.

   e.g. If $x \neq 0$, state the value of x^0.

   $x^0 = 1$ if $x \neq 0$

Skills check:

1. Simplify

   a) $\dfrac{x^8}{x^3}$ **b)** $\sqrt{x^{10}}$ **c)** $(x^7)^{\frac{1}{3}}$.

2. Simplify $(2^a)^b$.

2.1 Continuous exponential growth and decay

In P1 you met geometric progressions of the form ar^{n-1}, where n is an integer and r is the common ratio between terms. Such a progression models exponential growth when $r > 1$ and exponential decay when $r < 1$. When we model exponential change with a geometric progression of this form, we assume that the change takes place in discrete steps. Situations involving exponential growth or decay are not restricted to discrete steps, but often occur in continuous time. We sometimes simplify a phenomenon, modelling it as changing in discrete steps to make the mathematics easier. We might look at the change over a unit of time (an hour, a day, a year) and treat it as though it were a discrete step process – for example, where an investment has the interest added to the principal once a year. Shortening the units of time will give a better approximation to reality. As we consider change over shorter and shorter time divisions, we can consider the limit of that process as an exponential function $f(t) = a \times r^t$, where t can take any real value rather than just integer values (in practice, t is mostly restricted to any non-negative value).

> A function $f(t) = a \times r^t$, where t is real,
>
> gives exponential growth $(r > 1)$ or decay $(r < 1)$.

We have defined r^t only for rational values of t so far, but any irrational value of t can be approximated above and below by a pair of rational numbers which differ by as small an amount as you wish – giving a process to define $f(t)$ for all t.

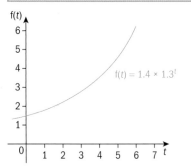

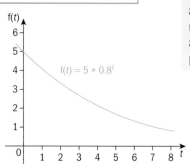

$f(t) = 1.4 \times 1.3^t$ The initial value of the function, a, is 1.4. We also see that $r = 1.3 > 1$, meaning this function describes exponential growth.

$f(t) = 5 \times 0.8^t$ The initial value of the function, a, is now 5 and $r = 0.8 < 1$, meaning this function describes exponential decay.

Did you know?

Examples of exponential growth include the growth of cells, nuclear chain reactions, epidemics, feedback in an amplification signal, and computer processing power.

Examples of exponential decay include radioactive decay (used in carbon dating), the rates of some chemical reactions (which depend on the concentration of one or more of the reactants), and light intensity in an absorbent medium.

One of the characteristics of exponential decay is that the graph will get closer to the horizontal axis as time increases, but will never reach it.

Exponential cell growth

Carbon dating relies on the fact that radioactive material decays exponentially.

Logarithms and exponential functions   **19**

Example 1

The mass in grams of a decaying radioactive material is given by $M(t) = 200(0.8)^t$ after t days. Draw a graph of M and use your graph to estimate the number of days it takes for M to fall to

a) 100 g   b) 50 g   c) 25 g.

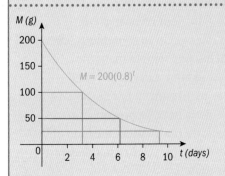

$M = 200(0.8)^t$

a) $M(3.1) = 100 \Rightarrow$ after 3.1 days, $M \approx 100$ g

b) $M(6.2) = 50 \Rightarrow$ after 6.2 days, $M \approx 50$ g

c) $M(9.3) = 25 \Rightarrow$ after 9.3 days, $M \approx 25$ g

The same length of time is taken for the amount remaining to halve each time. The 'half-life' of this radioactive material is approximately 3.1 days.

Example 2

The value V in dollars of a car which is t years old is $V(t) = 35\,000(0.75)^t$.
a) Find its value after 2 years.
b) Show that the value goes below $15\,000 before it is 3 years old.

a) $V(2) = \$35\,000(0.75)^2 = \$19\,687.50 = \$19\,700$ (3 s.f.)

b) $V(3) = \$35\,000(0.75)^3 = \$14\,765.63 < \$15\,000$ so the value goes below $15\,000 before the car is 3 years old.

Example 3

In a container, the number of cells doubles every six hours. At 10 am on Monday a new culture is started with 20 cells and placed in a container which will hold 10 million cells. In which six-hour period will the container be filled up?

During the 19th period (10 pm Friday – 4 am Saturday) it goes from 5 242 880 to 10 485 760 cells.

On your calculator if you do $20 \times 2 =$ and then $\times 2 =$ repeatedly, you can count the number of periods until it goes over 10 million.

Exercise 2.1

1. Calculate

   a) $0.6 \times (4)^5$ **b)** $400\,000 \times (0.3)^5$ **c)** $63.2 \times (1.03)^8$ **d)** $5.1 \times (0.08)^4$.

2. Air is escaping from a balloon. The volume V, in ml, of the balloon after t minutes is given by $V = 1500 \times (0.7)^t$. What is the volume after

   a) 5 minutes   **b)** a quarter of an hour?

3. A company has an accounting policy that equipment is depreciated at 20% of current book value each financial year, and the initial book value is set at the price paid for the equipment. The company buys equipment for $3 million in 2012, and makes a further purchase in 2013 for $2 million. What is the book value of the equipment in 2015?

> **Did you know?**
>
> 'Depreciation' is the loss in value over a period of time for equipment, and 'book value' is the value at any time in the company accounts.

4. The amount of a drug (in mg) in a patient's bloodstream t hours after the first injection is $100 \times (0.8)^t$. If injections are to be repeated at 12-hourly intervals, what should the dose be (to the nearest mg) for the second and subsequent injections to bring the total in the bloodstream back to 100 mg each time?

5. At 4 pm one afternoon, Jeremy's car bottoms out when going over a speed bump too quickly and oil starts leaking from an engine sump. The volume, in ml, remaining after t hours is $3000 \times (0.98)^t$. If the oil level falls below 2 litres there is a risk of permanently damaging the engine. The next morning at 9 am Jeremy notices the oil underneath his parked car. Show that there is still enough oil for him to drive the car to the garage for repair.

2.2 The logarithmic function

The graph shows the function $y = 2^x$ in blue, and the inverse of this function in green (recall that the inverse function is the mirror image of the function in the line $y = x$).

This inverse function is known as the **logarithmic function**.

> The logarithmic function is the inverse of the exponential function to the same base.

The green graph shown here is the logarithm to base 2, because the blue graph shows 2^x.

The exponential function has domain the set of all real numbers, with range the positive real numbers, so it follows that the logarithmic function is defined on a domain of the positive real numbers and its range is the set of all real numbers.

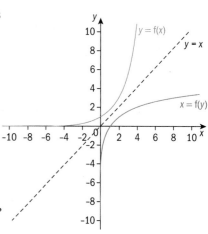

Formally, we define the logarithmic function by

$$y = b^x \iff x = \log_b y, \text{ where } x \in \mathbb{R},\ y \in \mathbb{R},\ y > 0$$

These two special cases are worth remembering:
For any b, $\log_b 1 = 0$.
For any n, $\log_b b^n = n$.

Example 4

Find the value of

a) $\log_2 16$ b) $\log_3 243$ c) $\log_{10} 1000$ d) $\log_7 343$.

....

a) $16 = 2^4 \Rightarrow \log_2 16 = 4$ b) $243 = 3^5 \Rightarrow \log_3 243 = 5$

c) $1000 = 10^3 \Rightarrow \log_{10} 1000 = 3$ d) $343 = 7^3 \Rightarrow \log_7 343 = 3$

> We identify the logarithm by thinking of what power we have to raise the base to.

Example 5

Find the value of

a) $\log_2 \dfrac{1}{8}$ b) $\log_8 2$ c) $\log_{\frac{1}{2}} 4$.

....

a) $\dfrac{1}{8} = 2^{-3} \Rightarrow \log_2 \dfrac{1}{8} = -3$ b) $2 = \sqrt[3]{8} = (8)^{\frac{1}{3}} \Rightarrow \log_8 2 = \dfrac{1}{3}$

c) $4 = \left(\dfrac{1}{2}\right)^{-2} \Rightarrow \log_{\frac{1}{2}} 4 = -2$

Example 6

a) Write these in the form $y = b^x$.

   i) $\log_2 64 = 6$ ii) $\log_k m = p$

b) Write these in the form $x = \log_b y$.

   i) $5^{-3} = 0.008$ ii) $h^5 = R$

....

a) i) $64 = 2^6$ ii) $m = k^p$

b) i) $\log_5 0.008 = -3$ ii) $\log_h R = 5$

Example 7

On the same axes sketch the graphs of

a) $y = 0.7^x$ **b)** $y = 1.2^x$ **c)** $y = 1.5^x$.

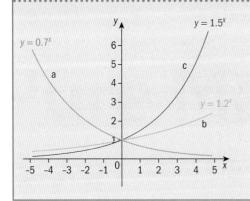

In part (a) the base value (0.7) is less than one so the function decreases as x gets larger. In parts (b) and (c) the base values are larger than 1 (1.2, 1.5) and so these functions increase as x gets larger.

Note also that (c) is steeper than (b) because (c) has a greater base value.

Since the logarithm function is the inverse of the exponential (or power) function, the rules of indices give rise to corresponding properties of logarithms.

$\log (xy) = \log x + \log y$

$\log\left(\dfrac{x}{y}\right) = \log x - \log y$

$\log (x^n) = n \log x$

These results are true for logarithms in any base.

Example 8

Express these as a single logarithm or number.

a) $\log_5 4 + \log_5 7$ **b)** $\log_6 4 + \log_6 9$

a) $\log_5 4 + \log_5 7 = \log_5 28$ **b)** $\log_6 4 + \log_6 9 = \log_6 36 = \log_6 (6^2) = 2$

Example 9

Express these in terms of $\log x$, $\log y$, and $\log z$.

a) $\log xy$ **b)** $\log\left(\dfrac{xy^2}{z}\right)$ **c)** $\log \sqrt{\left(\dfrac{x^3 y^2}{z^4}\right)}$

a) $\log xy = \log x + \log y$ **b)** $\log\left(\dfrac{xy^2}{z}\right) = \log x + 2 \log y - \log z$

c) $\log \sqrt{\left(\dfrac{x^3 y^2}{z^4}\right)} = \log\left(x^{\frac{3}{2}} y z^{-2}\right) = \dfrac{3}{2}\log x + \log y - 2 \log z$

Identify the power of each variable first.

Exercise 2.2

1. Evaluate the following logarithms.

a) $\log_3 \left(\dfrac{1}{81} \right)$ **b)** $\log_9 27$ **c)** $\log_4 128$ **d)** $\log_3 6561$

e) $\log_6 216$ **f)** $\log_8 \dfrac{1}{32}$ **g)** $\log_{16} 32$ **h)** $\log_4 \left(\dfrac{1}{16} \right)$

i) $\log_9 \left(\dfrac{1}{3} \right)$ **j)** $\log_{\frac{1}{2}} \left(\dfrac{1}{32} \right)$ **k)** $\log_{\frac{1}{3}} 27$ **l)** $\log_{\frac{1}{4}} 8$

2. Write in the form $y = b^x$.

a) $\log_2 256 = 8$ **b)** $\log_3 \left(\dfrac{1}{27} \right) = -3$ **c)** $\log_4 32 = 2.5$

d) $\log_3 (9\sqrt{3}) = 2.5$ **e)** $\log_a x = q$ **f)** $\log_s t = u$

3. Write in the form $x = \log_b y$.

a) $7^{-3} = \dfrac{1}{343}$ **b)** $10^9 = 1\,000\,000\,000$ **c)** $2^{-4} = \dfrac{1}{16}$

d) $5^{2.5} = 25\sqrt{5}$ **e)** $t^{-3} = v$ **f)** $p^x = m$

4. Evaluate, without using a calculator, the following logarithms.

a) $\log_{10} 100$ **b)** $\log_{10} \sqrt{10}$ **c)** $\log_{10} \dfrac{10}{\sqrt[3]{10}}$ **d)** $\log_{10} \left(\dfrac{1}{\sqrt{10}} \right)$

e) $\log_{10} (100\,000\sqrt{10})$ **f)** $\log_{10} \left(\dfrac{1}{100\sqrt[3]{10}} \right)$ **g)** $\log_{10} 1$

5. On the same axes, sketch the graphs of

a) $y = 0.5^x$ **b)** $y = 1.3^x$ **c)** $y = 1^x$.

6. Express the following as a single logarithm or number.

a) $\log_5 3 + \log_5 8$ b) $\log_6 7 + \log_6 10$ c) $\log_5 4 + \log_5 6.25$

d) $2\log_5 3 + \log_5 4$ e) $2\log_5 6 - \log_5 8 + \log_5 12$ f) $2\log_{10} 4 - 4\log_{10} 2$

g) $2\log_p 4 - 4\log_p 2$ h) $\log_{10} 12 - \dfrac{1}{2}\log_{10} 9 + \log_{10} 25$

7. Express as a single logarithm.

a) $\log_5 3 + 2$ b) $3 + \log_6 10$ c) $2\log_{10} x - \log_{10} 8 + \log_{10}(x+3)$

d) $3 + 2\log_5 x$ e) $\log_{10} x + 3$ f) $2 + \log_{10} 8 - \log_{10} x$

8. Express in terms of $\log x$, $\log y$, and $\log z$.

a) $\log\left(\dfrac{xy}{z^2}\right)$ b) $\log\left(x^2\left(xy^2z\right)^3\right)$ c) $\log\sqrt{\left(\dfrac{x^4y^2}{z^4}\right)}$

9. Given that $\log_a 2 = x$, $\log_a 5 = y$, and $\log_a 6 = z$, express the following in terms of x, y, and z.

a) $\log_a 50$ b) $\log_a 75$ c) $\log_a 0.75$ d) $\log_a 0.001$

2.3 e^x and logarithms to base e

You have seen that you can use any positive real number as a base for a power or logarithmic function, but in practice two bases are very commonly used, and they are available on your calculator.

- The button marked $\boxed{\log}$ uses base 10 since our number system is based on powers of 10.

- The $\boxed{\ln}$ button uses base e (= 2.71828 ...). These are sometimes called natural logarithms because of some special properties the number represented by 'e' has when used in an exponential function.

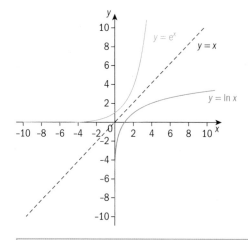

The blue graph is $y = e^x$ and the green graph is $y = \log_e x$ (also known as $y = \ln x$). These are inverses and are similar to the graphs you saw earlier at the beginning of section 2.2. One feature to notice about $y = e^x$ is that as x increases, not only does e^x also increase, but its rate of increase (the gradient of the graph) gets larger.

In x or $\log_e x$ is the inverse function of e^x.

Did you know?

Consider the following results, which you encountered in section 2.2.

$$\log(xy) = \log x + \log y \quad (1)$$

$$\log\left(\frac{x}{y}\right) = \log x - \log y \quad (2)$$

$$\log(x^n) = n \log x \quad (3)$$

Before hand-held electronic calculators became commonly available in the 1980s, scientists and students used these results to simplify numerical calculations. Using them, we are able to transform a calculation and so merely add instead of multiply (using (1)), subtract instead of divide (using (2)), and multiply instead of raising something to a power (using (3)).

Using base 10 made those new operations very simple, since (for example)

$$257.2 = 2.572 \times 100 \Rightarrow \log_{10} 257.2 = \log_{10} 2.572 + \log_{10} 100$$
$$= \log_{10} 2.572 + 2$$
$$= 2.4102\ldots$$

Check on your calculator that $\log_{10} 257.2 = 2 + \log_{10} 2.572$

When they did not have calculators, scientists and students used books of logarithm values. However, as they were able to transform the logarithm of a big number into the logarithm of a small one, as shown above, the tables only needed to show the logarithms of numbers x satisfying $1 \le x < 10$.

An earthquake under 3 on the Richter scale will go unnoticed by most people going about their normal business, but 3 will be noticed and 4 will cause anxiety amongst those unfamiliar with the feeling of earthquakes. Once it gets up to 5, buildings may be damaged, and 6 is likely to cause injuries, so the effects increase enormously for small increases in the reported measurement.

Photo of devastation to buildings/roads caused by a recent high-magnitude earthquake, e.g. Mexico 2017

When logarithmic scales are used with natural phenomena, it is because the range of measurement is so large that the numbers are not easily comprehensible, but it is then very important to remember that the scale is logarithmic when making comparisons. An earthquake which measures 4.0 on the Richter scale creates waves recorded on a seismograph which have ten times the amplitude of the waves an earthquake measuring 3.0 creates, and releases around 31.7 times as much energy.

Other commonly used logarithmic scales include the pH scale for measuring acidity, noise levels measured in decibels and the stellar magnitude scale for the brightness of stars.

Example 11

Use your calculator to find the values of

a) $\log_{10} 16$ **b)** $\log_{10} 0.01457$ **c)** $\ln 10$

d) $\ln 2$ **e)** $\ln 1$ **f)** $\ln 0.8$.

Give your answers to 3 decimal places.

a) 1.204   **b)** −1.837   **c)** 2.303

d) 0.693   **e)** 0   **f)** −0.223

Remember:
- 1 has logarithm 0 in any base
- numbers below 1 will have a negative logarithm
- anything between 1 and the base will have a positive logarithm less than 1.

Example 12

Find the values of

a) $\dfrac{\log_{10} 7}{\log_{10} 4}$ **b)** $\dfrac{\log_{10} 3.52}{\log_{10} 7.94}$ **c)** $\dfrac{\ln 8.2}{\ln 6.5}$.

Give your answers to 3 decimal places.

a) 1.404   **b)** 0.607   **c)** 1.124

Often when you solve logarithmic equations your solution will be in a form involving a quotient. Make sure you can enter these correctly in your calculator – often you need to remember to insert parentheses separately around both the numerator and the denominator.

Example 13

An earthquake of magnitude 3.0 on the Richter scale releases the energy equivalent of 480 kg of TNT. How much energy will be released by an earthquake of the following magnitudes?

An increase of 0.2 on the Richter scale roughly doubles the amount of energy released by an earthquake. An increase of 1.0 gives a multiplying factor of 31.7 (very close to 2^5).

a) 3.2   **b)** 4.0   **c)** 7.5

a) approximately 1 tonne of TNT   ← $0.48 \times 31.7^{0.2} = 0.958$

b) approximately 15 tonnes of TNT   ← $0.48 \times 31.7 = 15.2$

c) approximately $0.48 \times (31.7)^{4.5} = 2.7$ million tonnes (2.7 megatonnes) of TNT

Example 14

A radioactive isotope has a half-life of 20 days. Initially it has a mass of 60 grams. How much will there be after

a) 20 days   **b)** 30 days?

a) 30 grams

b) 21.2 grams ($= 60 \times (0.5)^{1.5}$) since 30 days = 1.5 half-lives

> **Did you know?**
> The 'half-life' is the length of time it will take for half the mass to decay.

Exercise 2.3

1. Use your calculator to find the value of these. Give your answers to 3 decimal places.

   a) $\log_{10} 23.2$ **b)** $\log_{10} 0.0232$

   c) $\log_{10} 2.7183$ (which is e)   **d)** $\ln 6.5$

   e) $\ln 0.0065$

2. Find the value of these. Give your answers to 3 decimal places.

   a) $\dfrac{\log_{10} 70}{\log_{10} 40}$ **b)** $\dfrac{\log_{10} 352}{\log_{10} 794}$ **c)** $\dfrac{\ln 0.82}{\ln 0.65}$

   > Compare your answers to those in Example 12. Make sure you never think of cancelling in an expression like this.

3. An earthquake of magnitude 3.0 on the Richter scale releases the energy equivalent of 480 kg of TNT.

   > Remember from Example 13 that an increase of 1.0 gives a multiplying factor of 31.7.

   a) How much energy will be released by an earthquake of magnitude

   i) 2   **ii)** 6?

   b) How much energy will be released by six earthquakes of magnitude 5?

4. The radioactive isotope carbon-14 has a half-life of 5730 years. Once a living organism dies, the decay process starts and can be used to date how old a fossil is.

   An archeologist finds a skull which he thinks is from an animal which became extinct 4000 years ago. For the size of the skull, he knows it would have had a mass of 28 mg of carbon-14 when the creature died. If the skull is 4000 years old, how much carbon-14 should be in the skull now?

2.4 Equations and inequalities using logarithms

You can use the laws of logarithms to transform equations in order to make their solutions accessible.

Example 15

Solve the equation $4^x = 16$ by taking logarithms of both sides.

$4^x = 16 \Rightarrow x \log 4 = \log 16 \Rightarrow x = \dfrac{\log 16}{\log 4} = 2$ (using any base)

Example 16

Solve the equation $5^x = 3^{2x+1}$.

$5^x = 3^{2x+1} \Rightarrow x \log 5 = (2x + 1) \log 3 \Rightarrow x(\log 5 - 2 \log 3) = \log 3$

$\Rightarrow x = \dfrac{\log 3}{\log 5 - 2\log 3} = -1.869$

Since x appears in more than one power, you need to collect x terms together to solve for x.

One important feature of the power (and logarithmic) functions is that they are strictly monotonic – that is to say, they are always increasing (if the base is > 1) or always decreasing (if the base is < 1). So if the logarithms are equal it follows that the expressions are equal. Examples 17 and 21 express both sides as a single logarithm and use this feature to remove the logarithm to enable the equation to be solved.

Example 17

Solve the equation $\log_{10} (2 + x) = 2 + \log_{10} x$.

$\log_{10} (2 + x) = 2 + \log_{10} x \Rightarrow \log_{10} (2 + x) = \log_{10} 100 + \log_{10} x$

$\Rightarrow \log_{10}(2 + x) = \log_{10} 100x \Rightarrow 2 + x = 100x \Rightarrow x = \dfrac{2}{99}$

You need to express the right-hand side as a single logarithm.

Another important consequence of the strictly monotonic property is that inequalities can be solved by considering the case of equality and then selecting the values above or below that critical value as appropriate.

Example 18

Solve the following inequalities.

a) $5^x \leq 13.3$ **b)** $(0.4)^x < 0.0001$

a) $5^x = 13.3 \Rightarrow x \log 5 = \log 13.3 \Rightarrow x = \dfrac{\log 13.3}{\log 5} = 1.608$ ⟵   Powers of 5 give a monotonically increasing function, since $5 > 1$.

so $5^x \leq 13.3$ when $x \leq 1.608$

b) $(0.4)^x = 0.0001 \Rightarrow x \log 0.4 = \log 0.0001$

$\Rightarrow x = \dfrac{\log 0.0001}{\log 0.4} = 10.1$ ⟵   Powers of 0.4 give a monotonically decreasing function, since $0.4 < 1$, so the direction of the inequality must be reversed.

so $(0.4)^x < 0.0001$ when $x > 10.1$

Example 19

How many terms of the geometric series $2 + 2 \times (1.1) + 2 \times (1.1)^2 + 2 \times (1.1)^3 + \dots$ must be taken for the sum to exceed one thousand?

If you solve for the sum of n terms of this GP to be 1000 it will (almost certainly) give a non-integer value. The solution will be the next integer above that value.

$1000 = \dfrac{2(1.1^n - 1)}{1.1 - 1} \Rightarrow 50 = 1.1^n - 1 \Rightarrow 1.1^n = 51$

Then taking logs gives $n = \dfrac{\log 51}{\log 1.1} = 41.25\dots$ ⟵   Remember $S_n = \dfrac{a(r^n - 1)}{r - 1}$ is the sum of n terms of a GP with first term a and common ratio r.

So 42 terms are needed.

Example 20

Solve the equation $\log_a (2 - x) - 2\log_a x = \log_a 3$.

$\log_a (2 - x) - 2\log_a x = \log_a 3 \Rightarrow \log_a \dfrac{(2-x)}{x^2} = \log_a 3$

$\Rightarrow \dfrac{(2-x)}{x^2} = 3 \Rightarrow 2 - x = 3x^2 \Rightarrow 3x^2 + x - 2 = 0$ ⟵   Having equated the expressions which were logarithms you need to check the 'solutions' actually gave a valid equation in log form.

This is now a standard quadratic equation, with solutions $x = \dfrac{2}{3}, -1$.

However, only $x = \dfrac{2}{3}$ is valid because the logarithm of a negative number is undefined.

Example 21

Solve the equation $\log_{10}(2-x) - 2\log_{10} x = 1$.

$\log_{10}(2-x) - 2\log_{10} x = 1 \Rightarrow \log_{10}\dfrac{(2-x)}{x^2} = \log_{10} 10$

$\Rightarrow \dfrac{(2-x)}{x^2} = 10 \Rightarrow 2 - x = 10x^2 \Rightarrow 10x^2 + x - 2 = 0$

This is now a standard quadratic equation, with solutions $x = \dfrac{2}{5}, -\dfrac{1}{2}$

but only $x = \dfrac{2}{5}$ is valid.

> You need to express both sides as a single logarithm in order to use this technique, so writing 1 as $\log_{10} 10$ is a critical step.

Example 22

A culture contains 300 000 bacteria at 9 am on Monday and increases at a rate of 20% every hour. At what time will there be

a) 600 000 bacteria   b) 2 million bacteria?

The number of bacteria t hours after 9 am on Monday is given by $300\,000 \times (1.2)^t$.

a) $300\,000 \times (1.2)^t = 600\,000 \Rightarrow (1.2)^t = 2 \Rightarrow t\log 1.2 = \log 2$

$\Rightarrow t = 3.8017 \ldots$ so at 12:48 pm.

b) $300\,000 \times (1.2)^t = 2\,000\,000 \Rightarrow (1.2)^t = \dfrac{20}{3} \Rightarrow t\log 1.2 = \log\left(\dfrac{20}{3}\right)$

$\Rightarrow t = 10.405 \ldots$ so at 7:24 pm.

Did you know?

The time for a bacteria population to double in size is known as the generation time and typically varies from 10 minutes to about 20 hours, though bacteria deep below the earth's surface may have a generation time of several thousand years.

Exercise 2.4

For this exercise, give your answers to 3 significant figures (unless they are exact).

1. Solve for x.

a) $5^x = 100$

b) $6^x = 3$

c) $0.4^x = 0.1$

d) $0.4^x = 2.5$

e) $7^{2x} = 12$

f) $7^{3x} = 64$

g) $4^{2x-1} = 7$

h) $7^{\frac{1}{2}x} = 10$

i) $2^{4x-1} = 128$

j) $3^{2x+1} = \dfrac{1}{27}$

k) $3^{\frac{1}{2}x} = 27$

l) $\left(\dfrac{1}{2}\right)^{2x-1} = 128$

2. Solve for x.

a) $2^{2x} = 3^{x+2}$

b) $5^{x+1} = 4^{2x-3}$

c) $3^{2x-1} = 7^{x+1}$

d) $4^{3x-1} = 2^{2x-6}$

e) $3^{x+4} = 5^{x+1}$

f) $7^{2x} \times 4^x = 3^{x+5}$

3. Solve for x.

   a) $3^x \geq 40$

   b) $6^x \leq 0.8$

   c) $0.4^x < 0.2$

   d) $0.4^x \leq 2$

   e) $7^{2x} \leq 5$

   f) $7^{\frac{1}{3}x} < 25$

4. Solve for x, v, and z.

   a) $7^x = 5$

   b) $7^v = 12$

   c) $7^z = 60$

   > Since $5 \times 12 = 60$ you should find that $x + v = z$.

5. Solve for x and y.

   a) $5^x = 30$

   b) $5^{2y} = 60$

   > Since question 5 shows us that $x \neq y$, we conclude that, in general, $5^{2x} \neq 2 \times 5^x$.

6. How many terms of the geometric series $1 + 3 + 9 + 27 + 81 + \ldots$ must be taken for the sum to exceed one million?

7. How many terms of the geometric series $1 + 3 + 9 + 27 + 81 + \ldots$ must be taken for the sum to exceed one billion? (1 billion = 1000 million)

8. How many terms of the geometric series $1 + 5 + 25 + 125 + 625 + \ldots$ must be taken for the sum to exceed one million?

9. How many terms of the geometric series $3 + 6 + 12 + 24 + 48 + \ldots$ must be taken for the sum to exceed ten thousand?

10. How many terms of the geometric series $10 + 5 + 2.5 + 1.25 + 0.625 + \ldots$ must be taken for the sum to exceed 19.999?

11. How many terms of the geometric series $8 + 4 + 2 + 1 + 0.5 + \ldots$ must be taken for the sum to get within 10^{-4} of its sum to infinity?

12. How many terms of the geometric series $1 + \dfrac{1}{3} + \dfrac{1}{9} + \dfrac{1}{27} + \dfrac{1}{81} + \ldots$ must be taken for the sum to get to within 0.01% of its sum to infinity?

13. Solve for x.

    a) $\log_a (2 - x) = \log_a x + \log_a 5$

    b) $\log_{10} (4 + x) = \log_{10} x + \log_{10} 7$

    c) $\ln (8x - 1) = \ln x + \ln 6$

    d) $\log_{10} (4 + x) = \log_{10} x + 2$

    e) $\ln (1 - x) = \ln x + 1$ (Give your answer in exact form, i.e. in terms of e.)

14. Solve for x.

    a) $\ln (6 - 13x) = 2 \ln x + \ln 5$

    b) $\log_{10} (13x - 6) = 2 \log_{10} x + \log_{10} 6$

    c) $\ln (17x - 6) = \ln x + \ln 12$

    d) $\log_{10} (13x - 4) = 2 \log_{10} x + 1$

15. There were 80 mg of a radioactive material stored at the start of 1950. The material has a half-life of 12 years.

    a) How much radioactive material will there be at the start of 2050?

    b) When will there be 1 mg left?

16. An accident occurs while transporting radioactive waste, and results in the area being contaminated by material with a half-life of 8 years. Experts say that the area should be quarantined until the radioactive material has reduced to 10% of its original level. How long are the experts recommending the quarantine should be in place for?

17. A colony of bacteria has a generation time (the time in which the population doubles) of 30 minutes under ideal conditions. Sunil conducts an experiment in which he grows one culture of the bacteria under ideal conditions and another under conditions he thinks will produce a generation time of 40 minutes. If he starts with equal sizes of culture in both cases, after how long would he expect one to be three times the size of the other?

2.5 Using logarithms to reduce equations to linear form

Many scientific, economic, and social science quantities can be described (at least approximately) by relationships which follow either an exponential growth or decay law, or else a power law. If we take logarithms of an equation which follows either of these two laws, we transform it into a linear expression. This allows values of unknown constants to be estimated from observational data.

Exponential growth or decay:

$y = ab^t \Rightarrow \log y = \log a + t \log b$

The graph of '$\log y$' against 't' has intercept '$\log a$' and gradient '$\log b$'.

Power law:

$y = ax^n \Rightarrow \log y = \log a + n \log x$

The graph is of '$\log y$' against '$\log x$' and has intercept '$\log a$' and gradient n.

> You know some examples of power laws in geometrical formulae for areas and volumes.

Example 23

By taking logarithms, transform these relationships between the two stated variables into a linear relationship between two new variables, and state the new variables.

a) y and t are related by $y = 7b^t$.

b) y and x are related by $y = ax^3$.

....

a) $\log y = \log 7 + t \log b$ is linear with variables '$\log y$' and 't'.

b) $\log y = \log a + 3 \log x$ is linear with variables '$\log y$' and '$\log x$'.

Example 24

For the following linear equations involving logarithms, find the relationship between the unknown variables, giving your answer in a form not involving logarithms.

a) $\log_{10} y = 2 + 3\log_{10} x$ **b)** $\log_{10} A = \log_{10} \pi + 2\log_{10} r$ **c)** $\log_{10} y = 0.3 + 0.7x$

a) $\log_{10} y = 2 + 3\log_{10} x \Rightarrow \log_{10} y = \log_{10} 100 + \log_{10} x^3 = \log_{10} 100\, x^3$

$\Rightarrow y = 100\, x^3$

b) $\log_{10} A = \log_{10} \pi + 2\log_{10} r = \log_{10} \pi r^2 \Rightarrow A = \pi r^2$

c) $\log_{10} y = 0.3 + 0.7x = \log_{10}(10^{0.3}) + \log_{10}(10^{0.7x})$

$\qquad = \log_{10}(10^{0.3}) + \log_{10}((10^{0.7})^x)$

$\qquad = \log_{10}(10^{0.3} \times (10^{0.7})^x)$

$\Rightarrow y = 1.995 \times 5.012^x$

> This step, using one of the laws of indices, is the vital one in transforming this into a power law relationship between y and x.

Example 25

The metabolic rate of mammals (M) and their body mass (B) are thought to obey a power law in the form $M = aB^k$. Data on four species of mammals are given in the table. Plot a graph of $\log M$ against $\log B$ and use it to estimate the values of a and k.

B	1.6	8	700	9000
M	1.9	6.4	143	1100

$M = aB^k \Rightarrow \log M = \log a + k \log B$

$\log B$	0.204	0.903	2.845	3.954
$\log M$	0.279	0.806	2.155	3.041

> Taking logarithms of the equation shows that the gradient is k and the intercept is $\log a$.

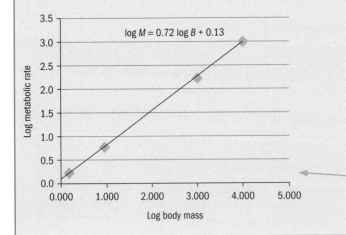

$\log M = 0.72 \log B + 0.13$

Log metabolic rate (y-axis)

Log body mass (x-axis)

> Plotting the points on the graph and drawing a line of best fit by eye gives $k = 0.72$ and $a = 10^{0.13} = 1.3$.

Example 26

The variables x and y satisfy the relation $5^y = 3^{2x+1}$. By taking logarithms, show that the graph of y against x is a straight line, and find the exact values of the gradient and the intercept.

· ·

$5^y = 3^{2x+1} \Rightarrow y\log 5 = (2x+1)\log 3$

$$\Rightarrow y = \left(\frac{2\log 3}{\log 5}\right)x + \left(\frac{\log 3}{\log 5}\right)$$

This equation describes a straight line with gradient $\dfrac{2\log 3}{\log 5}$ and intercept $\dfrac{\log 3}{\log 5}$.

Exercise 2.5

1. By taking logarithms, transform these relationships between the two stated variables into a linear relationship between two new variables, and state the new variables.

   a) p and t are related by $p = 3b^t$.

   b) y and x are related by $y = Kx^{-2}$.

   c) y and x are related by $y = \alpha\sqrt{x}$.

2. For the following linear equations involving logarithms find the relationship between the unknown variables in a form which does not involve logarithms.

   a) $\log_{10} y = 1 + 2\log_{10} x$

   b) $\log_{10} V = \log_{10}\left(\dfrac{4\pi}{3}\right) + 3\log_{10} r$

   c) $\log_{10} y = 0.1 + 1.3x$

3. Two variable quantities are related by the equation $y = Ax^n$ where A and n are constants. The graph shows the results of plotting $\log_{10} y$ against $\log_{10} x$ for four pairs of values of x and y. Use the diagram to estimate the values of A and n.

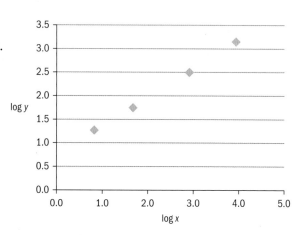

4. Two variable quantities are related by the equation $p = Ak^{-q}$ where A and k are constants. The graph shows the results of plotting $\ln p$ against q for four pairs of values of p and q. Use the diagram to estimate the values of A and k.

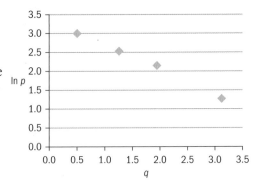

5. Two variable quantities are related by the equation $y = Ax^n$ where A and n are constants. The table shows four pairs of values of x and y. Estimate the values of A and n.

x	62	150	400	921
y	352	232	141	88.1

6. The resistance R (in newtons) to the motion of a train is measured at a number of speeds v (in m s^{-1}) and is shown in the table. If the resistance follows the relationship $R = kv^n$, find the values of k and n.

v	25.7	45.3	71	102.3
R	3562	4635	6037	7324

7. The distance D (in millions of km) from the Sun of the first four planets in the solar system and the time T (in days) for each planet to orbit the Sun are given in the table. If the orbit time follows the relationship $T = kD^n$, find the values of k and n.

	Mercury	Venus	Earth	Mars
D	58	108	150	228
T	88	225	365	687

8. The population P of a colony of bacteria is thought to obey an exponential growth law in the form $P = kr^t$ where P is thousands of bacteria and t is hours since the experiment started. Plot an appropriate graph and use it to estimate the values of k and r.

t	2	5	8	12
P	919	2049	4475	12817

9. The variables x and y satisfy the relation $7^y = 4^{3x-2}$. By taking logarithms show that the graph of y against x is a straight line, and find the exact values of the gradient and the intercept.

Summary exercise 2

1. Find the values of
   a) $\log_2\left(\dfrac{1}{32}\right)$ b) $\log_4 8$ c) $\log_{81} 3$.

2. Write these in the form $y = b^x$.
   a) $\log_4 256 = 4$ b) $\log_2\left(\dfrac{1}{32}\right) = -5$
   c) $\log_2\left(32\sqrt{2}\right) = 5.5$

3. Write these in the form $x = \log_b y$.
   a) $6^{-3} = \dfrac{1}{216}$ b) $10^7 = 10\,000\,000$

4. Find the values of
   a) $\log_{10} 10\,000$ b) $\log_{10} \sqrt{1000}$
   c) $\log_{10} \dfrac{100}{\sqrt[4]{10}}$.

5. Express as a single logarithm or number.
   a) $\log_4 2 + \log_4 7$
   b) $\log_{10} 17 + \log_{10} 12$
   c) $\ln 3 + \ln 4.5$
   d) $2 \log_6 3 + \log_6 4$
   e) $\log_{10} 24 - \log_{10} 3 + \log_{10} 12.5$
   f) $2 + 3\log_5 x$
   g) $3 \ln (2x) - \ln 8 + \ln (2x + 3)$
   h) $2 + \ln 4 - 2\ln x$

6. Express $\log\left(\dfrac{\left(xy^2 z\right)^5}{x^2}\right)$ in terms of $\log x$, $\log y$, and $\log z$.

7. Given $\log_a 2 = x$, $\log_a 5 = y$, and $\log_a 6 = z$, express $\log_a 0.15$ in terms of x, y, and z.

8. Solve
   a) $7^x = 83$ b) $5^x = 0.3$
   c) $0.2^x = 0.8$.

9. Solve
   a) $4^{2x} = 8^{x+2}$ b) $3^{2x+1} = 5^{3x-1}$
   c) $7^{2x-1} = 3^{x+1}$.

10. Solve
    a) $5^x \geq 43.2$ b) $5^x \leq 0.7$
    c) $0.7^x < 0.3$.

11. How many terms of the geometric series $2 + 10 + 50 + 250 + 1250 + \ldots$ must be taken for the sum to exceed one million?

12. How many terms of the geometric series $5 + 2.5 + 1.25 + 0.625 + \ldots$ must be taken for the sum to get within 10^{-5} of its sum to infinity?

13. Solve
    a) $\log_a (3 - 2x) = \log_a x + \log_a 7$
    b) $\log_{10}(3 + x) = \log_{10} x + 2$
    c) $\ln (5x - 2) = \ln x + \ln 4$
    d) $\ln (3 - 2x) = \ln x + 2$ (leave your answer in terms of e).

EXAM-STYLE QUESTION

14. There were 150 mg of a radioactive material stored at the start of the year 2000. The material has a half-life of 15 years.
    a) How much radioactive material will there be at the start of 2040?
    b) When will there be 1 mg of radioactive material left?

15. A type of bacteria has a generation time (the time in which the population doubles) of 30 minutes. Henry conducts an experiment in which he grows a culture of the bacteria starting with a colony of 10 000 bacteria, in a container which will hold 10 million bacteria. When will the container be filled?

16. Warren is an investor. He puts $A into a fund. Its value, V, T years later is expected to follow the equation $V = Ar^T$. The value of the fund on four occasions is given in the table. Find estimates of A and r and hence estimate the average annual return Warren gets on his investment in the fund.

T	2	4	6	9
V	3572	4000	4440	5106

17. The population P of a colony of bacteria is thought to obey an exponential growth law in the form $P = kr^t$ where P is thousands of bacteria and t is the number of hours since the experiment started. Plot an appropriate graph and use it to estimate the values of k and r.

t	1.2	2.5	4.2	6.2
P	12.3	17.2	26.9	45.9

18. The variables x and y satisfy the relation $5^y = 6^{2x-5}$. By taking logarithms show that the graph of y against x is a straight line, and find the exact values of the gradient and the intercept.

19. Two variable quantities are related by the equation $y = Ak^x$ where A and k are constants. The graph shows the results of plotting $\ln y$ against x for four pairs of values of x and y. Use the diagram to estimate the values of A and k.

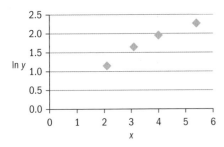

20. Use logarithms to solve the equation $3^{x+2} = 11^{x-1}$ giving the answer correct to 3 significant figures.

21. Use logarithms to solve the equation $e^{2x} = 5^{x-3}$, giving your answer correct to 3 decimal places.

22. Use logarithms to solve the equation

$$e^{x-1} = 5^{x+3}$$

giving the answer correct to 3 significant figures.

Chapter summary

Exponential growth/decay and logarithmic functions

- A function $f(t) = a \times r^t$, where t is real, describes exponential growth when $r > 1$, and describes exponential decay when $r < 1$.
- $y = b^x \Leftrightarrow x = \log_b y$, where $x \in \mathbb{R}$, $y \in \mathbb{R}$, $y > 0$.
- The logarithmic function is the inverse of the exponential function to the same base.

Properties of logarithms

$\log(xy) = \log x + \log y$

$\log\left(\dfrac{x}{y}\right) = \log x - \log y$

$\log(x^n) = n \log x$

Special bases

In practice, two bases are very commonly used for logarithms and both are available on your calculator.

- The button marked $\boxed{\log}$ uses base 10 since our number system is based on powers of 10.
- The $\boxed{\ln}$ button uses base e ($= 2.71828 \ldots$). These are sometimes called natural logarithms because of some special properties the number represented by 'e' has when used in an exponential function.

Equations and inequalities using logarithms

- You can use the laws of logarithms to transform equations to make their solutions accessible.
- Power (and logarithmic) functions are strictly monotonic, so if the logarithms are equal it follows that the expressions are equal. This also allows inequalities to be solved by solving for the critical value at which equality holds.

Using logarithms to reduce equations to linear form

- Exponential growth/decay: $y = ab^t \Rightarrow \log y = \log a + t \log b$. The graph of '$\log y$' against '$t$' has intercept '$\log a$' and gradient '$\log b$'.
- Power law: $y = ax^n \Rightarrow \log y = \log a + n \log x$. The graph of '$\log y$' against '$\log x$' has intercept '$\log a$' and gradient '$n$'.

3 Trigonometry

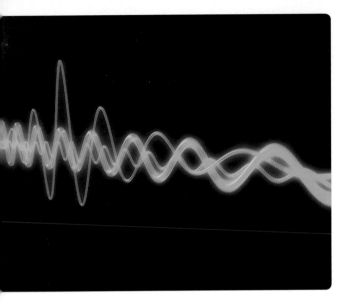

Trigonometry plays a major role in musical theory and production. A musical note can be represented by a sine curve, and a chord can be represented by multiple sine curves. A graphical representation of music allows computers to create and understand sounds. It also allows sound engineers to visualise sound waves so that they can adjust volume, pitch and other elements to create the desired sound. Trigonometry also plays an important role in speaker placement, since the angles of sound waves hitting our ears can influence the quality of the sound.

Objectives

- Understand the relationship of the secant, cosecant, and cotangent functions to cosine, sine, and tangent, and use properties and graphs of all six trigonometric functions for angles of any magnitude.
- Use trigonometric identities for the simplification and exact evaluation of expressions in the course of solving equations, and select an identity or identities appropriate to the context, showing familiarity in particular with the use of
  ○ $\sec^2\theta \equiv 1 + \tan^2\theta$ and $\csc^2\theta \equiv 1 + \cot^2\theta$
  ○ the expansions of $\sin(A \pm B)$, $\cos(A \pm B)$, and $\tan(A \pm B)$
  ○ the formulae for $\sin 2A$, $\cos 2A$, and $\tan 2A$
  ○ the expressions of $a\sin\theta + b\cos\theta$ in the forms $R\sin(\theta \pm \alpha)$ and $R\cos(\theta \pm \alpha)$.

Before you start

You should know how to:

1. Use exact trigonometric ratios,

   e.g. $\sin 30° = \dfrac{1}{2}$ $\qquad \cos 45° = \dfrac{1}{\sqrt{2}}\left(=\dfrac{\sqrt{2}}{2}\right)$

   $\tan 60° = \sqrt{3}$ $\qquad \sin\dfrac{\pi}{4} = \dfrac{1}{\sqrt{2}}\left(=\dfrac{\sqrt{2}}{2}\right)$

   $\cos\dfrac{\pi}{3} = \dfrac{1}{2}$ $\qquad \tan\dfrac{3\pi}{4} = -1$

Skills check:

1. State the exact values of
   a) $\sin 0°$
   b) $\cos 30°$
   c) $\tan 150°$
   d) $\sin\dfrac{2\pi}{3}$
   e) $\cos\dfrac{\pi}{6}$
   f) $\tan\dfrac{7\pi}{4}$.

2. Use inverse trigonometric functions, **e.g.** write down the principal values of

$$\tan^{-1}\left(\frac{1}{\sqrt{3}}\right), \cos^{-1} 0 \text{ and } \sin^{-1}\left(-\frac{\sqrt{3}}{2}\right).$$

$$\tan^{-1}\left(\frac{1}{\sqrt{3}}\right) = \frac{\pi}{6} (= 30°)$$

$$\cos^{-1} 0 = \frac{\pi}{2} (= 90°)$$

$$\sin^{-1}\left(-\frac{\sqrt{3}}{2}\right) = -\frac{\pi}{3} (= -60°)$$

3. Solve trigonometric equations using the identities $\tan\theta \equiv \dfrac{\sin\theta}{\cos\theta}$ and $\sin^2\theta + \cos^2\theta \equiv 1$.

e.g. Solve $\sqrt{2}\sin x - \cos x = 0$

for $0 \le x \le 360°$.

$$\sqrt{2}\sin x = \cos x$$

$$\tan x = \frac{1}{\sqrt{2}}$$

$$x = 35.3°, 215.3°$$

2. **a)** State the principal value, in radians, of

   i) $\sin^{-1}\left(\frac{1}{2}\right)$ **ii)** $\tan^{-1}(-1)$.

   b) State the principal value, in degrees, to 1 decimal place, of

   i) $\cos^{-1}(0.85)$ **ii)** $\sin^{-1}(-0.2)$.

3. Solve

   a) $4\sin\theta = 2\cos\theta$ for $0° \le \theta < 360°$

   b) $2(\sin^2\theta - \cos^2\theta) = 1$ for $0 \le \theta \le 2\pi$.

3.1 Secant, cosecant, and cotangent

We are familiar with the trigonometric functions sine, cosine, and tangent. In P1, we studied these functions together with their graphs, their inverse functions, and the trigonometric identities $\tan\theta \equiv \dfrac{\sin\theta}{\cos\theta}$ and $\sin^2\theta + \cos^2\theta \equiv 1$. We also studied how to use these identities to solve trigonometric equations.

We now introduce three other trigonometric functions: secant, cosecant and cotangent. These functions, usually abbreviated to sec, cosec, and cot, are closely related to sin, cos, and tan. They are defined as follows:

$$\sec\theta = \frac{1}{\cos\theta} \qquad \csc\theta = \frac{1}{\sin\theta} \qquad \cot\theta = \frac{1}{\tan\theta} = \left(\frac{\cos\theta}{\sin\theta}\right)$$

These functions are the reciprocals of cos θ, sin θ, and tan θ. Remember, for example, that $\csc\theta = \dfrac{1}{\sin\theta}$. Be careful not to confuse these functions with the inverse functions $\sin^{-1}\theta$, $\cos^{-1}\theta$, and $\tan^{-1}\theta$.

In order to sketch the graphs of sec θ, cosec θ, and cot θ we can consider the reciprocals of the functions cos θ, sin θ, and tan θ.

For example, to sketch the graph of $y = \sec \theta$, we consider the graph of $y = \cos \theta$.

As $y = \cos \theta$ is zero at $x = 90°, 270°, 450°, ...,$ we know that sec θ will not be defined at each of these points, and instead there will be vertical asymptotes on the graph of $y = \sec \theta$ at these points.

The graph can then be sketched by considering the value of $\dfrac{1}{\cos \theta}$ at different points.

We see that, just as the graph of $y = \cos \theta$ is periodic with period 360°, so is the graph of $y = \sec \theta$.

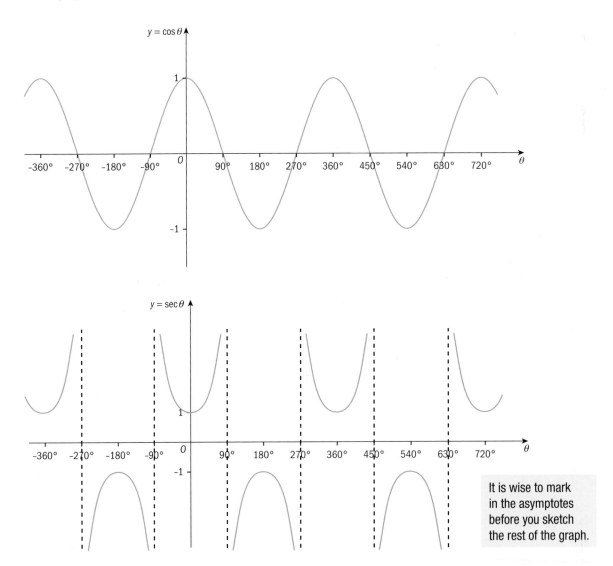

It is wise to mark in the asymptotes before you sketch the rest of the graph.

The graph of $y = \mathrm{cosec}\,\theta$ can be obtained from the graph of $y = \sin\theta$ in a similar way.

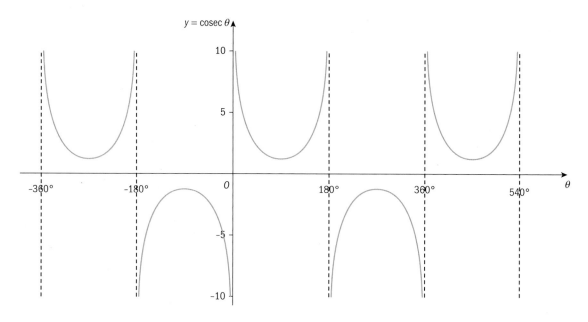

The graph of $y = \cot\theta$ can be obtained from the graph of $y = \tan\theta$.

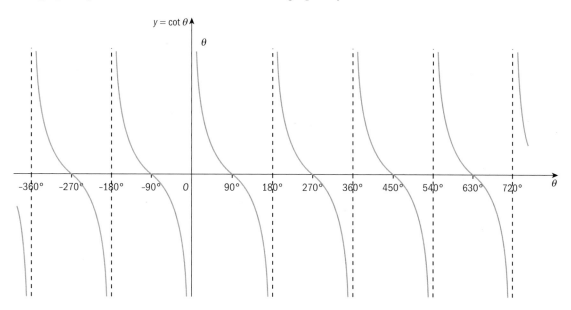

Example 1

Without using a calculator, write down the values of

a) $\sec 0°$ b) $\sec 60°$ c) $\operatorname{cosec} 120°$ d) $\cot 270°$ e) $\operatorname{cosec}\left(-\dfrac{\pi}{4}\right)$ f) $\cot\left(\dfrac{5\pi}{4}\right)$.

a) $\sec 0° = \dfrac{1}{\cos 0°} = 1$

Use $\cos 0° = 1$.

b) $\sec 60° = \dfrac{1}{\cos 60°} = 2$

Use $\cos 60° = \dfrac{1}{2}$.

c) $\operatorname{cosec} 240° = \dfrac{1}{\sin 240°} = -\dfrac{2}{\sqrt{3}}\left(\text{or } -\dfrac{2\sqrt{3}}{3}\right)$

Use $\sin 240° = -\dfrac{\sqrt{3}}{2}$ and take the reciprocal.

d) $\cot 270° = \dfrac{1}{\tan 270°} = \dfrac{\cos 270°}{\sin 270°} = \dfrac{0}{-1} = 0$

Note that $\tan 270°$ is infinitely large so $\cot 270° = 0$.

e) $\operatorname{cosec}\left(-\dfrac{\pi}{4}\right) = \dfrac{1}{\sin\left(-\dfrac{\pi}{4}\right)} = -\dfrac{2}{\sqrt{2}} \text{ (or } -\sqrt{2})$

$\sin\left(-\dfrac{\pi}{4}\right) = -\sin\left(\dfrac{\pi}{4}\right) = -\dfrac{\sqrt{2}}{2}$

f) $\cot\left(\dfrac{5\pi}{4}\right) = \dfrac{\cos\left(\dfrac{5\pi}{4}\right)}{\sin\left(\dfrac{5\pi}{4}\right)} = \left(-\dfrac{\sqrt{2}}{2}\right) \div \left(-\dfrac{\sqrt{2}}{2}\right) = 1$

You can use the reciprocal of $\tan\left(\dfrac{5\pi}{4}\right)$ or use $\cot\theta = \dfrac{\cos\theta}{\sin\theta}$.

We can use the definitions of $\sec\theta$, $\operatorname{cosec}\theta$, and $\cot\theta$ to help us solve some simple trigonometric equations.

Example 2

Solve for $0° \le \theta \le 360°$.

a) $\sec\theta = 2$ b) $\cot^2\theta = 3$ c) $11 + 3\operatorname{cosec} 2\theta = 1$

a) $\sec\theta = \dfrac{1}{\cos\theta} = 2$

Replace $\sec\theta$ with $\dfrac{1}{\cos\theta}$.

$\cos\theta = \dfrac{1}{2}$

Rearrange to obtain the value of $\cos\theta$.

$\theta = 60°, 300°$

Solve the equation for $0° \le \theta \le 360°$.

▶ Continued on the next page

b) $\cot\theta = \pm\sqrt{3}$ ←——— Write down the two values, $\pm\sqrt{3}$.

$\tan\theta = \pm\dfrac{1}{\sqrt{3}}\left(\text{or} \pm\dfrac{\sqrt{3}}{3}\right)$ ←——— Use $\cot\theta = \dfrac{1}{\tan\theta}$ to write down the values of tan θ.

$\theta = 30°, 210°$

or $\theta = 150°, 330°$ ←——— Solve each of the equations $\tan\theta = \dfrac{1}{\sqrt{3}}$ and $\tan\theta = -\dfrac{1}{\sqrt{3}}$.

$\theta = 30°, 150°, 210°, 330°$

c) $3\operatorname{cosec}2\theta = -10$ ←——— Rearrange to obtain the value of cosec 2θ.

$\operatorname{cosec}2\theta = -\dfrac{10}{3}$

$\sin 2\theta = -0.3$ ←——— Use $\operatorname{cosec}2\theta = \dfrac{1}{\sin 2\theta}$ find the value of sin 2θ.

$2\theta = 180° + 17.45°, 360° - 17.45°,$ ←——— Find all values of 2θ in the range $0°$ to $720°$.
$\quad\quad 540° + 17.45°, 720° - 17.45°$

$\theta = 98.8°, 171.3°, 278.8°, 351.3°$ ←——— Give each solution for θ correct to 1 decimal place.

We can also sketch the graphs of composite functions involving sec θ, cosec θ, and cot θ in the same way that we sketched the graphs of composite functions involving sin θ, cos θ, and tan θ in P1. You can practise this by answering questions **13** and **14** in Exercise 3.1.

Exercise 3.1

1. Find the exact values of
   a) $\sec 30°$
   b) $\operatorname{cosec}30°$
   c) $\cot 60°$
   d) $\sec 45°$
   e) $\sec 210°$
   f) $\operatorname{cosec}135°$
   g) $\cot 300°$
   h) $\sec 150°$.

2. Find the exact values of
   a) $\sec\dfrac{3\pi}{4}$
   b) $\sec\dfrac{\pi}{3}$
   c) $\cot\dfrac{\pi}{6}$
   d) $\operatorname{cosec}\dfrac{\pi}{2}$
   e) $\operatorname{cosec}\dfrac{4\pi}{3}$
   f) $\operatorname{cosec}\dfrac{3\pi}{2}$
   g) $\sec\dfrac{5\pi}{6}$
   h) $\cot\dfrac{7\pi}{4}$.

3. State whether each of the following values are defined or undefined.
   a) $\operatorname{cosec}90°$
   b) $\operatorname{cosec}130°$
   c) $\sec 180°$
   d) $\cot(-180°)$
   e) $\sec\dfrac{3\pi}{2}$
   f) $\cot\dfrac{7\pi}{6}$
   g) $\operatorname{cosec}(-\pi)$
   h) $\cot\dfrac{5\pi}{2}$

Trigonometry 45

4. Use your calculator to find these values correct to 3 significant figures.

a) $\cot 32°$

b) $\operatorname{cosec} 70°$

c) $\sec 215°$

d) $\sec(-22°)$

e) $\operatorname{cosec} \dfrac{\pi}{5}$

f) $\cot \dfrac{7\pi}{10}$

g) $\sec(0.6\pi)$

h) $\sec \dfrac{7\pi}{5}$

5. Solve each of these equations for $0° < \theta < 360°$.

a) $\cot \theta = 1$

b) $\sec \theta = \sqrt{2}$

c) $\operatorname{cosec} \theta = \dfrac{2}{\sqrt{3}}$

d) $\operatorname{cosec} \theta = 2$

e) $\cot \theta = -\sqrt{3}$

f) $\sec \theta = -1$

6. Solve each of these equations for $0 \le \theta \le 2\pi$. Give your answers in exact form.

a) $\operatorname{cosec} \theta = 1$

b) $\sec \theta = 1$

c) $\cot \theta = -1$

d) $\sec \theta = \dfrac{2}{\sqrt{3}}$

e) $\cot \theta = 0$

f) $\operatorname{cosec} \theta = -\dfrac{2}{\sqrt{3}}$

7. Solve each of these equations for $0° \le x < 360°$. Give your answers correct to 1 decimal place.

a) $\sec x = 3$

b) $\operatorname{cosec} x = 4$

c) $\cot x = 0.9$

d) $\cot x = 5$

e) $\operatorname{cosec} x = -\dfrac{9}{7}$

f) $\sec x = -\sqrt{6}$

8. a) Solve the equation $\sec x = 7$ for $0° < x < 360°$.

b) Solve the equation $\sec^2 x = 49$ for $0° < x < 360°$.

9. Solve the equation $8 \cot x - 5 = 7$ for $0° < x < 360°$.

10. Solve the equation $7 + 2 \operatorname{cosec} \theta = 4$ for $0 < \theta < 2\pi$.

11. Solve each of these equations for $0° \le x \le 360°$.

a) $\sec 2x = 1$

b) $\operatorname{cosec} 2x = 4$

c) $6 \cot 3x + 5 = 13$

12. Solve each of these equations for $0° < x < 360°$.

a) $\operatorname{cosec}(2x + 90°) = 1$

b) $4 - 3 \cot \dfrac{1}{2}x = 3$

c) $2 \sec 3x - 5 = \sqrt{2}$

13. Sketch these graphs on separate axes.

a) $y = \operatorname{cosec} 3x$ for $0° \le x \le 360°$

b) $y = \sec 2x$ for $0° \le x \le 360°$

c) $y = \cot \dfrac{1}{2}x$ for $-360° \le x \le 360°$

d) $y = 2 \operatorname{cosec} x$ for $0° \le x \le 180°$

e) $y = \cot(x + 90°)$ for $0° \le x \le 360°$

f) $y = 3 \sec 2x$ for $0° \le x \le 360°$

14. Sketch, on separate sets of axes, the graphs of the following functions for $-2\pi \le x \le 2\pi$.

a) $y = 6 \operatorname{cosec} x$

b) $y = \cot\left(x - \dfrac{\pi}{4}\right)$

c) $y = \dfrac{1}{2} \sec\left(x + \dfrac{\pi}{2}\right)$

d) $y = \operatorname{cosec}(2x - \pi)$

3.2 Further trigonometric identities

In P1, you used the identities $\tan\theta = \dfrac{\sin\theta}{\cos\theta}$ and $\sin^2\theta + \cos^2\theta \equiv 1$. We now turn our

attention to finding and using two new identities involving $\sec\theta$, $\csc\theta$, and $\cot\theta$.

Using the identity

$$\sin^2\theta + \cos^2\theta \equiv 1$$

We divide through by $\cos^2\theta$ to give $\quad \dfrac{\sin^2\theta}{\cos^2\theta} + \dfrac{\cos^2\theta}{\cos^2\theta} \equiv \dfrac{1}{\cos^2\theta}$

$$\Rightarrow \tan^2\theta + 1 \equiv \sec^2\theta$$

Similarly, dividing $\sin^2\theta + \cos^2\theta \equiv 1$ by $\sin^2\theta$ gives

$$\dfrac{\sin^2\theta}{\sin^2\theta} + \dfrac{\cos^2\theta}{\sin^2\theta} \equiv \dfrac{1}{\sin^2\theta}$$

$$\Rightarrow 1 + \cot^2\theta \equiv \csc^2\theta$$

$$1 + \tan^2\theta \equiv \sec^2\theta$$
$$1 + \cot^2\theta \equiv \csc^2\theta$$

Remember that we should use '$\equiv$' rather than '=' in identities to show that the statement is true for all values of θ. Although, don't worry if you use the '=' sign.

These identities are useful in helping us to simplify expressions, prove identities, and solve trigonometric equations.

Example 3

Prove the identity $(\tan\theta + \cot\theta)^2 \equiv \sec^2\theta + \csc^2\theta$.

$(\tan\theta + \cot\theta)^2 \equiv \tan^2\theta + 2\tan\theta\cot\theta + \cot^2\theta$ ← Start with the left-hand side and expand the brackets.

$\equiv \tan^2\theta + 2\tan\theta\dfrac{1}{\tan\theta} + \cot^2\theta$ ← Replace $\cot\theta$ with $\dfrac{1}{\tan\theta}$.

$\equiv \tan^2\theta + 2 + \cot^2\theta$

$\equiv \sec^2\theta - 1 + 2 + \csc^2\theta - 1$ ← Use the identities to replace $\tan^2\theta$ with $\sec^2\theta - 1$ and $\cot^2\theta$ with $\csc^2\theta - 1$ and then simplify.

$\equiv \sec^2\theta + \csc^2\theta$

Example 4

Solve the equation $\sec^2 \theta + \tan \theta - 1 = 0$ for $0° \le \theta \le 360°$.

$1 + \tan^2 \theta + \tan \theta - 1 = 0$ Substitute $1 + \tan^2 \theta$ for $\sec^2 \theta$ to get a quadratic equation in $\tan \theta$.

$$\tan^2 \theta + \tan \theta = 0$$

$\tan \theta (\tan \theta + 1) = 0$ Factorise the equation and solve for $\tan \theta$.

$$\tan \theta = 0 \text{ or } \tan \theta + 1 = 0$$

$$\tan \theta = 0 \text{ or } \tan \theta = -1$$

$$\theta = 0°, 135°, 180°, 315°, 360°$$

Exercise 3.2

1. Simplify the following expressions.

   a) $\sec^2 \theta - \tan^2 \theta$

   b) $\dfrac{\sin \theta}{\cosec \theta - \cos \theta \cot \theta}$

   c) $\dfrac{\tan \theta}{\sqrt{1 + \tan^2 \theta}}$

   d) $\tan \theta \cot \theta$

   e) $\dfrac{\sin \theta}{1 + \cot^2 \theta}$

   f) $\dfrac{1 + \tan^2 \theta}{1 + \cot^2 \theta}$

2. Solve the equation $2 \sec^2 \theta + 3 \tan \theta - 4 = 0$ for $0° \le \theta \le 360°$.

3. Solve the equation $\sec \theta \tan \theta = 1$ for $0 < \theta < 2\pi$.

4. Solve the equation $\tan^2 \theta - \sec \theta - 5 = 0$ for $0° \le x \le 360°$.

5. Prove the identity $\cosec \theta + \cot \theta \equiv \dfrac{1}{\cosec \theta - \cot \theta}$.

   > **Hint:** Start with multiplying the left-hand side by $\cosec \theta - \cot \theta$ and then consider the right-hand side.

6. Prove the identity $\dfrac{\tan \theta}{1 + \sec \theta} + \dfrac{1 + \sec \theta}{\tan \theta} \equiv \dfrac{2}{\sin \theta}$.

7. Prove the identity $(\sec \theta - \tan \theta)^2 \equiv \dfrac{1 - \sin \theta}{1 + \sin \theta}$.

8. Prove the identity $\dfrac{\cot \theta}{1 + \cot^2 \theta} \equiv \sin \theta \cos \theta$.

9. Prove the identity $\sec^2 \theta + \cosec^2 \theta \equiv \sec^2 \theta \cosec^2 \theta$.

10. Solve the equation $4 \cot^2 \theta - 2 \cot \theta = 3 \cosec^2 \theta$ for $0° \le \theta \le 360°$.

11. Solve the equation $\sec \theta = 3 \cos \theta + 1$ for $0° \le \theta \le 360°$.

12. Solve the equation $\operatorname{cosec}^2 \theta + 2 \cot \theta = 0$ for $0 \le \theta \le 2\pi$.

13. Prove the identity $\tan \theta + \cot \theta \equiv \sec \theta \operatorname{cosec} \theta$.

14. Solve the equation $3 \cot \theta + 2 \tan \theta = 5$ for $0° \le \theta \le 360°$.

15. Solve the equation $\tan^2 \theta = \sec \theta$ for $0° \le \theta \le 360°$.

16. Prove the identity $\cot^2 \theta \cos^2 \theta - \sin^2 \theta = \cot^2 \theta - 1$.

A geometrical interpretation of the trigonometric functions

All of the trigonometric functions of an angle can be constructed geometrically, in terms of a unit circle centre O, as shown in the diagram.

In P1 we saw that the term 'cosine (of θ)' originates from 'the sine of the complementary angle (to θ)'.

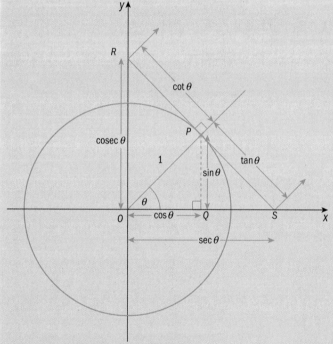

Can you show how the diagram may be used to help explain the following identities?

$\cos \theta \equiv \sin(90° - \theta)$ and $\sin \theta \equiv \cos(90° - \theta)$

$\cot \theta \equiv \tan(90° - \theta)$ and $\tan \theta \equiv \cot(90° - \theta)$

$\operatorname{cosec} \theta \equiv \sec(90° - \theta)$ and $\sec \theta \equiv \operatorname{cosec}(90° - \theta)$

$\tan \theta \equiv \dfrac{\sin \theta}{\cos \theta}$

$\sin^2 \theta + \cos^2 \theta \equiv 1$

$1 + \tan^2 \theta \equiv \sec^2 \theta$

$1 + \cot^2 \theta \equiv \operatorname{cosec}^2 \theta$

3.3 Addition formulae

In this section we will learn how to use formulae for $\sin(A \pm B)$, $\cos(A \pm B)$, and $\tan(A \pm B)$.

It is important to recognise that $\sin(A + B) \neq \sin A + \sin B$.

We can see this if we let $A = 30°$ and $B = 60°$, for example.

Then

$$\sin(A + B) = \sin(30° + 60°) = \sin 90° = 1$$

and

$$\sin A + \sin B = \sin 30° + \sin 60° = \frac{1}{2} + \frac{\sqrt{3}}{2} \neq 1.$$

So $\qquad \sin(A + B) \neq \sin A + \sin B$

We will now establish the formulae for $\sin(A + B)$ and $\cos(A + B)$.

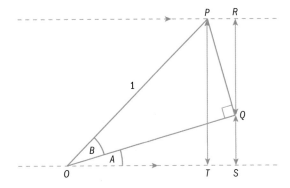

In the diagram, OPQ is a right-angled triangle with $OP = 1$ unit.

From triangle OPT,

$$PT = 1 \times \sin(A + B) = \sin(A + B) \qquad\qquad (1)$$

From triangle OPQ,

$$PQ = 1 \times \sin B = \sin B \qquad \text{and} \qquad OQ = 1 \times \cos B = \cos B$$

We know that $\qquad$ angle $OQS = 90° - A$

and $\qquad$ angle $PQR = A$

We also know that $\quad RQ = PQ \cos A = \sin B \cos A \qquad (2)$

and $\qquad\qquad\quad QS = OQ \sin A = \cos B \sin A \qquad (3)$

Now

$$PT = RQ + QS$$

So, using equations (1), (2) and (3),

$$\sin(A + B) = \sin B \cos A + \cos B \sin A$$

This is usually written as

$$\sin(A + B) \equiv \sin A \cos B + \cos A \sin B$$

Similarly, from triangle OPT,

$$OT = 1 \times \cos(A + B) = \cos(A + B)$$

but

$$OT = OS - TS = OS - PR$$

and

$$OS = \cos B \cos A$$
$$PR = \sin B \sin A$$

so

$$\cos(A + B) = \cos B \cos A - \sin A \sin B$$

or

$$\cos(A + B) \equiv \cos A \cos B - \sin A \sin B$$

Replacing B with $-B$ in $\sin(A + B) = \sin A \cos B + \cos A \sin B$, we obtain

$$\sin(A - B) = \sin A \cos(-B) + \cos A \sin(-B)$$

or

$$\sin(A - B) \equiv \sin A \cos B - \cos A \sin B$$

because

$$\cos(-B) = \cos B \quad \text{and} \quad \sin(-B) = -\sin B.$$

Similarly, replacing B with $-B$ in $\cos(A + B) = \cos A \cos B - \sin A \sin B$, we obtain

$$\cos(A - B) = \cos A \cos(-B) + \sin A \sin(-B)$$

or

$$\cos(A - B) \equiv \cos A \cos B + \sin A \sin B$$

We now turn our attention to obtaining expressions for $\tan(A + B)$ and $\tan(A - B)$ by using the identity $\tan \theta \equiv \dfrac{\sin \theta}{\cos \theta}$.

$$\tan(A + B) = \frac{\sin(A + B)}{\cos(A + B)} = \frac{\sin A \cos B + \cos A \sin B}{\cos A \cos B - \sin A \sin B}$$

Dividing the numerator and denominator by $\cos A \cos B$,

$$\tan(A + B) = \frac{\dfrac{\sin A \cos B}{\cos A \cos B} + \dfrac{\cos A \sin B}{\cos A \cos B}}{\dfrac{\cos A \cos B}{\cos A \cos B} - \dfrac{\sin A \sin B}{\cos A \cos B}}$$

$$\equiv \frac{\tan A + \tan B}{1 - \tan A \tan B}$$

Replacing B with $-B$, we obtain

$$\tan(A - B) \equiv \frac{\tan A - \tan B}{1 + \tan A \tan B}$$

$$\sin(A + B) \equiv \sin A \cos B + \cos A \sin B$$

$$\sin(A - B) \equiv \sin A \cos B - \cos A \sin B$$

$$\cos(A + B) \equiv \cos A \cos B - \sin A \sin B$$

$$\cos(A - B) \equiv \cos A \cos B + \sin A \sin B$$

$$\tan(A + B) \equiv \frac{\tan A + \tan B}{1 - \tan A \tan B}$$

$$\tan(A - B) \equiv \frac{\tan A - \tan B}{1 + \tan A \tan B}$$

These formulae complete our set of six **addition formulae** (or compound angle formulae).

Example 5

By considering $\sin(45° + 30°)$, prove that $\sin 75° = \dfrac{1 + \sqrt{3}}{2\sqrt{2}}$.

$\sin(A + B) = \sin A \cos B + \cos A \sin B$

$\sin(45° + 30°) = \sin 45° \cos 30° + \cos 45° \sin 30°$

Put $A = 45°$ and $B = 30°$ into the addition formula for $\sin(A + B)$.

$$= \frac{1}{\sqrt{2}} \times \frac{\sqrt{3}}{2} + \frac{1}{\sqrt{2}} \times \frac{1}{2}$$

Substitute exact values for $\sin 30°$, $\sin 45°$, $\cos 30°$ and $\cos 45°$.

$$\sin 75° = \frac{1 + \sqrt{3}}{2\sqrt{2}}$$

Example 6

Prove that $2 \cos\left(\theta - \dfrac{\pi}{3}\right) \equiv \cos\theta + \sqrt{3} \sin\theta$.

$$2 \cos\left(\theta - \frac{\pi}{3}\right) \equiv 2\left(\cos\theta\cos\frac{\pi}{3} + \sin\theta\sin\frac{\pi}{3}\right)$$

Use the formula
$\cos(A - B) = \cos A \cos B + \sin A \sin B$.

$$\equiv 2\left(\cos\theta \times \frac{1}{2} + \sin\theta \times \frac{\sqrt{3}}{2}\right)$$

Substitute $\cos\dfrac{\pi}{3} = \dfrac{1}{2}$ and $\sin\dfrac{\pi}{3} = -\dfrac{\sqrt{3}}{2}$.

$$\equiv \cos\theta + \sqrt{3} \sin\theta$$

Example 7

Solve the equation $\cos(\theta + 60°) = 2\sin(\theta - 45°)$ for $0° \le \theta \le 360°$.

$\cos\theta\cos 60° - \sin\theta\sin 60° = 2(\sin\theta\cos 45° - \cos\theta\sin 45°)$ ← Expand both expressions.

$\dfrac{1}{2}\cos\theta - \dfrac{\sqrt{3}}{2}\sin\theta = 2\left(\dfrac{1}{\sqrt{2}}\sin\theta - \dfrac{1}{\sqrt{2}}\cos\theta\right)$ ← Substitute values for $\sin 45°$, $\cos 45°$, $\sin 60°$ and $\cos 60°$.

$\left(\dfrac{1}{2} + \dfrac{2}{\sqrt{2}}\right)\cos\theta = \left(\dfrac{2}{\sqrt{2}} + \dfrac{\sqrt{3}}{2}\right)\sin\theta$ ← Collect terms in $\sin\theta$ and $\cos\theta$.

$\dfrac{\sqrt{2}+4}{2\sqrt{2}}\cos\theta = \dfrac{4 + \sqrt{3}\sqrt{2}}{2\sqrt{2}}\sin\theta$

$\dfrac{\sqrt{2}+4}{4+\sqrt{3}\sqrt{2}} = \tan\theta$ ← Rearrange to find $\tan\theta$.

$\tan\theta = 0.8394\ldots$ ← Find all the angles.

$\theta = 40.0°, 220.0°$

We have worked in exact terms until the final step, but you may wish to use a calculator at an earlier stage in the solution.

Exercise 3.3

1. Use the addition formulae to find, in exact form, the following expressions.

   a) $\cos 75°$ b) $\cos 15°$ c) $\tan 105°$ d) $\dfrac{1 + \sin 15°}{1 - \sin 15°}$

2. a) Prove that $\tan(\theta + 45°) \equiv \dfrac{1 + \tan\theta}{1 - \tan\theta}$.

   b) Show that $\tan 75° \equiv \dfrac{\sqrt{3} + 1}{\sqrt{3} - 1}$.

3. Given that $\sin\alpha = \dfrac{4}{5}$ and $\sin\beta = \dfrac{5}{13}$, where α and β are acute angles, without using calculators find the exact values of

   a) $\cos\alpha$ and $\cos\beta$

   b) $\sin(\alpha + \beta)$, $\sin(\alpha - \beta)$, $\cos(\alpha + \beta)$, and $\cos(\alpha - \beta)$

   c) $\tan(\alpha + \beta)$ and $\tan(\alpha - \beta)$.

4. Evaluate, in exact form,

   a) $\sin\dfrac{5\pi}{12}$ b) $\cos\dfrac{7\pi}{12}$ c) $\tan\dfrac{\pi}{12}$ d) $\sin\left(-\dfrac{\pi}{12}\right)$.

5. Evaluate, in exact form,

   a) $\cos 10° \cos 20° - \sin 10° \sin 20°$

   b) $\sin 75° \cos 45° + \cos 75° \sin 45°$

   c) $\dfrac{\tan 103° - \tan 58°}{1 + \tan 103° \tan 58°}$

   d) $\dfrac{\tan 75° + 1}{\tan 75° - 1}$.

6. Simplify

   a) $\sin(\alpha + \beta) + \sin(\alpha - \beta)$

   b) $\cos(\alpha + \beta) + \cos(\alpha - \beta)$

   c) $\sin(\alpha - \beta)\cos\alpha - \cos(\alpha - \beta)\sin\alpha$.

7. Use the addition formulae to show that

   a) $\sin\left(\theta + \dfrac{\pi}{2}\right) \equiv \cos\theta$ b) $\sin(90° - \theta) \equiv \cos\theta$

   c) $\cos(90° - \theta) \equiv \sin\theta$ d) $\cos(180° + \theta) \equiv -\cos\theta$

   e) $\tan(\theta + 180°) \equiv \tan\theta$ f) $\tan(\pi - \theta) \equiv -\tan\theta$.

8. Prove the identity $\operatorname{cosec}(\theta + \varphi) \equiv \dfrac{\operatorname{cosec}\theta \ \operatorname{cosec}\varphi}{\cot\theta + \cot\varphi}$.

9. Express $\cot(\theta + \varphi)$ in terms of $\cot\theta$ and $\cot\varphi$.

10. Prove that $\tan(\alpha + \beta + \gamma) = \dfrac{\tan\alpha + \tan\beta + \tan\gamma - \tan\alpha\tan\beta\tan\gamma}{1 - \tan\alpha\tan\beta - \tan\beta\tan\gamma - \tan\alpha\tan\gamma}$.

    Hint: Consider $\tan\{\alpha + (\beta + \gamma)\}$.

11. In the diagram, find the exact values of

    a) the sine of angle PSR

    b) the tangent of angle PSR

    c) the secant of angle PQR.

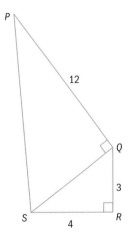

12. Solve the equation $\tan\theta = 2\tan(45° - \theta)$ for $0° \leq \theta \leq 360°$.

13. Solve the equation $\cos\left(\theta - \dfrac{\pi}{6}\right) = \cos\left(\theta + \dfrac{\pi}{6}\right)$ for $0 \leq \theta \leq 2\pi$.

14. Solve the equation $5\cos(\theta + 45°) = \sin(\theta - 45°)$ for $0° \leq \theta \leq 360°$.

3.4 Double angle formulae

We can substitute $B = A$ into the addition formulae (section 3.3) to obtain three new identities.

$$\sin(A + B) \equiv \sin A \cos B + \cos A \sin B$$
$$\Rightarrow \sin(A + A) \equiv \sin A \cos A + \cos A \sin A$$
$$\Rightarrow \sin 2A \equiv 2 \sin A \cos A$$

$$\cos(A + B) \equiv \cos A \cos B - \sin A \sin B$$
$$\Rightarrow \cos(A + A) \equiv \cos A \cos A - \sin A \sin A$$
$$\Rightarrow \cos 2A \equiv \cos^2 A - \sin^2 A$$

$$\tan(A + B) \equiv \frac{\tan A + \tan B}{1 - \tan A \tan B}$$

$$\Rightarrow \tan(A + A) \equiv \frac{\tan A + \tan A}{1 - \tan A \tan A}$$

$$\Rightarrow \tan 2A \equiv \frac{2 \tan A}{1 - \tan^2 A}$$

Further, using the identity $\cos^2 A + \sin^2 A \equiv 1$ to rewrite the identity for $\cos 2A$ in terms of $\sin A$ or $\cos A$ only, we obtain

$$\cos 2A \equiv \cos^2 A - (1 - \cos^2 A) = 2 \cos^2 A - 1$$

and $\cos 2A \equiv (1 - \sin^2 A) - \sin^2 A = 1 - 2 \sin^2 A$

$\sin 2A \equiv 2 \sin A \cos A$	$\cos 2A \equiv \cos^2 A - \sin^2 A$ $\equiv 2 \cos^2 A - 1$ $\equiv 1 - 2 \sin^2 A$	$\tan 2A \equiv \dfrac{2 \tan A}{1 - \tan^2 A}$

These results are called the **double angle formulae**.

It is important to know all three versions of the formula for $\cos 2A$.

Example 8

Solve the following equations for $0° \leq \theta \leq 360°$.

a) $4 \sin 2\theta = \sin \theta$ b) $\cos 2\theta = \cos \theta$

▶ Continued on the next page

a) $4 \times 2 \sin \theta \cos \theta = \sin \theta$ → Write the equation in terms of $\sin \theta$ and $\cos \theta$ only.

$8 \sin \theta \cos \theta - \sin \theta = 0$ → Write in the form $f(\theta) = 0$.

$\sin \theta (8 \cos \theta - 1) = 0$ → Factorise the LHS and solve the equation.

$\sin \theta = 0$ or $\cos \theta = \dfrac{1}{8}$

$\theta = 0°, 180°, 360°$ or $\theta = 82.8°, 360° - 82.8°$

$\theta = 0°, 82.8°, 180°, 277.2°, 360°$

b) $2 \cos^2 \theta - 1 = \cos \theta$ → Write the equation in terms of $\cos \theta$ only.

$2 \cos^2 \theta - \cos \theta - 1 = 0$

$(2 \cos \theta + 1)(\cos \theta - 1) = 0$ → Factorise and solve the resulting quadratic equation in $\cos \theta$.

$\cos \theta = -\dfrac{1}{2}$ or $\cos \theta = 1$

$\theta = 120°, 240°$ or $\theta = 0°, 360°$

$\theta = 0°, 120°, 240°, 360°$

Example 9

Express $\cos 3\theta$ in terms of $\cos \theta$.

$\cos 3\theta \equiv \cos(2\theta + \theta)$ → Express $\cos 3\theta$ in a way that the addition formulae may be used.

$\equiv \cos 2\theta \cos \theta - \sin 2\theta \sin \theta$ → Use the double angle formulae.

$\equiv (2 \cos^2 \theta - 1) \times \cos \theta - 2 \sin \theta \cos \theta \times \sin \theta$

$\equiv 2 \cos^3 \theta - \cos \theta - 2 \sin^2 \theta \cos \theta$

$\equiv 2 \cos^3 \theta - \cos \theta - 2(1 - \cos^2 \theta)\cos \theta$ → Ensure that the expression is expressed in terms of $\cos \theta$ only.

$\equiv 4 \cos^3 \theta - 3 \cos \theta$

Exercise 3.4

1. Solve these equations for $0° \le \theta \le 360°$.

   a) $\sin 2\theta = \cos \theta$

   b) $\sin 2\theta - \sqrt{3} \cos \theta = 0$

   c) $\cos 2\theta + 3 \cos \theta - 1 = 0$

   d) $4 \cos 2\theta + 2 \sin \theta - 1 = 0$

   e) $\tan 2\theta + \tan \theta = 0$

   f) $\sin 2\theta = \tan \theta$

2. Solve these equations for $0 \le \theta \le 2\pi$.

   a) $2 \tan 2\theta = 5 \tan \theta$

   b) $5 \sin 2\theta = 2 \sin \theta$

   c) $2 \cos 2\theta = 1 - 4 \cos \theta$

3. An acute angle θ is such that $\sin \theta = \dfrac{3}{5}$. Find the exact value of

   a) $\cos 2\theta$

   b) $\tan 2\theta$

   c) $\tan 4\theta$.

4. For the triangle ABC shown in the diagram, calculate the exact values of

   a) $\sin 2A$

   b) $\tan 2A$

   c) $\sec 2A$.

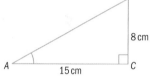

5. If $\tan \theta = 3$, where $180° \le \theta \le 270°$, find the exact values of

   a) $\sin \theta$

   b) $\cos \theta$

   c) $\tan 2\theta$

   d) $\cos 2\theta$.

6. a) Express $(\sin x - \cos x)^2$ in terms of $\sin 2x$.

   b) Express $(\cos^4 x - \sin^4 x)(\cos^2 x - \sin^2 x)$ in terms of $\cos 2x$.

7. Express $\sin 4\theta \sin \theta$ in terms of $\cos \theta$.

8. a) i) By considering $\sin(2\theta + \theta)$, express $\sin 3\theta$ in terms of $\sin \theta$.

      ii) Solve the equation $\sin 3\theta = \sin \theta$ for $0° \le \theta \le 360°$.

   b) Express $\tan 3\theta$ in terms of $\tan \theta$.

9. Express $\cot 2\theta$ in terms of $\cot \theta$.

10. Prove that   a) $\dfrac{\sin 2\theta}{1 + \cos 2\theta} = \tan \theta$ b) $\dfrac{\sin 2\theta}{1 - \cos 2\theta} = \cot \theta$.

11. Prove that $\dfrac{\sin \alpha}{\cos \beta} + \dfrac{\cos \alpha}{\sin \beta} = \dfrac{2 \cos(\alpha - \beta)}{\sin 2\beta}$.

12. Prove that $\sec 2\theta - \tan 2\theta = \dfrac{\cos \theta - \sin \theta}{\cos \theta + \sin \theta}$.

3.5 Expressing $a \sin\theta + b \cos\theta$ in the form $R\sin(\theta \pm \alpha)$ or $R\cos(\theta \pm \alpha)$

We saw in P1 how to solve equations such as $2\sin(\theta + 30°) = 1$.
In this section, we will learn how to write expressions of the form
$a \sin\theta + b \cos\theta$ as a single sine or cosine and hence be able to solve
equations such as $3\sin\theta + 4\cos\theta = 2$.

If we draw the graph of $y = 3\sin\theta + 4\cos\theta$, we obtain a single wave
(see the graph on the right). This wave is the same as the graph we obtain
if we draw $y = 5\sin(\theta + 53.1°)$, so the expressions $3\sin\theta + 4\cos\theta$ and
$5\sin(\theta + 53.1°)$ are equivalent.

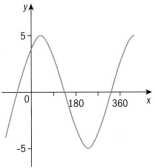

In order to write $a \sin\theta + b\cos\theta$ $(a, b > 0)$ in the form $R\sin(\theta + \alpha)$,
let $a \sin\theta + b\cos\theta \equiv R\sin(\theta + \alpha)$ where R and α are constants to be determined.

Using addition formulae,

$$a \sin\theta + b\cos\theta \equiv R(\sin\theta\cos\alpha + \cos\theta\sin\alpha)$$
$$\equiv R\sin\theta\cos\alpha + R\cos\theta\sin\alpha$$

We require the identity to be true for all values of θ, so we must equate the
coefficients of $\sin\theta$ and $\cos\theta$.

$$R\cos\alpha = a$$
$$R\sin\alpha = b$$

Squaring and adding these equations gives

$$R^2\cos^2\alpha + R^2\sin^2\alpha = a^2 + b^2$$
$$R^2(\cos^2\alpha + \sin^2\alpha) = a^2 + b^2$$
$$R^2 = a^2 + b^2$$
$$R = \sqrt{a^2 + b^2}, \text{ taking } R > 0$$

> **Remember:**
> $\cos^2\alpha + \sin^2\alpha = 1$

Also,

$$\frac{R\sin\alpha}{R\cos\alpha} = \frac{b}{a}$$
$$\tan\alpha = \frac{b}{a}$$
$$\alpha = \tan^{-1}\frac{b}{a}$$

> **Remember:**
> $\dfrac{\sin\alpha}{\cos\alpha} \equiv \tan\alpha$

$$a \sin\theta + b\cos\theta \equiv R\sin(\theta + \alpha), \quad \text{where } R = \sqrt{a^2 + b^2}$$
$$\text{and } \alpha = \tan^{-1}\frac{b}{a}$$

Examples 10 and 11 will demonstrate how we can use this technique and how we
can extend it to writing expressions in one of the other three possible forms:
$R\sin(\theta - \alpha)$, $R\cos(\theta + \alpha)$, or $R\cos(\theta - \alpha)$.

$a \sin \theta + b \cos \theta \equiv R \cos(\theta - \alpha)$ where $R = \sqrt{a^2 + b^2}$

   and $\alpha = \tan^{-1} \dfrac{a}{b}$

$a \sin \theta - b \cos \theta \equiv R \sin(\theta - \alpha)$ where $R = \sqrt{a^2 + b^2}$

   and $\alpha = \tan^{-1} \dfrac{b}{a}$

$a \cos \theta - b \sin \theta \equiv R \cos(\theta + \alpha)$ where $R = \sqrt{a^2 + b^2}$

   and $\alpha = \tan^{-1} \dfrac{b}{a}$

Note: It is important that you learn the process of how to find the values of R and α as outlined in Examples 10 and 11, rather than relying on learning these formulae.

Example 10

a) Express $2 \sin \theta + \cos \theta$ in the form $R \sin(\theta + \alpha)$ where $R > 0$ and $0° \le \alpha \le 90°$, giving the exact value of R and the value of α correct to 2 decimal places.

b) Hence solve the equation $2 \sin \theta + \cos \theta = 2$ for $0° \le \theta \le 360°$.

a) $2 \sin \theta + \cos \theta \equiv R \sin(\theta + \alpha)$

   $\equiv R \sin \theta \cos \alpha + R \cos \theta \sin \alpha$ ⟵ Expand the expression.

   $R \cos \alpha = 2$ ⟵ Compare the coefficients of $\sin \theta$ and $\cos \theta$.
   $R \sin \alpha = 1$

   $R^2 \cos^2 \alpha + R^2 \sin^2 \alpha = 2^2 + 1^2$ ⟵ Square and add to find the exact value of R.
   $R^2 (\cos^2 \alpha + \sin^2 \alpha) = 5$

   $R = \sqrt{5}$

Also,

   $\dfrac{R \sin \alpha}{R \cos \alpha} = \dfrac{1}{2}$ ⟵ Divide to find $\tan \alpha$ and then find α to the required accuracy.

   $\tan \alpha = \dfrac{1}{2}$

   $\alpha = \tan^{-1} \dfrac{1}{2} = 26.57°$

Thus, $2 \sin \theta + \cos \theta = \sqrt{5} \sin(\theta + 26.57°)$

b) $2 \sin \theta + \cos \theta = 2$

   $\sqrt{5} \sin(\theta + 26.57°) = 2$ ⟵ Substitute $\sqrt{5} \sin(\theta + 26.57°)$ for $2 \sin \theta + \cos \theta$.

   $\sin(\theta + 26.57°) = \dfrac{2}{\sqrt{5}}$

   $\theta + 26.57° = \sin^{-1} \dfrac{2}{\sqrt{5}}$ ⟵ Divide both sides by $\sqrt{5}$.

   $\theta + 26.57° = 63.43°, \; 180° - 63.43°, \; 360° + 63.43°$ ⟵ Find all the values for $\theta + 26.57°$.

   $\theta = 36.9°, \; 90°, \; 396.9°$

   $\theta = 36.9°, \; 90°$ for $0° \le \theta \le 360°$ ⟵ Subtract 26.57° from each value and check you only include values in the required interval.

Example 11

a) Express $3\cos\theta + \sqrt{3}\sin\theta$ in the form $R\cos(\theta - \alpha)$ where $R > 0$ and $0° \leq \alpha \leq 90°$, stating the exact values of R and α.

b) Determine the greatest and least possible values of $[(3\cos\theta + \sqrt{3}\sin\theta)^2 - 5]$ as θ varies.

a) $3\cos\theta + \sqrt{3}\sin\theta = R\cos(\theta - \alpha)$

$\phantom{3\cos\theta + \sqrt{3}\sin\theta} \equiv R\cos\theta\cos\alpha + R\sin\theta\sin\alpha$ ←   Expand the expression.

$R\cos\alpha = 3$

$R\sin\alpha = \sqrt{3}$ ←   Equate coefficients of $\sin\theta$ and $\cos\theta$.

$R^2\cos^2\alpha + R^2\sin^2\alpha = 3^2 + (\sqrt{3})^2$

$R = \sqrt{12}\ (\text{or } 2\sqrt{3})$

Also,

$\tan\alpha = \dfrac{\sqrt{3}}{3}$

$\alpha = 30°$

$3\cos\theta + \sqrt{3}\sin\theta = \sqrt{12}\cos(\theta - 30°)$ ←   Write out the alternative form of the expression.

b) $3\cos\theta + \sqrt{3}\sin\theta = \sqrt{12}\cos(\theta - 30°)$ ←   Use the expression found in part (a).

but $-1 \leq \cos(\theta - 30°) \leq 1$

so $-\sqrt{12} \leq \sqrt{12}\cos(\theta - 30°) \leq \sqrt{12}$ ←   Use the fact that the cosine of an angle of any size is between -1 and 1.

$-\sqrt{12} \leq (3\cos\theta + \sqrt{3}\sin\theta) \leq \sqrt{12}$

$0 \leq (3\cos\theta + \sqrt{3}\sin\theta)^2 \leq 12$ ←   Remember that the square of an expression cannot be negative.

$-5 \leq (3\cos\theta + \sqrt{3}\sin\theta)^2 - 5 \leq 7$

Greatest value of $[(3\cos\theta + 3\sin\theta)^2 - 5]$ is 7

Least value of $[(3\cos\theta + 3\sin\theta)^2 - 5]$ is -5

Exercise 3.5

1. Express each of the following in the form $R\sin(\theta + \alpha)$, where $R > 0$ and $0° \leq \alpha \leq 90°$, giving the exact value of R and the value of α correct to 2 decimal places.

 a) $3\sin\theta + 2\cos\theta$ **b)** $10\sin\theta + 7\cos\theta$

 c) $2\sin\theta + \cos\theta$ **d)** $40\sin\theta + 9\cos\theta$

   Expressing $a\sin\theta + b\cos\theta$ in the form $R\sin(\theta \pm \alpha)$ or $R\cos(\theta \pm \alpha)$

2. **a)** Express each of the following in the form $R \sin(\theta - \alpha)$, where $R > 0$ and $0° \leq \alpha \leq 90°$, giving the exact value of R and the value of α correct to 2 decimal places.

   i) $5 \sin \theta - 8 \cos \theta$ **ii)** $7 \sin \theta - 24 \cos \theta$

   iii) $2 \sin \theta - \cos \theta$ **iv)** $\sqrt{3} \sin \theta - 4 \cos \theta$

   b) Find the greatest and least possible values of each of the expressions in part **(a)**.

3. Express each of the following in the form $R \cos(\theta + \alpha)$, where $R > 0$ and $0° \leq \alpha \leq 90°$, giving the exact value of R and the value of α correct to 2 decimal places.

   a) $\cos \theta - \sin \theta$ **b)** $2 \cos \theta - \sqrt{2} \sin \theta$

   c) $4 \cos \theta - 3 \sin \theta$ **d)** $\sqrt{5} \cos \theta - \sin \theta$

4. **a)** Express each of the following in the form $R \cos(\theta - \alpha)$, where $R > 0$ and $0° \leq \alpha \leq 90°$, giving the exact value of R and the value of α correct to 2 decimal places.

   i) $3 \cos \theta + 4 \sin \theta$ **ii)** $5 \cos \theta + 12 \sin \theta$

   iii) $\sqrt{2} \cos \theta + \sin \theta$ **iv)** $\sin \theta + \cos \theta$

   b) Determine the greatest and least possible values of each of the expressions in part **(a)**.

   c) For each of the expressions in part **(a)**, find a value of θ for which the expression has its greatest value.

5. **a)** Express $\sin \theta + 2 \cos \theta$ in the form $R \sin(\theta + \alpha)$, where $R > 0$ and $0° \leq \alpha \leq 90°$, giving the exact value of R and the value of α correct to 2 decimal places.

   b) Hence solve the equation $\sin \theta + 2 \cos \theta = 1$ for $0° \leq \theta \leq 360°$.

6. **a)** Express $\sqrt{3} \sin \theta - \cos \theta$ in the form $R \sin(\theta - \alpha)$, where $R > 0$ and $0° \leq \alpha \leq 90°$, giving the exact values of R and α.

   b) Hence solve the equation $\sqrt{3} \sin \theta - \cos \theta = 1$ for $0° \leq \theta \leq 360°$.

   c) Find the greatest and least possible values of $(\sqrt{3} \sin \theta - \cos \theta)^2$ as θ varies.

7. **a)** Express $2 \cos \theta - 2 \sin \theta$ in the form $R \cos(\theta + \alpha)$, where $R > 0$ and $0° \leq \alpha \leq 90°$, giving the exact values of R and α.

   b) Hence solve the equation $\cos \theta - \sin \theta = \dfrac{1}{2}$ for $0° \leq \theta \leq 360°$.

8. a) Express $4\cos\theta + 6\sin\theta$ in the form $R\cos(\theta - \alpha)$, where $R > 0$ and $0° \le \alpha \le 90°$, giving the exact value of R and the value of α correct to 2 decimal places.

   b) Hence solve the equation $4\cos\theta + 6\sin\theta = 5$ for $0° \le \theta \le 360°$.

   c) Find the greatest and least possible values of $[(4\cos\theta + 6\sin\theta)^2 + 5]$ as θ varies.

9. Solve the following equations for $0° \le \theta \le 360°$.

   a) $4\cos\theta - 6\sin\theta = 5$

   b) $7\sin\theta - 24\cos\theta = 25$

   c) $7\cos\theta = 5 - \sin\theta$

   d) $4\cos\theta = 2 + 3\sin\theta$

   Hint: You may use any appropriate form of a single sine or single cosine to help you.

10. a) Express $2\sin 2x + \cos 2x$ in the form $R\sin(2x + \alpha)$, where $R > 0$ and $0° \le \alpha \le 90°$, giving the exact value of R and the value of α correct to 2 decimal places.

    b) Hence solve the equation $2\sin 2x + \cos 2x = 1$ for $0° \le x \le 360°$.

    c) Find the greatest and least possible values of $10 - (2\sin 2x + \cos 2x)$ as x varies.

11. a) Express $\sin x + 4\cos x$ in the form $R\sin(x + \alpha)$, where $R > 0$ and $0 \le \alpha \le \frac{\pi}{2}$, giving the exact value of R and the value of α correct to 3 decimal places.

    b) Hence solve the equation $\sin x + 4\cos x = 3$ for $0 \le x \le 2\pi$.

    c) Determine the greatest and least possible values of $[(\sin x + 4\cos x)^2 - 1]$ as x varies.

12. Solve the equation $15\cos 2\theta + 20\sin 2\theta + 7 = 0$ for $0° \le \theta \le 180°$.

Summary exercise 3

EXAM-STYLE QUESTIONS

1. Solve the equation
   $\cos(45° - \theta) = \sin(30° + \theta)$ for $0 \le \theta \le 360°$.

2. Solve the equation $6\cos\theta - 2\sec\theta = 1$ for $0 \le \theta \le 2\pi$.

3. Solve the equation $5\cos 2\theta - 11\sin\theta + 1 = 0$ for $0° \le x \le 360°$.

4. Prove the identity
   $\tan\left(\frac{\pi}{4} + \theta\right) - \tan\left(\frac{\pi}{4} - \theta\right) \equiv 2\tan 2\theta$.

5. Prove the identity $\csc 2\theta + \cot 2\theta \equiv \cot \theta$.

6. a) Prove that $\dfrac{1 + \sin 2\theta - \cos 2\theta}{1 + \sin 2\theta + \cos 2\theta} \equiv \tan \theta$.

b) Hence find the exact value of $\tan 22.5°$.

7. Prove that $(\sec \theta - \tan \theta)^2 \equiv \dfrac{1 - \sin \theta}{1 + \sin \theta}$.

8. Acute angles θ and α are such that $\tan \theta = \dfrac{1}{7}$ and $\tan \alpha = \dfrac{1}{3}$.

Find the exact value of $\tan(\theta + 2\alpha)$.

9. Show that

$$\cos \theta + \cos\left(\theta + \frac{2\pi}{3}\right) + \cos\left(\theta + \frac{4\pi}{3}\right) = 0.$$

10. a) Express $5 \sin \theta + 12 \cos \theta$ in the form $R \sin(\theta + \alpha)$, where $R > 0$ and $0 \le \alpha \le 90°$, giving the exact value of R and the value of α correct to 2 decimal places.

b) Hence

i) write down the greatest value of $5 \sin \theta + 12 \cos \theta$ as θ varies

ii) find a value of θ at which this greatest value occurs.

11. a) Express $4 \cos \theta - 5 \sin \theta$ in the form $R \cos(\theta + \alpha)$, where $R > 0$ and $0 \le \alpha \le 90°$, giving the exact value of R and the value of α correct to 2 decimal places.

b) Hence solve the equation $4 \cos \theta - 5 \sin \theta = -2$ for $0 \le \theta \le 360°$.

c) Find the greatest possible value of $20 - (4 \cos \theta - 5 \sin \theta)^2$ as θ varies.

12. a) Express $7 \sin \theta + 24 \cos \theta$ in the form $R \sin(\theta + \alpha)$, where $R > 0$ and $0 \le \alpha \le 90°$,

giving the exact value of R and the value of α correct to 2 decimal places.

b) Hence solve the inequality $7 \sin \theta + 24 \cos \theta < 15$ for $0 \le \theta \le 360°$.

13. If $\sin(\theta + \alpha) = 2 \sin(\theta - \alpha)$, prove that $\tan \theta = 3 \tan \alpha$.

14. Solve the equation $\tan(\theta + 45°) = 1 - 4 \tan \theta$ for $0° \le \theta \le 360°$.

15. Solve the equation

$$\sin 60° + \sin(60° + x) + \sin(60° + 2x) = 0$$

for $0 \le x \le 360°$.

16. If $p = \csc \theta - \sin \theta$ and $q = \sec \theta - \cos \theta$, show that $p^2 q^2 (p^2 + q^2 + 3) = 1$.

17. a) Given that $\tan \dfrac{\theta}{2} = t$, find expressions for

$\sin \dfrac{\theta}{2}$ and $\cos \dfrac{\theta}{2}$ in terms of t.

b) Show that $\sin \theta = \dfrac{2t}{1 + t^2}$ and $\cos \theta = \dfrac{1 - t^2}{1 + t^2}$.

c) Hence solve the equation $4 \sin \theta = 3 + 2 \cos \theta$ for $0° \le \theta \le 360°$.

18. Solve the equation $6 \sec^2 2x + 5 \tan 2x = 12$ giving all solutions in the interval $0° \le x \le 180°$

19. a) Prove the identity $\sin 3x \equiv 3 \sin x - 4 \sin^3 x$

b) Hence, solve the equation $\sin 3x = 2 \sin x$ for $0° \le x \le 180°$

20. Solve the equation $\sin(2x + \dfrac{\pi}{3}) = \cos(2x - \dfrac{\pi}{6})$ giving all solutions in the interval $0° \le x \le 360°$

Chapter summary

Secant, cosecant and cotangent

- $\sec \theta = \dfrac{1}{\cos \theta}$ $\qquad$ $\text{cosec}\, \theta = \dfrac{1}{\sin \theta}$ $\qquad$ $\cot \theta = \dfrac{1}{\tan \theta}\left(= \dfrac{\cos \theta}{\sin \theta}\right)$

- $1 + \tan^2 \theta \equiv \sec^2 \theta$

- $1 + \cot^2 \theta \equiv \text{cosec}^2 \theta$

Graphs of sec θ, cosec θ and cot θ

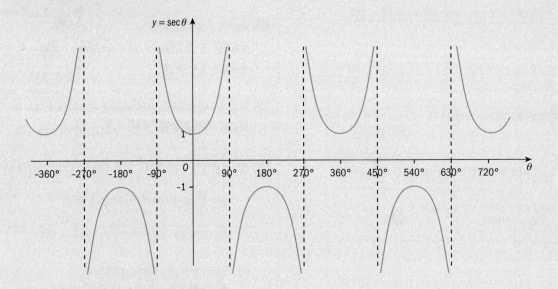

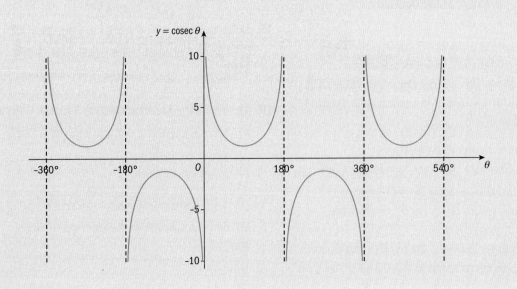

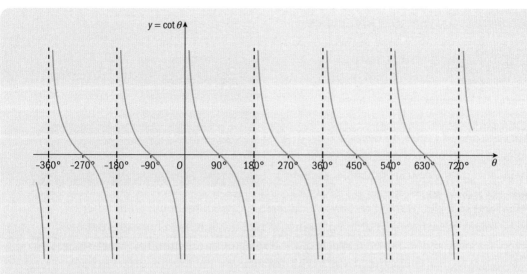

Complementary angle identities

- $\sin(90° - \theta) \equiv \cos\theta$
- $\tan(90° - \theta) \equiv \cot\theta$
- $\sec(90° - \theta) \equiv \operatorname{cosec}\theta$

- $\cos(90° - \theta) \equiv \sin\theta$
- $\cot(90° - \theta) \equiv \tan\theta$
- $\operatorname{cosec}(90° - \theta) \equiv \sec\theta$

Addition formulae

- $\sin(A + B) \equiv \sin A \cos B + \cos A \sin B$
- $\cos(A + B) \equiv \cos A \cos B - \sin A \sin B$
- $\tan(A + B) \equiv \dfrac{\tan A + \tan B}{1 - \tan A \tan B}$

- $\sin(A - B) \equiv \sin A \cos B - \cos A \sin B$
- $\cos(A - B) \equiv \cos A \cos B + \sin A \sin B$
- $\tan(A - B) \equiv \dfrac{\tan A - \tan B}{1 + \tan A \tan B}$

Double angle formulae

- $\sin 2A \equiv 2\sin A \cos A$
- $\tan 2A \equiv \dfrac{2\tan A}{1 - \tan^2 A}$

- $\cos 2A \equiv \cos^2 A - \sin^2 A$
  $\equiv 2\cos^2 A - 1$
  $\equiv 1 - 2\sin^2 A$

Expressing $a\sin\theta + b\cos\theta$ in the form $R\sin(\theta \pm \alpha)$ or $R\cos(\theta \pm \alpha)$

- $a\sin\theta + b\cos\theta \equiv R\sin(\theta + \alpha)$ where $R = \sqrt{a^2 + b^2}$ and $\alpha = \tan^{-1}\dfrac{b}{a}$

- $a\sin\theta + b\cos\theta \equiv R\cos(\theta - \alpha)$ where $R = \sqrt{a^2 + b^2}$ and $\alpha = \tan^{-1}\dfrac{a}{b}$

- $a\sin\theta - b\cos\theta \equiv R\sin(\theta - \alpha)$ where $R = \sqrt{a^2 + b^2}$ and $\alpha = \tan^{-1}\dfrac{b}{a}$

- $a\cos\theta - b\sin\theta \equiv R\cos(\theta + \alpha)$ where $R = \sqrt{a^2 + b^2}$ and $\alpha = \tan^{-1}\dfrac{b}{a}$

Maths in real-life

Predicting tidal behaviour

The surface level of water in oceans all over the world is seen to rise and fall periodically – known as high and low tides. To develop a good mathematical model of what is happening, you need to understand the causes of the tidal motion.

Water is fluid and so it can move in response to different gravitational pulls from other bodies in the solar system acting at different points. While the Sun has a huge gravitational pull, it is very far away and so the tidal effects of the Sun's gravity are minimal. Most of the tidal effects we see are caused by the Moon, as the distances on the Earth's surface are a much smaller proportion of the total distance between the Moon and the Earth, and so the gravitational field difference across the Earth is very small.

The Earth rotates on its axis every 24 hours and the Moon orbits around the Earth over a period of 27.3 days. Both of these cause points on the Earth's surface to move further away from (and closer to) the Moon; however, the Earth's rotation happens over a much shorter time period than the lunar cycle, and so the effect is much greater.

The first model would be a 24 hour cycle with two high and low tides – as shown here, a wave function in the form $h = a + b \sin 30t$ gives a starting point.

In fact, because of the relative motion of the Moon, the tidal cycle is actually about 24 hours and 50 minutes. In this time, two high and low tides of not quite equal sizes are seen. The side of the Earth closest to the Moon experiences the biggest pull so there is a difference in the size of the two high tides in a day – however, this is hard to model. Changing the period to 12 hours 25 minutes is easy ($h = a + b \sin 29t$).

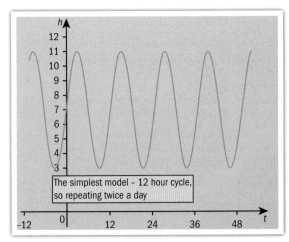

The simplest model – 12 hour cycle, so repeating twice a day

Once the period is adjusted we can add in the effect of the lunar cycle. When the Sun, Moon and Earth are directly in line with one another, the smaller effect of the Sun is added to that of the Moon – this happens twice in the cycle: once when the Moon is closer to the Sun than the Earth is (at new moon), and once when it is further away (at full moon) – and the highest tides occur.
These are known as spring tides and the corresponding lowest high tides are known as neap tides.

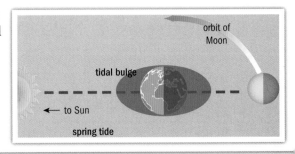

orbit of Moon

tidal bulge

← to Sun

spring tide

This diagram shows an improved model. The small effect of the Moon's orbit is shown by the blue line at the bottom, while the purple line shows the tidal pattern taking into account both the effect of the Earth's rotation and the Moon's orbit. The phenomenon of the spring and neap tides can be seen.

The model for this graph is
$y = 7 + 4 \sin(696x) + 0.4 \sin(26.4x)$, where
x is measured in days. The mean depth at this position is 7 metres, with a main tidal effect of 4 metres, and the Moon's effect is 0.4 metres. (Note that $696 = 360 \times \dfrac{24}{12.42}$ gives the period of the main tidal effect of 12 hours 25 minutes, and $26.4 = \dfrac{720}{27.3}$ gives the period of 27.3 days for the lunar effect to complete two cycles.)

Very high tides can cause untold damage when floods occur. A further regular perturbation occurs at the two equinoxes – where the axis of the Earth's rotation is perpendicular to the line between the Earth and the Sun. This is when the spring tides have their greatest variation between high and low tides. This is often a time at which stormy meteorological conditions occur (perhaps because of the way the bodies are moving relative to one another), so the risk of abnormally high tides and flooding is raised.

By understanding tides we can gain important insight into other problems, such as how oil spills will behave – whether as the result of an accident in a drilling operation or some mishap to a tanker transporting oil. The picture here shows an oil-drenched pelican being cleaned after the Deepwater Horizon oil spill in the Gulf of Mexico in 2010. Experts suggested that the spill was being pulled southeast by the loop current and would be extremely difficult to contain. Being able to predict the behaviour of pollutants – where they will go, how much they will become diluted, etc. – is a crucial component of making good decisions about what environmental measures are needed in order to contain the damage caused by toxic materials.

4 Differentiation

In real-world applications, differentiation is often complex, or involves trigonometric, exponential or logarithmic functions. For example, differentiation is used to maximise the electrical power produced by a certain source, to minimise the stress on a hollow tube, and to find the acceleration of a rocket which moves such that the only force on it is due to gravity and where its mass is decreasing at a constant rate.

Objectives
- Use the derivatives of e^x, $\ln x$, $\sin x$, $\cos x$, $\tan x$, $\tan^{-1} x$, together with constant multiples, sums, differences, and composites.
- Differentiate products and quotients.
- Find and use the first derivative of a function which is defined parametrically or implicitly.

Before you start

You should know how to:

1. Differentiate expressions that can be simplified to single terms in x,

e.g. find $\dfrac{dy}{dx}$ when

a) $y = \dfrac{x^6 - 12x}{3x^3}$ **b)** $y = 2x\left(4\sqrt{x} + \dfrac{1}{\sqrt{x}}\right)$.

....

a) $y = \dfrac{1}{3}x^3 - 4x^{-2}$, $\dfrac{dy}{dx} = x^2 + 8x^{-3}$

b) $y = 8x^{\frac{3}{2}} + 2x^{\frac{1}{2}}$, $\dfrac{dy}{dx} = 12x^{\frac{1}{2}} + x^{-\frac{1}{2}}$

2. Differentiate composite functions using the chain rule,

e.g. find the rate of change of y with respect to x when

a) $y = \dfrac{6}{2x - 3}$ **b)** $y = \dfrac{8}{\sqrt{5x + 1}}$.

....

a) $y = 6(2x - 3)^{-1}$,

$\dfrac{dy}{dx} = -6(2x - 3)^{-2}\,(2) = -12(2x - 3)^{-2}$

b) $y = 8(5x + 1)^{-\frac{1}{2}}$

$\dfrac{dy}{dx} = -4(5x + 1)^{-\frac{3}{2}}\,(5) = -20(5x + 1)^{-\frac{3}{2}}$

Skills check:

1. Find $\dfrac{dy}{dx}$ when

a) $y = \dfrac{6x^8 + x^2}{2x^4}$

b) $y = (4\sqrt{x} - 1)(2\sqrt{x} + 5)$

c) $y = \dfrac{x^{10} - 2x^5 + 10x}{5x^5}$

d) $y = \dfrac{1}{3}\sqrt{x}\left(\dfrac{6}{\sqrt{x}} - \sqrt[3]{x^6}\right)$.

2. Find the rate of change of y with respect to x when

a) $y = \dfrac{5}{4 - x}$ **b)** $y = \dfrac{7}{\sqrt{2x + 3}}$

c) $y = \dfrac{1}{x^2 - 3x + 7}$ **d)** $y = \dfrac{4}{\sqrt{(5x + 1)^3}}$.

4.1 Differentiating the exponential function

In Chapter 2, you learned about the natural exponential function e^x. This function is the only function whose derivative is the same as the function itself.
Consider the graph of $y = e^x$. Let δx be a small increase in x.

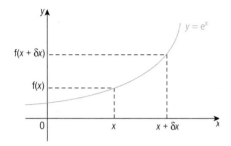

$$\frac{dy}{dx} = \lim_{\delta x \to 0} \frac{f(x + \delta x) - f(x)}{(x + \delta x) - x} = \lim_{\delta x \to 0} \frac{e^{x + \delta x} - e^x}{(x + \delta x) - x} = \lim_{\delta x \to 0} \frac{e^x e^{\delta x} - e^x}{\delta x} = \lim_{\delta x \to 0} \frac{e^x (e^{\delta x} - 1)}{\delta x}$$

Consider $\dfrac{(e^{\delta x} - 1)}{\delta x}$ with values of δx getting smaller and smaller.

δx	0.1	0.01	0.001	0.0001
$\dfrac{(e^{\delta x} - 1)}{\delta x}$	1.05170925	1.00501672	1.00050023	1.00005006

We can see that as $\delta x \to 0$, $\dfrac{(e^{\delta x} - 1)}{\delta x} \to 1$

Thus $\dfrac{dy}{dx} \to e^x (1) = e^x$

Note: You are not expected to know this proof for the examination.

> If $y = e^x$, then $\dfrac{dy}{dx} = e^x$.

If we want to differentiate $y = e^{kx}$ where k is a constant, we can use the chain rule:
Let $u = kx \Rightarrow y = e^u$.

Thus $\dfrac{du}{dx} = k$ and $\dfrac{dy}{du} = e^u$.

$\dfrac{dy}{dx} = \dfrac{dy}{du} \times \dfrac{du}{dx} = e^u \times k = e^{kx} \times k = ke^{kx}$

> If $y = e^{kx}$, then $\dfrac{dy}{dx} = ke^{kx}$.

In general we can say:

If $y = e^{f(x)}$, then $\dfrac{dy}{dx} = f'(x) \, e^{f(x)}$.

Example 1

Differentiate

a) e^{7x} **b)** e^{5x-4} **c)** e^{x^3}.

a) $\dfrac{d}{dx} e^{7x} = 7 \, e^{7x}$ ← Substitute for $k = 7$ in $k \, e^{kx}$.

b) $\dfrac{d}{dx} e^{5x-4} = 5 \, e^{5x-4}$ ← Use $\dfrac{dy}{dx} = f'(x) \, e^{f(x)}$ where $f'(x) = 5$.

c) $\dfrac{d}{dx} e^{x^3} = 3x^2 \, e^{x^3}$ ← Use $\dfrac{dy}{dx} = f'(x) \, e^{f(x)}$ where $f'(x) = 3x^2$.

Example 2

Find $\dfrac{dy}{dx}$ when

a) $y = 4 \, e^{2x}$ **b)** $y = -2 \, e^{\sqrt{x}}$ **c)** $y = 3 \, e^{\frac{1}{x}}$.

a) $\dfrac{dy}{dx} = 4 \, e^{2x} \,(2) = 8 \, e^{2x}$ ← Write down $4 \, e^{2x}$ and then multiply by $\dfrac{d}{dx} 2x$.

b) $\dfrac{dy}{dx} = -2 \, e^{\sqrt{x}} \left(\dfrac{1}{2} x^{-\frac{1}{2}} \right)$ ← Write down $-2 \, e^{\sqrt{x}}$ and then multiply by $\dfrac{d}{dx} x^{\frac{1}{2}}$.

$\qquad = -\dfrac{1}{\sqrt{x}} \, e^{\sqrt{x}}$

c) $\dfrac{dy}{dx} = 3 \, e^{\frac{1}{x}} (-1x^{-2})$ ← Write down $3 \, e^{\frac{1}{x}}$ and then multiply by $\dfrac{d}{dx} x^{-1}$.

$\qquad = -\dfrac{3 \, e^{\frac{1}{x}}}{x^2}$

Exercise 4.1

1. Differentiate

a) e^{2x}

b) e^{-5x}

c) e^{3x+9}

d) e^{8-x}

e) $-7e^x$

f) $5e^{x^2}$

g) $3e^{2x-1}$

h) $-4e^{6+x^2}$

i) e^{5x^2}

j) $e^{\frac{3}{x}}$

k) $6e^{\sqrt{x}}$

l) $\dfrac{1}{e^{2x}}$.

2. Find $\dfrac{dy}{dx}$ when

a) $y = e^{-2x}$

b) $y = e^{5+x}$

c) $y = 9e^{3x}$

d) $y = -8e^{\sqrt{x}}$

e) $y = 3e^x - 2e^{4x}$

f) $y = \dfrac{1}{2}e^{-8x}$

g) $y = e^{4x} + e^{-x}$

h) $y = 6e^{-x} - 4e^{3x}$

i) $y = (e^x + 2)(e^x - 1)$

Hint for part (i): First expand the brackets.

j) $y = (2e^{3x} - 1)^2$

k) $y = \dfrac{1+e^x}{e^x}$

l) $y = e^{-4x+2} + 7e^{5-2x}$.

3. Find the gradient of the tangent to the curve $y = 5e^{2x}$ at the point $(0, 5)$.

4. Find the exact value of the gradient of the tangent to the curve $y = e^x - 6\sqrt{x}$ when $x = 1$.

5. Find the coordinates of the minimum point on the curve with equation $y = e^x - x$.

6. Find the coordinates of the turning point on the curve $y = 2x - e^{2x} + 3$ and determine whether this point is a maximum or a minimum point.

4.2 Differentiating the natural logarithmic function

In Chapter 2, you learned about the natural logarithmic function $\log_e x$. This is also known as $\ln x$.

If $y = \log_e x$ (or $y = \ln x$), then $e^y = x$.

Since $x = e^y$, $\dfrac{dx}{dy} = e^y$.

As $\dfrac{dy}{dx} = \dfrac{1}{\dfrac{dx}{dy}}$, we can say $\dfrac{dy}{dx} = \dfrac{1}{e^y} = \dfrac{1}{x}$.

If $y = \ln x$, then $\dfrac{dy}{dx} = \dfrac{1}{x}$.

Example 3

Find $\dfrac{dy}{dx}$ when   **a)** $y = \ln 5x$ **b)** $y = \ln(3x^2 - 2)$.

a) Let $u = 5x$, then $\dfrac{du}{dx} = 5$

$y = \ln u$, so $\dfrac{dy}{du} = \dfrac{1}{u}$

Use the chain rule.

$\dfrac{dy}{dx} = \dfrac{dy}{du} \times \dfrac{du}{dx} = \dfrac{1}{u} \times 5 = \dfrac{5}{u}$

Substitute $u = 5x$ to have your answer in terms of x.

Thus $\dfrac{dy}{dx} = \dfrac{5}{5x} = \dfrac{1}{x}$

Note: Alternatively, we could say $y = \ln 5x = \ln 5 + \ln x$, so $\dfrac{dy}{dx} = 0 + \dfrac{1}{x} = \dfrac{1}{x}$ (since $\ln 5$ is a constant).

b) Let $u = 3x^2 - 2$, then $\dfrac{du}{dx} = 6x$

$y = \ln u$, so $\dfrac{dy}{du} = \dfrac{1}{u}$

$\dfrac{dy}{dx} = \dfrac{dy}{du} \times \dfrac{du}{dx} = \dfrac{1}{u} \times 6x = \dfrac{6x}{u}$

Use the chain rule.

Thus $\dfrac{dy}{dx} = \dfrac{6x}{3x^2 - 2}$

Eliminate u to present your answer in terms of x.

If $y = \ln[f(x)]$, then $\dfrac{dy}{dx} = \dfrac{f'(x)}{f(x)}$.

Example 4

Differentiate

a) $\ln(9x + 2)$ **b)** $\ln(e^x - 3x)$ **c)** $4\ln 8x$ **d)** $\ln(5x - 1)^7$.

a) $\dfrac{dy}{dx} = \dfrac{f'(x)}{f(x)} = \dfrac{9}{9x + 2}$

$f(x) = (9x + 2),\ f'(x) = 9$

b) $\dfrac{dy}{dx} = \dfrac{e^x - 3}{e^x - 3x}$

$f(x) = (e^x - 3x),\ f'(x) = e^x - 3$

c) $\dfrac{dy}{dx} = 4 \times \dfrac{f'(x)}{f(x)} = 4 \times \dfrac{8}{8x} = \dfrac{4}{x}$

$f(x) = 8x,\ f'(x) = 8$

d) $\ln(5x - 1)^7 = 7\ln(5x - 1)$

$\dfrac{dy}{dx} = 7 \times \dfrac{f'(x)}{f(x)} = 7 \times \dfrac{5}{5x - 1} = \dfrac{35}{5x - 1}$

It is best to use laws of logarithms (covered in Chapter 2) to first simplify the expression.

Differentiating the natural logarithmic function

Exercise 4.2

1. Differentiate

   a) $\ln 4x$

   b) $\ln(1 - x^2)$

   c) $\ln(3 + 5x)$

   d) $\ln(x^3 + 2)$

   e) $\ln(e^x - 7)$

   f) $\ln\sqrt{x}$

   g) $\ln(6x - 3)^4$

   h) $\ln(4x^2 + 2x)$

   i) $\ln\dfrac{1}{x^2}$

   j) $3\ln 4x$

   k) $2\ln\dfrac{1}{x}$

   l) $6\ln\dfrac{9x - 2}{3}$.

2. Find $\dfrac{dy}{dx}$ when

   a) $y = 8\ln 3x$

   b) $y = \ln 2x^3$

   c) $y = \ln(2 - 4x)$

   d) $y = 6\ln\dfrac{x}{2}$

   e) $y = \ln\sqrt{(2x+1)}$

   f) $y = \ln(5x)^{10}$

   g) $y = \ln(3e^x - 2x)$

   h) $y = 9\ln(x^2 + 3x)^2$

   i) $y = 5\ln\left(7 - \dfrac{2}{x}\right)$

   j) $y = 4\ln\sqrt{(x^2 - 8x)}$

   k) $y = \ln e^{5x}$

   l) $y = \ln\dfrac{7}{(x + 3)^6}$.

3. Find the gradient of the tangent to the curve $y = 4 + \ln x$ at the point $(1, 4)$.

4. Find the gradient of the tangent to the curve $y = \ln(x^2 + 1)$ when $x = 3$.

5. Find the exact coordinates of the point on the curve $y = \ln 3x$ where the gradient is $\dfrac{1}{2}$.

6. Find the coordinates of the turning point on the curve $y = x - \ln x$, and determine whether this point is a maximum or a minimum point.

4.3 Differentiating products

Consider $y = uv$ where u and v are functions of x.
Let δx be a small increase in x. Let δu, δv, and δy be the corresponding increases in u, v, and y.

If $y = uv$

$$y + \delta y = (u + \delta u)(v + \delta v)$$

$$uv + \delta y = uv + u\,\delta v + v\,\delta u + \delta u\,\delta v$$

$$\delta y = u\,\delta v + v\,\delta u + \delta u\,\delta v$$

$$\frac{\delta y}{\delta x} = u\frac{\delta v}{\delta x} + v\frac{\delta u}{\delta x} + \delta u\frac{\delta v}{\delta x}$$

As $\delta x \to 0$, $\delta u \to 0$, $\dfrac{\delta y}{\delta x} \to \dfrac{dy}{dx}$, $\dfrac{\delta u}{\delta x} \to \dfrac{du}{dx}$, and $\dfrac{\delta v}{\delta x} \to \dfrac{dv}{dx}$.

Thus, $\dfrac{dy}{dx} = u\dfrac{dv}{dx} + v\dfrac{du}{dx} + 0\dfrac{dv}{dx}$ | If $y = uv$, then $\dfrac{dy}{dx} = u\dfrac{dv}{dx} + v\dfrac{du}{dx}$. |

This is known as the product rule.

Example 5

Find $\dfrac{dy}{dx}$ when $y = e^{2x}(x^3 - 3)$.

Let $y = uv$ where $u = e^{2x}$ and $v = (x^3 - 3)$.   ←——— y is a product of two functions.

$\dfrac{du}{dx} = 2\,e^{2x}$ and $\dfrac{dv}{dx} = 3x^2$ ←——— Differentiate u and differentiate v.

$\dfrac{dy}{dx} = u\dfrac{dv}{dx} + v\dfrac{du}{dx}$ ←——— Use the product rule.

$\dfrac{dy}{dx} = (e^{2x})(3x^2) + (x^3 - 3)(2\,e^{2x})$ ←——— Substitute for each of the four unknowns.

$= e^{2x}(2x^3 + 3x^2 - 6)$ ←——— Factorise by taking out the common e^{2x}.

Example 6

Find $\dfrac{dy}{dx}$ when $y = (2x + 1)(x - 5)^4$.

Let $y = uv$ where $u = (2x + 1)$ and $v = (x - 5)^4$.   ←——— y is a product of two functions.

$\dfrac{du}{dx} = 2$ and $\dfrac{dv}{dx} = 4(x - 5)^3(1)$ ←——— Differentiate u and differentiate v.

$\dfrac{dy}{dx} = u\dfrac{dv}{dx} + v\dfrac{du}{dx}$ ←——— Use the product rule.

$\dfrac{dy}{dx} = (2x + 1)[4(x - 5)^3] + [(x - 5)^4](2)$ ←——— Substitute for each of the four unknowns.

$= 2(x - 5)^3(4x + 2 + x - 5)$

$= 2(x - 5)^3(5x - 3)$ ←——— Factorise by taking out the common factor $2(x - 5)^3$.

Example 7

Show that $\dfrac{dy}{dx} = \dfrac{4x(5x+2)}{\sqrt{2x+1}}$ when $y = 4x^2\sqrt{2x+1}$.

Let $y = uv$, where $u = 4x^2$ and $v = \sqrt{2x+1}$. ← y is a product of two functions.

$\dfrac{du}{dx} = 8x$ ← Differentiate u.

$\dfrac{dv}{dx} = \dfrac{1}{2}(2x+1)^{-\frac{1}{2}}(2) = \dfrac{1}{\sqrt{2x+1}}$ ← Differentiate $v = (2x+1)^{\frac{1}{2}}$.

$\dfrac{dy}{dx} = u\dfrac{dv}{dx} + v\dfrac{du}{dx}$ ← Use the product rule.

$\dfrac{dy}{dx} = (4x^2)\left(\dfrac{1}{\sqrt{2x+1}}\right) + (\sqrt{2x+1})(8x)$ ← Substitute for each of the four unknowns.

$= \dfrac{4x^2 + (2x+1)(8x)}{\sqrt{2x+1}}$ ← $(2x+1) \div \sqrt{2x+1} = \sqrt{2x+1}$

$= \dfrac{4x^2 + 16x^2 + 8x}{\sqrt{2x+1}}$

$= \dfrac{20x^2 + 8x}{\sqrt{2x+1}}$

$= \dfrac{4x(5x+2)}{\sqrt{2x+1}}$

Thus $\dfrac{dy}{dx} = \dfrac{4x(5x+2)}{\sqrt{2x+1}}$

Exercise 4.3

1. Use the product rule to find $\dfrac{dy}{dx}$ when

a) $y = x\,e^x$

b) $y = (4x - 1)(x^2 - 5)$

c) $y = x^2\sqrt{x^2 + 1}$

d) $y = \sqrt{x}\,(\ln x)$

e) $y = x^2(2x + 5)^7$

f) $y = \sqrt{x}\,(x - 3)^2$

g) $y = e^{2x}(4x + 1)$

h) $y = x(x + 3)^4$

i) $y = x^2 \ln x$

j) $y = x^2(x + 3)^3$

k) $y = e^x(2x - 1)^4$

l) $y = \dfrac{x}{x + 1}$

> **Hint for part (l):** Let $y = x(x + 1)^{-1}$ and use the product rule.

m) $y = 6e^{\sqrt{x}}(x + 4)$

n) $y = 4x^3\,e^{2x}$

o) $y = 2e^x \ln 2x$

p) $y = \dfrac{x - 2}{x + 2}$

q) $y = \dfrac{2x + 3}{1 - 5x}$

r) $y = (x^2 + 1) \ln (x^2 + 1)$.

2. Find the gradient of the tangent to the curve $y = x^2 e^{-x}$ at the point $\left(1, \dfrac{1}{e}\right)$.

3. Find the equation of the tangent to the curve $y = x(1 + x^2)^3$ when $x = 2$.

4. Show that $\dfrac{dy}{dx} = \dfrac{2x^2 + 4x - 1}{\sqrt{(x^2 - 1)}}$ when $y = (x + 4)\sqrt{(x^2 - 1)}$.

5. Find the exact coordinates of the turning point on the curve $y = e^x(1 + x)$ and determine whether this point is a maximum or a minimum point.

4.4 Differentiating quotients

Consider $y = \dfrac{u}{v}$ where u and v are functions of x.

Let δx be a small increase in x. Let δu, δv, and δy be the corresponding increases in u, v, and y.

If $\qquad y = \dfrac{u}{v}$

$$y + \delta y = \dfrac{u + \delta u}{v + \delta v}$$

$$\delta y = \dfrac{u + \delta u}{v + \delta v} - y$$

$$= \dfrac{u + \delta u}{v + \delta v} - \dfrac{u}{v}$$

$$= \dfrac{v(u + \delta u) - u(v + \delta v)}{v(v + \delta v)}$$

$$= \dfrac{uv + v\delta u - uv - u\delta v}{v^2 + v\delta v}$$

$$\dfrac{\delta y}{\delta x} = \dfrac{v\dfrac{\delta u}{\delta x} - u\dfrac{\delta v}{\delta x}}{v^2 + v\delta v}$$

As $\delta x \to 0$, $v\,\delta v \to 0$, $\dfrac{\delta y}{\delta x} \to \dfrac{dy}{dx}$, $\dfrac{\delta u}{\delta x} \to \dfrac{du}{dx}$, and $\dfrac{\delta v}{\delta x} \to \dfrac{dv}{dx}$.

Thus, $\quad \dfrac{dy}{dx} = \dfrac{v\dfrac{du}{dx} - u\dfrac{dv}{dx}}{v^2}$

$$\boxed{\text{If } y = \dfrac{u}{v}, \text{ then } \dfrac{dy}{dx} = \dfrac{v\dfrac{du}{dx} - u\dfrac{dv}{dx}}{v^2}.}$$

This is known as the quotient rule.

Example 8

Find $\dfrac{dy}{dx}$ when $y = \dfrac{x^3}{x^2 + 1}$.

Let $y = \dfrac{u}{v}$ where $u = x^3$ and $v = x^2 + 1$. ← y is a quotient of two functions.

$\dfrac{du}{dx} = 3x^2$ and $\dfrac{dv}{dx} = 2x$ ← Differentiate u and differentiate v.

$\dfrac{dy}{dx} = \dfrac{v\dfrac{du}{dx} - u\dfrac{dv}{dx}}{v^2}$ ← Use the quotient rule.

$\dfrac{dy}{dx} = \dfrac{(x^2 + 1)(3x^2) - (x^3)(2x)}{(x^2 + 1)^2}$ ← Substitute for each of the four unknowns.

$= \dfrac{3x^4 + 3x^2 - 2x^4}{(x^2 + 1)^2}$

$= \dfrac{x^2(x^2 + 3)}{(x^2 + 1)^2}$ ← Factorise by taking out the common x^2.

Example 9

Find $\dfrac{dy}{dx}$ when $y = \dfrac{4x + 3}{\sqrt{(2x - 1)}}$.

Let $y = \dfrac{u}{v}$ where $u = 4x + 3$ and $v = \sqrt{(2x-1)}$.

$\dfrac{du}{dx} = 4$ ← Differentiate u and differentiate v.

$\dfrac{dv}{dx} = \dfrac{1}{2}(2x - 1)^{-\frac{1}{2}}(2) = (2x - 1)^{-\frac{1}{2}}$

$\dfrac{dy}{dx} = \dfrac{v\dfrac{du}{dx} - u\dfrac{dv}{dx}}{v^2}$ ← Use the quotient rule.

$\dfrac{dy}{dx} = \dfrac{(2x-1)^{\frac{1}{2}}(4) - (4x+3)(2x-1)^{-\frac{1}{2}}}{2x - 1}$ ← Substitute for each of the four unknowns.

$= \dfrac{(2x-1)^{-\frac{1}{2}}[4(2x-1) - (4x+3)]}{2x-1}$ Note: $(2x-1)^{-\frac{1}{2}} \times (2x-1)^1 = (2x-1)^{\frac{1}{2}}$

$= \dfrac{4x - 7}{(2x-1)^{\frac{3}{2}}}$ ← $\dfrac{(2x-1)^{-\frac{1}{2}}}{(2x-1)^1} = \dfrac{1}{(2x-1)^{\frac{3}{2}}}$

Exercise 4.4

1. Use the quotient rule to find $\dfrac{dy}{dx}$ when

   a) $y = \dfrac{2x}{x+3}$

   b) $y = \dfrac{e^{2x}}{x}$

   c) $y = \dfrac{x^2}{x+4}$

   d) $y = \dfrac{x}{1+x^2}$

   e) $y = \dfrac{3x}{2-x}$

   f) $y = \dfrac{3x^2}{2e^x - x}$

   g) $y = \dfrac{3x-4}{2x+1}$

   h) $y = \dfrac{3x+5}{(x+1)^2}$

   i) $y = \dfrac{6x^2}{2-x}$

   j) $y = \dfrac{4x}{(1-x)^3}$

   k) $y = \dfrac{x^2+1}{x-2}$

   l) $y = \dfrac{5x^2}{\ln x}$

   m) $y = \dfrac{x^3}{\sqrt{(1-2x^2)}}$

   n) $y = \dfrac{3x+x^4}{2x^2+1}$

   o) $y = \dfrac{x}{1+\sqrt{x}}$

   p) $y = \dfrac{x^2+6}{2x-7}$

   q) $y = \dfrac{\sqrt{(x+1)^5}}{x}$

   r) $y = \dfrac{1-2x^2}{\sqrt{(1+2x^2)}}$.

2. If $y = \dfrac{x^2}{2x-3}$ find the values of x at the points where $\dfrac{dy}{dx} = 0$.

3. Find the equation of the normal to the curve $y = \dfrac{3x^2}{2x-1}$ at the point $(2, 4)$.

4. If $y = \dfrac{x}{x-1}$, find $\dfrac{d^2y}{dx^2}$.

5. Show that the coordinates of the turning point on the curve $y = \dfrac{x+3}{\sqrt{(1+x^2)}}$ are $\left(\dfrac{1}{3}, \sqrt{10}\right)$.

6. Show that if $y = \dfrac{7\ln x - x^3}{e^{3x}}$ then $\dfrac{dy}{dx} = \dfrac{7}{e^3}$ when $x = 1$.

4.5 Differentiating $\sin x$, $\cos x$, and $\tan x$

When we differentiate trigonometric functions, we assume the angle is measured in radians.

Consider $y = \sin x$ where x is measured in radians.

Let δx be a small increase in x. Let δy be the corresponding small increase in y.

If $\qquad y = \sin x$

$\qquad y + \delta y = \sin(x + \delta x)$

$\qquad\qquad = \sin x \cos \delta x + \cos x \sin \delta x$ ← $\sin(A + B) = \sin A \cos B + \cos A \sin B$

As $\delta x \to 0$, $\cos \delta x \to 1$ and $\sin \delta x \to \delta x$ ←   **Note:** The proof of this limit is beyond this course.

Thus, $\qquad y + \delta y \approx \sin x + \delta x \cos x$

$\qquad\qquad \sin x + \delta y \approx \sin x + \delta x \cos x$

$\qquad\qquad\qquad \delta y \approx \delta x \cos x$

$\qquad\qquad\qquad \dfrac{\delta y}{\delta x} \approx \cos x$

$\qquad\qquad \lim\limits_{\delta x \to 0}\left(\dfrac{\delta y}{\delta x}\right) = \dfrac{dy}{dx} = \cos x$

> If $y = \sin x$, then $\dfrac{dy}{dx} = \cos x$.

We can use a similar proof to differentiate $\cos x$. The result is stated below.

> If $y = \cos x$, then $\dfrac{dy}{dx} = -\sin x$.

Consider $y = \tan x$.

Therefore $y = \dfrac{\sin x}{\cos x}$

Using the quotient rule: $\dfrac{dy}{dx} = \dfrac{\cos x(\cos x) - \sin x(-\sin x)}{\cos^2 x} = \dfrac{\cos^2 x + \sin^2 x}{\cos^2 x} = \dfrac{1}{\cos^2 x} = \sec^2 x$

> If $y = \tan x$, then $\dfrac{dy}{dx} = \sec^2 x$.

Example 10

Find $\dfrac{dy}{dx}$ when

a) $y = \cos 2x$ **b)** $y = 5 \sin 3x^2$ **c)** $y = 7 \tan\left(4x + \dfrac{\pi}{2}\right)$.

a) Let $u = 2x \Rightarrow \dfrac{du}{dx} = 2$

$y = \cos u \Rightarrow \dfrac{dy}{du} = -\sin u = -\sin 2x$ ⟵ Substitute $u = 2x$.

$\dfrac{dy}{dx} = \dfrac{dy}{du} \times \dfrac{du}{dx}$ ⟵ Use the chain rule to help you.

$= (-\sin 2x)(2) = -2\sin 2x$ ⟵ Substitute for each of the two unknowns.

b) Let $u = 3x^2 \Rightarrow \dfrac{du}{dx} = 6x$ ⟵ Substitute $u = 3x^2$ and use the chain rule to help you differentiate.

$y = 5\sin u \Rightarrow \dfrac{dy}{du} = 5\cos u$

$\dfrac{dy}{dx} = (5\cos u)(6x)$ ⟵ Using the chain rule $\dfrac{dy}{dx} = \dfrac{dy}{du} \times \dfrac{du}{dx}$.

$= 30x \cos 3x^2$

c) Let $u = 4x + \dfrac{\pi}{2} \Rightarrow \dfrac{du}{dx} = 4$ ⟵ Substitute $u = 4x + \dfrac{\pi}{2}$.

$y = 7\tan u \Rightarrow \dfrac{dy}{du} = 7\sec^2 u$

$\dfrac{dy}{dx} = (7\sec^2 u) \times 4$ ⟵ Using the chain rule $\dfrac{dy}{dx} = \dfrac{dy}{du} \times \dfrac{du}{dx}$.

$= 28\sec^2\left(4x + \dfrac{\pi}{2}\right)$

Exercise 4.5

1. Find $\dfrac{dy}{dx}$ when

a) $y = \sin 5x$

b) $y = 6\tan\dfrac{x}{2}$

c) $y = \cos(x^2 + 2x)$

d) $y = 4\tan^2 x$

e) $y = \sin(x^3 - 7)$

f) $y = 5\tan 4x$

g) $y = \cos^3 x$

h) $y = \sin 2x \cos x$

i) $y = x^2 \tan 3x$

j) $y = \dfrac{\cos 2x}{x^2}$

k) $y = \sin^3 3x \cos x$

l) $y = 4\tan\sqrt{x}$.

Differentiating $\sin x$, $\cos x$, and $\tan x$

2. **a)** If $y = (\cos 2x - \sin x)^2$, show that $\dfrac{dy}{dx} = -(\cos 2x - \sin x)(2 \sin 2x + \cos x)$.

   b) If $y = e^x \tan x$, show that $\dfrac{dy}{dx} = e^x (\sec^2 x + \tan x)$.

   c) If $y = \ln(\cos x)$, show that $\dfrac{dy}{dx} = -\tan x$.

   d) If $y = x\, e^{\sin x}$, show that $\dfrac{dy}{dx} = e^{\sin x}(x \cos x + 1)$.

3. Find the gradient of the tangent to the curve $y = 3 \sin 2x$ when $x = \dfrac{\pi}{3}$.

4. Find the equation of the normal to the curve $y = x - \sin x$ at the point $\left(\dfrac{\pi}{2}, \dfrac{\pi}{2} - 1\right)$.

5. Show that $\dfrac{d^2 y}{dx^2} = -\dfrac{1}{2} \sec^2 x$ when $y = \ln \sqrt{\cos x}$.

6. Find the coordinates of the turning points on the curve $y = 2 \sin x + \cos 2x$ for $0 < x < \pi$ and determine whether these points are maximum or minimum points.

7. Show that the maximum value of the curve $y = x - 2 \sin x$ for $0 \le x \le 2\pi$ is $\dfrac{5\pi}{3} + \sqrt{3}$, and find the minimum value of the curve.

8. If $y = \ln \sqrt{1 - \cos x}$, show that $\dfrac{d^2 y}{dx^2} = \dfrac{1}{2(\cos x - 1)}$.

4.6 Implicit differentiation

Functions of the form $y = f(x)$ are explicit functions as y is given explicitly in terms of x.

However, some functions that have two variables are not of this form – that is, sometimes we are not given one variable explicitly in terms of the other. We call these functions implicit functions.

Examples of implicit functions include:

(1) $x^2 + 3xy^2 - 3y = 0$

(2) $x \ln(y + 2) - x^2 = 2y$

(3) $\dfrac{4y^2}{y^2 + x} = 5$

How can we differentiate an implicit function?

> We use implicit differentiation when one of the variables is not given explicitly as a function of the other variable.

To illustrate the process of implicit differentiation, we will consider the example $x^2 + 3y = 6xy$.

Differentiating each term with respect to x,

$$\frac{d}{dx}x^2 + \frac{d}{dx}3y = \frac{d}{dx}6xy$$

Thus $2x + \frac{d}{dx}3y = 6x\frac{d}{dx}y + 6y\frac{d}{dx}x$

Using the product rule on the RHS.

Thus $2x + \left(\frac{d}{dy}3y\right)\left(\frac{dy}{dx}\right) = (6x)\left(\frac{d}{dy}y\right)\left(\frac{dy}{dx}\right) + 6y$

We can say $\frac{d}{dx} = \frac{d}{dy} \times \frac{dy}{dx}$.

Thus $2x + 3\frac{dy}{dx} = 6x\frac{dy}{dx} + 6y$

Rearranging gives:

Collect terms in $\frac{dy}{dx}$ on LHS.

$(3 - 6x)\frac{dy}{dx} = 6y - 2x$

Thus $\frac{dy}{dx} = \frac{6y - 2x}{3 - 6x}$

It is important to note that when differentiating an implicit equation to find $\frac{dy}{dx}$, we must use the chain rule to differentiate any expression involving y. For example, $\frac{d}{dx}(3y^2) = 6y\frac{dy}{dx}$.

Example 11

Find $\frac{dy}{dx}$ in terms of x and y when $x^2 - 3y^3 = 4y^2$.

$2x - 9y^2\frac{dy}{dx} = 8y\frac{dy}{dx}$

Differentiate each term with respect to x.

$2x = 8y\frac{dy}{dx} + 9y^2\frac{dy}{dx}$

Rearrange terms.

$\frac{dy}{dx}(8y + 9y^2) = 2x$

Factorise.

$\frac{dy}{dx} = \frac{2x}{8y + 9y^2}$

Rearrange for $\frac{dy}{dx}$.

Example 12

Find the equation of the normal to the curve $5x^2 + 6xy - y^2 = 10$ at the point $(1, 5)$.

$10x + 6x\dfrac{dy}{dx} + 6y - 2y\dfrac{dy}{dx} = 0$ ←   Use the product rule to differentiate $6xy$.

$(6x - 2y)\dfrac{dy}{dx} = -10x - 6y$ ←   Rearrange terms.

At $(1, 5) \Rightarrow (6 - 10)\dfrac{dy}{dx} = -10 - 30$ ←   Substitute for $x = 1$ and $y = 5$.

So $\dfrac{dy}{dx} = \dfrac{-40}{-4} = 10$

Thus gradient of tangent to curve at $(1, 5) = 10$

Gradient of normal to curve at $(1, 5) = -\dfrac{1}{10}$ ← $m_1 \times m_2 = -1$ for perpendicular lines.

Equation of normal at $(1, 5)$ is

$y - 5 = -\dfrac{1}{10}(x - 1)$ ← $(y - y_1) = m(x - x_1)$ is the equation of a straight line.

or $-10y + 50 = x - 1$

or $10y = 51 - x$ ←   Multiply both sides by -10.

Example 13

It is given that $y = \tan^{-1} x$. Find an expression for $\dfrac{dy}{dx}$.

$y = \tan^{-1} x$

$x = \tan y$ ←   Use what you learned in P1.

$1 = \sec^2 y \dfrac{dy}{dx}$ ←   Differentiate with respect to x implicitly.

$1 = (1 + \tan^2 y) \dfrac{dy}{dx}$ ←   Use $1 + \tan^2 y \equiv \sec^2 y$.

Rearranging:

$\dfrac{1}{1 + \tan^2 y} = \dfrac{dy}{dx}$

and substituting $x = \tan y$ gives

$\dfrac{dy}{dx} = \dfrac{1}{1 + x^2}$

$$\text{If } y = \tan^{-1} x, \text{ then } \frac{dy}{dx} = \frac{1}{1 + x^2}.$$

You are given this formula in the formula book.

If $y = \tan^{-1} ax$, then, using the chain rule with $u = ax$,

$$\frac{dy}{dx} = \frac{dy}{du} \times \frac{du}{dx} = \frac{1}{1 + (ax)^2} \times a$$

Exercise 4.6

1. Find $\dfrac{dy}{dx}$ in terms of x and y when

   a) $x^2 + 2y - y^2 = 5$

   b) $x \cos y = y^2 + x$

   c) $y = 5 \tan^{-1} 3x$

   d) $x^3 + xy^2 = 5x$

   e) $x^2 + y^2 = y$

   f) $2xy - 3y = y^2 - 7x$

   g) $x \ln y = 1 + x$

   h) $y = e^x \tan^{-1} x$

   i) $x^2 + y^2 + 2x - 4y + 4 = 0$

   j) $x^4 + x^2 y^2 - y^4 = 5$

   k) $x \sin y - y \sin x = 8$

   l) $e^{x+y} = 2x$

   m) $y = x \tan^{-1} x$

   n) $y + y^3 = x - x^2$

   o) $x e^y + 2y = 1$

   p) $y = x \ln y$

   q) $y = \dfrac{1}{4} \tan^{-1} 2x$

   r) $\sin y + x^2 y^3 = 2y + \cos x.$

2. Find the gradients of the tangents to the curve $x^2 + 6y^2 = 10$ when $x = 2$.

3. Find the equation of the tangent to the curve $x^3 + y^3 = 2xy$ at the point $(1, 1)$.

4. Show that $\dfrac{dy}{dx} = \dfrac{2y^2 - 2xy^3}{3x^2 y^2 - 4xy - 3}$ when $x^2 y^3 - 2xy^2 = 3y$.

5. Given that $\ln(xy) = x^2 + y^2$, show that $\dfrac{dy}{dx} = \dfrac{y(2x^2 - 1)}{x(1 - 2y^2)}$.

6. If $y = \tan^{-1}(1 - 2x)$, find, in its simplest form, an expression for $\dfrac{dy}{dx}$.

4.7 Parametric differentiation

When x and y are related via a third variable, t, then t is called a parameter.
Equations which state x and y in terms of t are called parametric equations.

Suppose $x = t^3$ and $y = t^2$.
We can look at values of x and y for particular values of t.

t	-2	-1	0	1	2
x	-8	-1	0	1	8
y	4	1	0	1	4

If we want to try to relate x and y directly, notice that
$x = t^3 \Rightarrow x^2 = t^6$, and $y = t^2 \Rightarrow y^3 = t^6$.

Therefore $x^2 = y^3$.

Using implicit differentiation, $2x = 3y^2 \dfrac{dy}{dx}$, thus $\dfrac{dy}{dx} = \dfrac{2x}{3y^2} = \dfrac{2t^3}{3t^4} = \dfrac{2}{3t}$.

We can also use the chain rule to produce the same result, but without
needing to eliminate t. Consider again $x = t^3$ and $y = t^2$.

Differentiating with respect to $t \Rightarrow \dfrac{dx}{dt} = 3t^2$ and $\dfrac{dy}{dt} = 2t$

Using the chain rule, $\dfrac{dy}{dx} = \dfrac{dy}{dt} \times \dfrac{dt}{dx} = 2t \times \dfrac{1}{3t^2} = \dfrac{2}{3t}$.

When differentiating parametric equations involving a parameter t,
$$\frac{dy}{dx} = \frac{dy}{dt} \div \frac{dx}{dt}$$
This is called **parametric differentiation**.

Example 14

a) If $x = 2t^3$ and $y = 5t^2 - 6$, find $\dfrac{dy}{dx}$ in terms of t.

b) Find the value of $\dfrac{dy}{dx}$ when $x = -16$.

. . . .

a) $\dfrac{dx}{dt} = 6t^2$, $\dfrac{dy}{dt} = 10t$ ←   Differentiate x and y with respect to t.

$\dfrac{dy}{dx} = \dfrac{10t}{6t^2} = \dfrac{5}{3t}$ ← $\dfrac{dy}{dx} = \dfrac{dy}{dt} \div \dfrac{dx}{dt}$

b) When $x = -16 \Rightarrow -16 = 2t^3$ ←   Use $x = -16$ to find the value of t.

$\Rightarrow t = -2$

$\dfrac{dy}{dx} = \dfrac{5}{-6} = -\dfrac{5}{6}$ ←   Substitute this value of t into the expression you found in (**a**).

When differentiating parametric equations where the parameter is an angle, θ,
$$\frac{dy}{dx} = \frac{dy}{d\theta} \div \frac{dx}{d\theta}, \text{ where } \theta \text{ is in radians}$$

Example 15

The parametric equations of a curve are given by $x = 2\cos\theta$, $y = \sin^2\theta$, $0 \le \theta \le 2\pi$.

Find the values of θ when the tangent to the curve is parallel to the x-axis.

$\dfrac{dx}{d\theta} = -2\sin\theta$, $\dfrac{dy}{d\theta} = 2\sin\theta\cos\theta$ ← Differentiate x and y with respect to θ.

$\dfrac{dy}{dx} = \dfrac{2\sin\theta\cos\theta}{-2\sin\theta} = -\cos\theta$ ← $\dfrac{dy}{dx} = \dfrac{dy}{d\theta} \div \dfrac{dx}{d\theta}$

When $-\cos\theta = 0$, ← $\dfrac{dy}{dx} = 0$ when tangent parallel to x-axis.

$\theta = \dfrac{\pi}{2}$ and $\dfrac{3\pi}{2}$ ← $0 \le \theta \le 2\pi$

Exercise 4.7

1. Find $\dfrac{dy}{dx}$ for the following parametric equations.

    Give your answer in its simplest form.

    a) $x = 2t^3$, $\quad y = 4t^2 + 1$

    b) $x = t^3 - 6t$, $\quad y = t + \dfrac{2}{t}$

    c) $x = 3\cos t$, $\quad y = 2\sin t$

    d) $x = 4t^2 - 7$, $\quad y = 7 + 8t$

    e) $x = t - \sin t$, $\quad y = 2 - \cos t$

    f) $x = \dfrac{1}{1-t}$, $\quad y = t^2 - 9$

    g) $x = t^2 - 3$, $\quad y = 2 + \sqrt{t}$

    h) $x = 3\cos^3\theta$, $\quad y = 3\sin^3\theta$

    i) $x = t^2 + 3t - 1$, $\quad y = 3(t - 2)$

    j) $x = \cos 2\theta$, $\quad y = 4\sin\theta$

    k) $x = a\cos\theta$, $\quad y = b\sin\theta$

    l) $x = 2 - t^2$, $\quad y = 5 + 4t - t^3$

2. The parametric equations of a curve are $x = t + \dfrac{1}{t}$, $y = t - \dfrac{1}{t}$. Show that $\dfrac{dy}{dx} = \dfrac{t^2 + 1}{t^2 - 1}$.

3. The parametric equations of a curve are $x = 4\sin t$, $y = \cos 2t$.

    Show that $\dfrac{dy}{dx} = -\sin t$.

4. The parametric equations of a curve are $x = 2 + t$, $y = \dfrac{1}{t^2} - 6$. Show that $\dfrac{dy}{dx} = -2t^{-3}$.

5. The parametric equations of a curve are $x = \cos 2\theta$, $y = 2\theta + \sin 2\theta$. Show that $\dfrac{dy}{dx} = -\cot\theta$.

6. The parametric equations of a curve are $x = 2t + 3$, $y = t^2 - 4$.
   Find the gradient of the curve at the point for which $x = 5$.

7. The parametric equations of a curve are $x = 3t$, $y = \dfrac{3}{t}$.

   Find the equation of the normal to the curve at the point $(-1, -9)$.
   Give your answer in the form $ax + by + c = 0$, where a, b, and c are integers.

8. Show that $\dfrac{dy}{dx} = \dfrac{\sin\theta}{1 + \cos\theta}$ when $x = a(\theta + \sin\theta)$ and $y = a(1 - \cos\theta)$, where a is a constant.

9. The parametric equations of a curve are $x = 2t$, $y = 4t^3 - 8t^2$.

   Find the two values of t for which $\dfrac{dy}{dx} = 0$.

10. The parametric equations of a curve are $x = 5\cos 3t$, $y = 2\sin 3t$, $0 < t < \dfrac{\pi}{2}$.

    a) Find $\dfrac{dy}{dx}$ in terms of t.

    b) Find the coordinates of the stationary point on the curve.

Summary exercise 4

1. Find $\dfrac{dy}{dx}$ when

   a) $y = e^{-5x}$
   b) $y = \ln(4x^2 + 5)$
   c) $y = \tan^{-1}(4x)$
   d) $y = \dfrac{x-1}{2x-3}$

   e) $y = \dfrac{\sin x}{e^x}$
   f) $y = (x+1)^2(x-2)^3$
   g) $y = \dfrac{x^2}{\sqrt{(x^2-1)}}$
   h) $y = 6\ln\sqrt{(3x-4)}$

   i) $y = xy^2 - x^3$
   j) $y = 7\ln e^{2x}$
   k) $y = \dfrac{xe^x}{\ln x}$
   l) $y = x^2\sqrt{(1+x^2)}$

   m) $x = 3t - 5$, $y = \dfrac{1}{t}$
   n) $y = -2e^{\sqrt{x}}$
   o) $2x^3 + 3xy^2 - y^3 = 6$
   p) $y = \dfrac{(2e^x - 3)^2}{e^x}$

   q) $y = e^{\tan^2 x}$
   r) $y = x^2 e^{3x-2}$
   s) $x = 1 + 2\cos\theta$, $y = 3 - 6\sin\theta$

   t) $y = \log_e(3x+1)$
   u) $x = (t+2)^4$, $y = 8t^3 - 2t$
   v) $y = x\ln x$

   w) $y = \dfrac{\sqrt{x}}{1-x^2}$
   x) $x^3 + xy^2 - y^3 = 1$
   y) $y = \dfrac{1}{5}\tan^{-1} 10x$.

2. Find the exact coordinates of the point on
   the curve $y = 3xe^{2x}$ at which $\dfrac{d^2y}{dx^2} = 0$.

3. Show that if $y = \ln(\sin\sqrt{x})$ then $\dfrac{dy}{dx} = \dfrac{\cot\sqrt{x}}{2\sqrt{x}}$.

4. The parametric equations of a curve are
$x = (1 + t)^2$, $y = (1 - t)^2$.

   a) Express $\dfrac{dy}{dx}$ in terms of t, simplifying your answer.

   b) Find the point on the curve where the gradient is parallel to the x-axis.

5. Find the equation of the tangent to the curve $y = \dfrac{x^2 + 2x}{x^2 - 3}$ at the point $(2, 8)$.

6. The equation of a curve is $x \ln y^3 = 6$. Find the gradient of the curve at the point $(2, e)$.

7. A curve has equation $2x^2 + y^2 + 5x - 3y = 1$. Find the equation of the normal to the curve at the point $(-1, 4)$. Give your answer in the form $ax + by + c = 0$, where a, b, and c are integers.

8. Show that if $y = 4 \tan^{-1}\left(\dfrac{x}{2}\right)$, then
$$\frac{dy}{dx} = \frac{8}{4 + x^2}$$

9. If $y = \tan^{-1} 2x$, find an expression for $\dfrac{d^2y}{dx^2}$.

10. The parametric equations of a curve are $x = t^2$, $y = t^3$.
Find the equation of the tangent to the curve at the point $(1, 1)$.

11. The curve with equation $y = \dfrac{\cos x}{e^x}$ has one stationary point in the interval $0 \le x \le \pi$. Find the exact x coordinate of this point.

12. If $y = 2 \cos(\ln x)$, show that
$$x^2 \frac{d^2y}{dx^2} + x\frac{dy}{dx} + y = 0$$

13. Given that $y = \dfrac{5x^2 - 10x + 9}{(x - 1)^2}$, $x \ne 1$, show that

$$\frac{dy}{dx} = -\frac{8}{(x - 1)^3}$$

14. The parametric equations of a curve are $x = t + 1$, $y = 4 - t^2$.
Find the equation of the normal to the curve at the point where $x = 2$.

15. If $y = \sqrt{5x^2 + 3}$, find the value of $\left(\dfrac{dy}{dx}\right)^2$ when $x = -1$.

16. Find the exact coordinates of the stationary point on the curve with equation $y = x \ln 3x$, giving your answer in terms of e.

17. Prove that there are no stationary points on the curve $y = \dfrac{ax + b}{cx + d}$ where a, b, c, and d are constants and $ad \ne bc$.

18. Prove that if $y = e^{-\frac{x}{2}} \sin 2x$, then
$$4\frac{d^2y}{dx^2} + 4\frac{dy}{dx} + 17y = 0$$

19. By differentiating $\dfrac{1}{\sin 2\theta}$, show that if
$y = \operatorname{cosec} 2\theta$ then $\dfrac{dy}{d\theta} = -2 \operatorname{cosec} 2\theta \cot 2\theta$.

20. Show that the curve with equation $y = \sin x - 8 \tan x$ cannot have any stationary values.

21. The equation of a curve is $x^2 - y^2 - 5x + 3y + 13 = 0$.

   a) Find $\dfrac{dy}{dx}$ in terms of x and y.

   b) Find the values of y at the stationary points.

22. The parametric equations of a curve are given by $x = (t^2 - 2)^2$, $y = t(t - 6)$.
Find the coordinates of the stationary point of the curve.

Chapter summary

Differentiating the exponential function

- If $y = e^x$, then $\dfrac{dy}{dx} = e^x$.

- If $y = e^{kx}$, then $\dfrac{dy}{dx} = k e^{kx}$.

- If $y = e^{f(x)}$, then $\dfrac{dy}{dx} = f'(x)\, e^{f(x)}$.

Differentiating the natural logarithmic function

- If $y = \ln x$, then $\dfrac{dy}{dx} = \dfrac{1}{x}$.

- If $y = \ln [f(x)]$, then $\dfrac{dy}{dx} = \dfrac{f'(x)}{f(x)}$.

Differentiating products

- If $y = uv$, then $\dfrac{dy}{dx} = u\dfrac{dv}{dx} + v\dfrac{du}{dx}$.

Differentiating quotients

- If $y = \dfrac{u}{v}$, then $\dfrac{dy}{dx} = \dfrac{v\dfrac{du}{dx} - u\dfrac{dv}{dx}}{v^2}$.

Differentiating sin x, cos x, and tan x

- If $y = \sin x$, then $\dfrac{dy}{dx} = \cos x$.

- If $y = \cos x$, then $\dfrac{dy}{dx} = -\sin x$.

- If $y = \tan x$, then $\dfrac{dy}{dx} = \sec^2 x$.

- If $y = \tan^{-1} x$, then $\dfrac{dy}{dx} = \dfrac{1}{1 + x^2}$.

Implicit differentiation

- We use implicit differentiation when one of the variables is not given explicitly as a function of the other variable.

- When differentiating an implicit equation to find $\dfrac{dy}{dx}$, we must use the chain rule to differentiate any expression involving y. For example, $\dfrac{d}{dx}(3y^2) = 6y\dfrac{dy}{dx}$.

Parametric differentiation

- When differentiating parametric equations involving a parameter t,

$$\frac{dy}{dx} = \frac{dy}{dt} \div \frac{dx}{dt}$$

- When differentiating parametric equations where the parameter is an angle, θ,

$$\frac{dy}{dx} = \frac{dy}{d\theta} \div \frac{dx}{d\theta}, \text{ where } \theta \text{ is in radians}$$

5 Integration

Integration is the basic building block for solving any equation which describes how things move or change. Such equations are called differential equations and you will meet them in Chapter 10. Integration is used to design and construct dams like this one, which need to be built to withstand extreme stresses caused by the water they hold back.

Objectives

- Extend the idea of 'reverse differentiation' to include the integration of e^{ax+b}, $\dfrac{1}{ax + b}$, $\sin(ax + b)$, $\cos(ax + b)$, and $\sec^2(ax + b)$ (knowledge of the general method of integration by substitution is not required).

- Use trigonometrical relationships (such as double angle formulae) to facilitate the integration of functions such as $\cos^2 x$.

- (Pure 2 only) Understand and use the trapezium rule to estimate the value of a definite integral, and use sketch graphs in simple cases to determine whether the trapezium rule gives an overestimate or an underestimate.

Before you start

You should know how to:

1. Differentiate standard functions,

e.g. a) $\dfrac{d}{dx}(e^{3x+2})$

$= 3\,e^{3x+2}$

e.g. b) $\dfrac{d}{dx}(\sin 3x + \tan 2x)$

$= 3\cos 3x + 2\sec^2 2x$

2. Use trigonometrical relationships,

e.g. show that

$\cos 4x = \cos^4 x + \sin^4 x - 6\sin^2 x \cos^2 x$.

$\cos 4x = \cos^2 2x - \sin^2 2x$

$= (\cos^2 x - \sin^2 x)^2 - (2\sin x \cos x)^2$

$= \cos^4 x + \sin^4 x - 2\sin^2 x \cos^2 x$

$\qquad\qquad - 4\sin^2 x \cos^2 x$

$= \cos^4 x + \sin^4 x - 6\sin^2 x \cos^2 x$

Skills check:

1. Differentiate

a) $y = e^{2x+1} + e^{-x}$

b) $y = \sin 4x - 3\cos 2x$

c) $y = \ln(3x + 7)$.

2. Show that $\sin\theta \tan\theta + \cos\theta = \sec\theta$.

In P1 you saw that $y = \int ax^n \, dx = \dfrac{ax^{n+1}}{n+1} + c$ for $n \neq -1$, and more generally that

$$\int (ax+b)^n \, dx = \frac{(ax+b)^{n+1}}{a(n+1)} + c \quad (n \neq -1)$$

In Chapter 4 we looked at how to differentiate a number of other functions. In this chapter we will extend the reverse process of differentiation – that is, integration – to cover many other functions.

5.1 Integration of e^{ax+b}

In section 4.1 you saw that $\dfrac{d}{dx}\left(e^{ax+b}\right) = ae^{ax+b}$. This means that

$$\int e^{ax+b} \, dx = \frac{1}{a}e^{ax+b} + c$$

> It is extremely important you do not forget to write the $+c$ at the end if you are not integrating between two definite limits.

Example 1

Find $\int e^{2x+3} \, dx$.

$$\int e^{2x+3} \, dx = \frac{1}{2}\left(e^{2x+3}\right)$$

> The coefficient of x in $2x + 3$ will be a multiplying factor when the (e^{2x+3}) term is differentiated, so you need to include the $\dfrac{1}{2}$ to compensate for this.

When the integral has no limits (called an **indefinite integral**) as in Example 1, the integral is determined only up to a constant (the '$+c$' term). However, all the functions you will learn to integrate in this chapter can appear either as **definite integrals** (where you integrate between two limits), or as **improper integrals** (where the function is not defined at one end of the interval). Example 2 looks at a definite integral, and Example 3 at an improper integral.

Example 2

Calculate $\displaystyle\int_0^1 e^{2x+3} \, dx$.

$$\int_0^1 e^{2x+3} \, dx = \left[\frac{1}{2}e^{2x+3}\right]_0^1$$

> The function is as in Example 1, but the $+c$ is not needed because we are integrating between two definite limits.

$$= \frac{1}{2}e^5 - \frac{1}{2}e^3$$

> Evaluate the expression at the top and bottom limits.

$$= \frac{1}{2}(e^5 - e^3)$$

> If we were to put the constant term $+c$ in the integral above, it would just cancel out when we evaluate the expression at the top and bottom limits. As such, we do not need to include the $+c$ when integrating any expression between two definite limits.

Example 3

Calculate $\int_{-\infty}^{1} e^{2x+3}\, dx$.

$$\int_{-\infty}^{1} e^{2x+3}\, dx = \left[\frac{1}{2}e^{2x+3}\right]_{-\infty}^{1}$$

> The integral is as in Example 2, but the lower limit is different.

$$= \frac{1}{2}e^5 - \lim_{a\to-\infty}\left(\frac{1}{2}e^{2a+3}\right)$$

> Remember that (formally) this is the justification for saying the lower limit is 0.

$$= \frac{1}{2}e^5 - 0$$

$$= \frac{1}{2}e^5$$

Example 4

Calculate $\int \sqrt{e^{x+1}}\, dx$.

$$\int \sqrt{e^{x+1}}\, dx \int \left(e^{x+1}\right)^{\frac{1}{2}} dx$$

> The function needs to be expressed with a linear power.

$$= \int e^{\frac{1}{2}x+1}\, dx$$

> The derivative of the power is $\frac{1}{2}$ so you need to multiply by 2 to compensate.

$$= 2e^{\frac{1}{2}(x+1)} + c$$

Example 5

Find the area bounded by the graph of $y = e^{2x}$, the x-axis, the y-axis, and the line $x = 1$.

$$\text{Area} = \int_0^1 e^{2x}\, dx$$

> Sketching the graphs helps identify the required integral.

$$= \left[\frac{1}{2}e^{2x}\right]_0^1$$

> Integrating and substituting the limits gives the area we're asked for.

$$= \frac{1}{2}e^2 - \frac{1}{2}$$

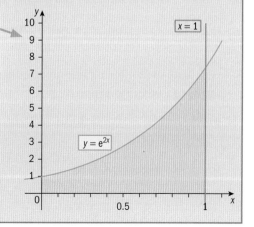

Exercise 5.1

1. Find the following indefinite integrals.

   a) $\displaystyle\int e^{3x-1}\,dx$

   b) $\displaystyle\int e^{4x+1}\,dx$

   c) $\displaystyle\int e^{0.5x+1}\,dx$

   d) $\displaystyle\int e^{\pi x+3}\,dx$

   e) $\displaystyle\int e^{4(x+1)}\,dx$

   f) $\displaystyle\int \sqrt{e^{x+1}}\,dx$

   g) $\displaystyle\int e^{2-x}\,dx$

   h) $\displaystyle\int e^{3-2x}\,dx$

   i) $\displaystyle\int \sqrt{e^{1-x}}\,dx$

2. Calculate the following definite integrals, leaving your answers in terms of e.

   a) $\displaystyle\int_0^1 e^{3x}\,dx$

   b) $\displaystyle\int_0^{0.25} e^{4x}\,dx$

   c) $\displaystyle\int_1^2 e^{x-1}\,dx$

   d) $\displaystyle\int_0^1 4e^{4x}\,dx$

   e) $\displaystyle\int_0^1 12e^{3x}\,dx$

   f) $\displaystyle\int_0^2 2e^{4x}\,dx$

   g) $\displaystyle\int_0^1 e^{-x}\,dx$

   h) $\displaystyle\int_0^2 2e^{-2x}\,dx$

   i) $\displaystyle\int_{-1}^1 e^{0.5x}\,dx$

3. Calculate the following integrals, giving your answers as exact values.

   a) $\displaystyle\int_0^{\ln 2} e^x\,dx$

   b) $\displaystyle\int_0^{2\ln 3} e^x\,dx$

   c) $\displaystyle\int_{\ln 2}^{\ln 3} e^x\,dx$

   d) $\displaystyle\int_0^{\ln 3} 4e^{4x}\,dx$

   e) $\displaystyle\int_0^{\ln 4} 2e^{2x}\,dx$

   f) $\displaystyle\int_0^{\ln 5} e^{-x}\,dx$

   g) $\displaystyle\int_0^1 e^{x\ln 3}\,dx$

   h) $\displaystyle\int_{-2}^1 e^{x\ln 2}\,dx$

   i) $\displaystyle\int_{-1}^4 e^{-x\ln 3}\,dx$

4. Find the area bounded by the graph of $y = e^x$, the x- and y-axes, and the line $x = 3$.

5. Find the area bounded by the graph of $y = e^{-2x}$, the x- and y-axes, and the line $x = 2$.

6. Find the area bounded by the graphs of $y = e^x$, $y = e^{-x}$, the y-axis, and the line $x = 2$.

7. Find $\displaystyle\int_0^k e^{-x}\,dx$ and hence find $\displaystyle\int_0^\infty e^{-x}\,dx$.

8. Calculate the following improper integrals, leaving your answers in terms of e where appropriate.

   a) $\displaystyle\int_0^\infty e^{-3x}\,dx$

   b) $\displaystyle\int_1^\infty e^{-4x}\,dx$

   c) $\displaystyle\int_1^\infty e^{1-x}\,dx$

   d) $\displaystyle\int_{-\infty}^0 4e^{4x}\,dx$

   e) $\displaystyle\int_{-\infty}^1 6e^{3x}\,dx$

   f) $\displaystyle\int_{-\infty}^2 \sqrt{e^x}\,dx$

9. The line $y = e^{-x}$ between $x = 0$ and $x = 3$ is rotated through $360°$ about the x-axis. Find the volume of the solid obtained.

5.2 Integration of $\dfrac{1}{ax+b}$

In section 4.2 you saw that $\dfrac{d}{dx}(\ln x) = \dfrac{1}{x}$. This means that $\displaystyle\int \dfrac{1}{x}\,dx = \ln x + c.$

However, the function $\ln x$ is only defined on the interval $0 < x < \infty$, which is written as $(0, \infty)$. Does this mean that the integral $\displaystyle\int \dfrac{1}{x}\,dx$ is also only defined on the interval $(0, \infty)$?

In actual fact, the integral $\displaystyle\int \dfrac{1}{x}\,dx$ is defined for all real x except $x = 0$, and we will explain how this is the case by using the graph on the right.

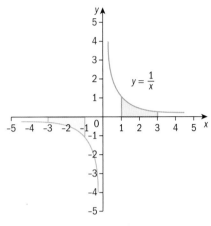

If you look at the graph of $y = \dfrac{1}{x}$ and consider the rotational symmetry about the origin, it is clear that the shaded areas are the same size. Therefore the integral $\displaystyle\int \dfrac{1}{x}\,dx$ from -3 to -1 should just be the negative of the integral from 1 to 3 (it's negative because it lies below the x-axis). The resolution of this dilemma is surprisingly simple – but the first step is to do something trivial:

write $\dfrac{1}{x}$ as $\dfrac{-1}{-x}$, which is obviously true provided $x \neq 0$.

Then $\displaystyle\int \dfrac{-1}{-x}\,dx = \ln(-x) + c.$

So $\displaystyle\int \dfrac{1}{x}\,dx = \ln x + c$ or $\displaystyle\int \dfrac{1}{x}\,dx = \ln(-x) + c$, which can be summarised as

$$\int \dfrac{1}{x}\,dx = \ln|x| + c$$

More generally you saw that $\dfrac{d}{dx}(\ln(ax+b)) = \dfrac{a}{ax+b}$, giving the general result

$$\int \dfrac{1}{ax+b}\,dx = \dfrac{1}{a}\ln|ax+b| + c$$

It is normal practice only to write the modulus sign in the case of definite integrals.

For indefinite integrals it is normal to restrict the function by the condition $ax + b > 0$.

Evaluating the two integrals above,

$$\int_1^3 \dfrac{1}{x}\,dx = \big[\ln|x|\big]_1^3 = \ln 3 - \ln 1 = \ln 3 \text{ and } \int_{-3}^{-1} \dfrac{1}{x}\,dx = \big[\ln|x|\big]_{-3}^{-1} = \ln 1 - \ln 3 = -\ln 3$$

so this not only gives the correct absolute value, but gives the correct sign for the function being below the axis.

Note:

- $\ln x$ gets large without bound as x tends to infinity and so there will not be any improper integrals involving logarithms.
- $\ln 0$ does not exist, which means that definite integrals in this form need to be defined on an interval (α, β) such that $ax + b \neq 0$ anywhere in the interval.
- It does not matter whether the end points of intervals are included or not, so definite integrals can be defined on an interval $[\alpha, \beta]$, i.e. $\alpha \leq x \leq \beta$, or (α, β), which is $\alpha < x < \beta$.

Example 6

Find $\displaystyle\int \frac{2}{2x-3}\,dx$ and state the values of x for which the answer is valid.

$\displaystyle\int \frac{2}{2x-3}\,dx = \ln(2x-3) + c; \text{ valid for } x > \frac{3}{2}$ ⟵ The derivative of the denominator is 2, so it is an exact integral, and $2x - 3 > 0$ when $x > \dfrac{3}{2}$.

Example 7

a) Sketch the graph of $y = \dfrac{1}{x-2}$.

b) Mark on your sketch the two areas represented by $\displaystyle\int_{-2}^{-1} \frac{1}{x-2}\,dx$ and $\displaystyle\int_{5}^{6} \frac{1}{x-2}\,dx$.

c) Explain, using the symmetry of the reciprocal function, why $\displaystyle\int_{-2}^{-1} \frac{1}{x-2}\,dx = -\int_{5}^{6} \frac{1}{x-2}\,dx$.

d) Calculate the two integrals in part (c) and verify the result.

---

a) and b)

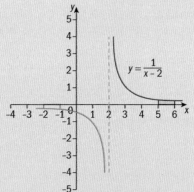

c) The vertical asymptote is at $x = 2$, so the two intervals $(-2, -1)$ and $(5, 6)$ are symmetrically placed on either side of the asymptote. Therefore the two shaded areas are equal.

d) $\displaystyle\int_{-2}^{-1} \frac{1}{x-2}\,dx = \left[\ln|x-2|\right]_{-2}^{-1} = \ln 3 - \ln 4 = \ln\left(\frac{3}{4}\right)$

$\displaystyle\int_{5}^{6} \frac{1}{x-2}\,dx = \left[\ln|x-2|\right]_{5}^{6} = \ln 4 - \ln 3 = \ln\left(\frac{4}{3}\right) = -\ln\left(\frac{3}{4}\right)$ as required

Example 8

Find $\displaystyle\int \frac{2}{5-x}\,dx$ and state the values of x for which the answer is valid.

$$\int \frac{2}{5-x}\,dx = -2\ln(5-x)+c; \quad x<5$$

The derivative of the denominator is -1, so the integral requires a multiplicative constant of -2 and the answer is valid when $5-x>0$, i.e. when $x<5$.

Example 9

a) Sketch the graph of $y = \dfrac{1}{x+2}$.

b) Mark on your sketch the area represented by $\displaystyle\int_{-3}^{1} \frac{1}{x+2}\,dx$.

c) Explain why this integral cannot be calculated.

a) and b)

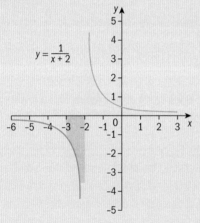

$y = \dfrac{1}{x+2}$

c) The vertical asymptote is at $x = -2$, so the interval $(-3, 1)$ includes the asymptote, and you cannot integrate across an asymptote as the area would be of infinite size.

Both the area above the x-axis and the area below the x-axis are infinite.

Example 10

Calculate $\displaystyle\int_{3}^{5} \frac{1}{3x-2}\,dx$, giving your answer as a single logarithm (i.e. in the form $\ln k$, where k is real).

$$\int_{3}^{5} \frac{1}{3x-2}\,dx = \left[\frac{1}{3}\ln|3x-2|\right]_{3}^{5}$$

Use the standard form of the logarithmic integral.

$$= \frac{1}{3}\ln 13 - \frac{1}{3\ln}7$$

Evaluate the expression at the top and bottom limits.

$$= \frac{1}{3}\ln\left(\frac{13}{7}\right) = \ln\left(\sqrt[3]{\frac{13}{7}}\right)$$

First combine the two logarithms, then use the laws of logarithms to take the $\frac{1}{3}$ into the logarithm as the cube root.

Example 11

Calculate $\displaystyle\int_{-2}^{1} \frac{1}{7-2x}\,dx$, giving your answer as a single logarithm (i.e. in the form $\ln k$,

where k is real).

$$\int_{-2}^{1} \frac{1}{7-2x}\,dx = \left[-\frac{1}{2}\ln|7-2x|\right]_{-2}^{1}$$

Use the standard from of the logarithmic integral, noting a is negative.

$$= -\frac{1}{2}\ln 5 - \left(-\frac{1}{2}\ln 11\right)$$

Evaluate the expression at the top and bottom limits.

$$= \frac{1}{2}\ln\frac{11}{5} = \ln\sqrt{2.2}$$

First combine the two logarithms, then use the laws of logarithms to take the $\frac{1}{2}$ into the logarithm as the square root.

Example 12

Calculate $\displaystyle\int_{4}^{6} \frac{1}{7-2x}\,dx$, giving your answer as a single logarithm (i.e. in the form $\ln k$,

where k is real).

$$\int_{4}^{6} \frac{1}{7-2x}\,dx = \left[-\frac{1}{2}\ln|7-2x|\right]_{4}^{6}$$

The integral is the same as in Example 11, but with different limits.

$$= -\frac{1}{2}\ln 5 - \left(-\frac{1}{2}\ln 1\right)$$

Evaluating the expression at the top and bottom limits now requires taking the absolute values of −5 and −1.

$$= \frac{1}{2}\ln\frac{1}{5} = \ln\sqrt{0.2}$$

As before, first combine the two logarithms, then use the laws of logarithms to take the $\frac{1}{2}$ into the logarithm as the square root.

When ln 1 occurs in calculations, it can either be replaced by 0, or you can combine it into a single logarithm using the standard laws.

Exercise 5.2

1. Find the following indefinite integrals and state the values of x for which the answer is valid.

a) $\displaystyle\int \frac{1}{2x-1}\,dx$

b) $\displaystyle\int \frac{2}{2x-1}\,dx$

c) $\displaystyle\int \frac{7}{7x+3}\,dx$

d) $\displaystyle\int \frac{10}{5-2x}\,dx$

e) $\displaystyle\int \frac{-6}{1-3x}\,dx$

f) $\displaystyle\int \frac{-2}{6x+5}\,dx$

g) $\displaystyle\int \frac{-3}{x+1}\,dx$

h) $\displaystyle\int \frac{1}{7-x}\,dx$

i) $\displaystyle\int \frac{2}{x+e}\,dx$

j) $\displaystyle\int \frac{-0.5}{2x+1}\,dx$

k) $\displaystyle\int \frac{-12}{9x+16}\,dx$

l) $\displaystyle\int \frac{-2}{10x-3}\,dx$

Integration of $\dfrac{1}{ax+b}$

2. Calculate the following definite integrals. Note that parts **(a)** to **(c)** use the same indefinite integrals as you calculated in parts **(a)** to **(c)** of question 1.

a) $\displaystyle\int_{2}^{3} \frac{1}{2x-1}\,dx$

b) $\displaystyle\int_{5}^{7} \frac{2}{2x-1}\,dx$

c) $\displaystyle\int_{0}^{1} \frac{7}{7x+3}\,dx$

d) $\displaystyle\int_{0}^{1} \frac{4}{5-2x}\,dx$

e) $\displaystyle\int_{0}^{1} \frac{9}{4-2x}\,dx$

f) $\displaystyle\int_{0}^{2} \frac{-7}{5-x}\,dx$

g) $\displaystyle\int_{2}^{4} \frac{2}{x-7}\,dx$

h) $\displaystyle\int_{-6}^{-4} \frac{-4}{7-2x}\,dx$

i) $\displaystyle\int_{-1}^{1} \frac{6}{5-3x}\,dx$

3. Find the area bounded by the graph of $y = \dfrac{1}{3+x}$, the x- and y-axes, and the line $x = 3$.

4. Find the area bounded by the graph of $y = \dfrac{-2}{x+3}$, the x- and y-axes, and the line $x = 2$.

5. Find the area bounded by the graphs of $y = \dfrac{1}{x-3}$, $y = \dfrac{1}{2x+3}$, the y-axis, and the line $x = 2$.

5.3 Integration of $\sin(ax+b)$, $\cos(ax+b)$, $\sec^2(ax+b)$

In section 4.5 you saw that

$$\frac{d}{dx}(\sin(ax+b)) = a\cos(ax+b)$$

$$\frac{d}{dx}(\cos(ax+b)) = -a\sin(ax+b)$$

$$\frac{d}{dx}(\tan(ax+b)) = a\sec^2(ax+b).$$

This means that the following are the standard trigonometric integral forms.

$$\int \sin(ax+b)\,dx = -\frac{1}{a}\cos(ax+b)+c$$

$$\int \cos(ax+b)\,dx = \frac{1}{a}\sin(ax+b)+c$$

$$\int \sec^2(ax+b)\,dx = \frac{1}{a}\tan(ax+b)+c$$

Remember: If you are using calculus with trigonometric functions, then x (or equivalent) is measured in radians and not in degrees.

Since all the trigonometric functions are periodic, there will not be a limit to which an integral approaches as x tends to infinity, so there will be no improper integrals of this sort, but care does need to be taken with the discontinuities that occur with $\tan x$ and also with all the reciprocal trigonometric functions.

Example 13

Find $\displaystyle\int \cos 2x\,dx$.

∙∙

$$\int \cos 2x\,dx = \frac{1}{2}\sin 2x + c$$

Remember to divide by the coefficient of x (i.e. 2).

Example 14

Find $\int \sin\left(3x + \frac{\pi}{4}\right)dx$.

....

$\int \sin\left(3x + \frac{\pi}{4}\right)dx = -\frac{1}{3}\cos\left(3x + \frac{\pi}{4}\right) + c$

Divide by the coefficient of x (i.e. 3).

Example 15

a) Sketch the graph of $y = \sec x$ and use it to sketch the graph of $y = \sec^2 x$.

b) On your sketch, shade the area corresponding to $\int_0^{\frac{\pi}{4}} \sec^2 x \, dx$ and estimate to 1 significant figure the value of the area.

c) Calculate $\int_0^{\frac{\pi}{4}} \sec^2 x \, dx$.

....

a) $y = \sec x$ $y = \sec^2 x$

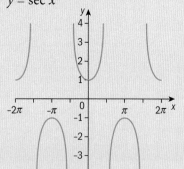 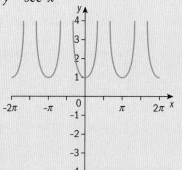

b) The area is approximately 0.8 wide (the exact width is $\frac{\pi}{4}$) and between 1 and 2 high, mostly closer to 1, so estimate is 1 correct to 1 significant figure.

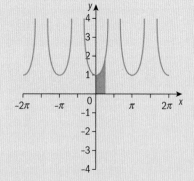

c) $\int_0^{\frac{\pi}{4}} \sec^2 x \, dx = [\tan x]_0^{\frac{\pi}{4}} = 1 - 0 = 1$

Integration of $\sin(ax + b)$, $\cos(ax + b)$, $\sec^2(ax + b)$

Example 16

By first sketching an appropriate graph, find the area bounded by the graph
of $y = \sin\left(x - \frac{\pi}{3}\right)$, the x- and y-axes and the line $x = \pi$.

The sketch shows that the area required is partly above
and partly below the x-axis, so the integral will need
to be done in two sections:

i) from 0 to $\frac{\pi}{3}$,   **ii)** from $\frac{\pi}{3}$ to π.

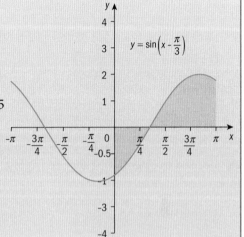

$$\int_0^{\frac{\pi}{3}} \sin\left(x - \frac{\pi}{3}\right)dx = \left[-\cos\left(x - \frac{\pi}{3}\right)\right]_0^{\frac{\pi}{3}} = -1 - (-0.5) = -0.5$$

so the first area is 0.5 (below the axis)

$$\int_{\frac{\pi}{3}}^{\pi} \sin\left(x - \frac{\pi}{3}\right)dx = \left[-\cos\left(x - \frac{\pi}{3}\right)\right]_{\frac{\pi}{3}}^{\pi} = 0.5 - (-1) = 1.5$$

Thus, the total area required is $0.5 + 1.5 = 2$.

Exercise 5.3

1. Find these integrals.

a) $\displaystyle\int \cos 2x \, dx$

b) $\displaystyle\int \sin\left(2x + \frac{\pi}{4}\right)dx$

c) $\displaystyle\int \sec^2\left(3x + \frac{\pi}{12}\right)dx$

d) $\displaystyle\int \cos(\pi - x)\,dx$

e) $\displaystyle\int \cos\left(\frac{\pi}{2} - \frac{x}{4}\right)dx$

f) $\displaystyle\int -\sin\left(2x - \frac{\pi}{4}\right)dx$

g) $\displaystyle\int \sec^2\left(\frac{\pi}{12} - 5x\right)dx$

h) $\displaystyle\int -\frac{1}{2}\sin\left(\frac{1}{2}x\right)dx$

i) $\displaystyle\int \cos\left(4 - \frac{1}{3}x\right)dx$

2. Calculate the following definite integrals. Note that parts **(a)** to **(c)** use the
same indefinite integral as you calculated in parts **(a)** to **(c)** of question 1.

a) $\displaystyle\int_0^{\frac{\pi}{4}} \cos 2x \, dx$

b) $\displaystyle\int_0^{\frac{\pi}{2}} \sin\left(2x + \frac{\pi}{4}\right)dx$

c) $\displaystyle\int_0^{\frac{\pi}{12}} \sec^2\left(3x + \frac{\pi}{12}\right)dx$

d) $\displaystyle\int_0^{\frac{\pi}{2}} \cos 2x \, dx$

e) $\displaystyle\int_0^{\pi} \cos\left(\frac{3\pi}{2} - \frac{x}{3}\right)dx$

f) $\displaystyle\int_0^{\frac{\pi}{2}} \sin\left(3x - \frac{\pi}{4}\right)dx$

g) $\displaystyle\int_{-\frac{\pi}{12}}^{\frac{\pi}{12}} \sec^2\left(\frac{\pi}{12} - 3x\right)dx$

h) $\displaystyle\int_{-\frac{\pi}{4}}^{\frac{\pi}{4}} -\frac{1}{3}\cos\left(\frac{1}{2}x\right)dx$

i) $\displaystyle\int_1^3 \cos\left(2 - \frac{1}{2}x\right)dx$

3. A curve passes through the point $\left(\frac{\pi}{4}, 2\right)$ and its gradient function is $\sec^2 x$.
Find the equation of the curve.

4. Find the area bounded by the graph of $y = \cos\left(x + \dfrac{\pi}{4}\right)$, the x- and y-axes, and the line $x = \pi$.

5. Find the area bounded by the graphs of $y = \cos\left(x + \dfrac{\pi}{4}\right)$, $y = \cos\left(x + \dfrac{\pi}{2}\right)$, the y-axis, and the line $x = \dfrac{\pi}{2}$.

6. Find the area bounded by the graph of $y = \sec^2\left(2x + \dfrac{\pi}{2}\right)$, the x-axis, and the lines $x = \dfrac{\pi}{4}$ and $x = \dfrac{\pi}{3}$.

5.4 Extending integration of trigonometric functions

In Chapter 3, you met a number of trigonometric identities which can be used to extend the range of trigonometric functions you can integrate. Those identities were

- $1 + \tan^2 \theta \equiv \sec^2 \theta$

- $1 + \cot^2 \theta \equiv \operatorname{cosec}^2 \theta$

- $\sin 2A \equiv 2 \sin A \cos A$

- $\cos 2A \equiv \cos^2 A - \sin^2 A$
  $\equiv 2 \cos^2 A - 1$
  $\equiv 1 - 2 \sin^2 A$

- $\tan 2A \equiv \dfrac{2 \tan A}{1 - \tan^2 A}$

These are formulae you should know well enough that if you are faced with an integral involving trigonometric functions which you cannot integrate, you can manipulate the expression into one which can be integrated using these formulae.

The double angle formula for $\cos 2A$ is particularly useful in the forms $\sin^2 A = \dfrac{1}{2}(1 - \cos 2A)$ and $\cos^2 A = \dfrac{1}{2}(1 + \cos 2A)$.

The other formulae developed in Chapter 3 will normally be hinted at, or given, if you need to use them in an integration problem.

Example 17

Find the integrals

a) $\displaystyle\int \cos^2 x \, dx$ b) $\displaystyle\int 12 \sin^2 x \, dx$.

a) $\displaystyle\int \cos^2 x \, dx = \int \dfrac{1}{2}(1 + \cos 2x) \, dx$

Use the double angle formula to get an integrable form.

$= \dfrac{1}{2}x + \dfrac{1}{4} \sin 2x + c$

Remember to integrate the number part as well as the trigonometric function, and include the $+ c$.

b) $\displaystyle\int 12 \sin^2 x \, dx = \int 12 \times \dfrac{1}{2}(1 - \cos 2x) \, dx$

$= 6x - 3 \sin 2x + c$

Again the double angle formula transforms this into an integrable form.

Example 18

Calculate the exact value of $\displaystyle\int_0^{\frac{\pi}{6}} 4\tan^2\theta\, d\theta$.

$$\int_0^{\frac{\pi}{6}} 4\tan^2\theta\, d\theta = \int_0^{\frac{\pi}{6}} 4(\sec^2\theta - 1)\, d\theta$$

> The use of the identity makes this a function you know how to integrate.

$$= \left[4\tan\theta - 4\theta\right]_0^{\frac{\pi}{6}}$$

> These are standard integrals.

$$= \left(\frac{4}{\sqrt{3}} - \frac{4\pi}{6}\right) - (0 - 0)$$

> Evaluate the expression at the top and bottom limits explicitly to show your working.

$$= \frac{4}{\sqrt{3}} - \frac{2\pi}{3}$$

Example 19

a) By writing $2x = 3x - x$ and $4x = 3x + x$, show that $\sin 3x \cos x = \frac{1}{2}(\sin 2x + \sin 4x)$.

b) Hence calculate $\displaystyle\int_0^{\frac{\pi}{6}} \sin 3x \cos x\, dx$.

a) $\sin 2x = \sin(3x - x) = \sin 3x \cos x - \cos 3x \sin x$

$\sin 4x = \sin(3x + x) = \sin 3x \cos x + \cos 3x \sin x$

Adding the two expressions

$\Rightarrow \sin 2x + \sin 4x = 2\sin 3x \cos x$

$\Rightarrow \sin 3x \cos x = \frac{1}{2}(\sin 2x + \sin 4x)$

> Using the compound angle formulae for $\sin x$ gives functions that you know how to integrate.

b) $\displaystyle\int_0^{\frac{\pi}{6}} \sin 3x \cos x\, dx = \int_0^{\frac{\pi}{6}} \frac{1}{2}(\sin 2x + \sin 4x)\, dx$

> Using part (a).

$$= \left[-\frac{1}{4}\cos 2x - \frac{1}{8}\cos 4x\right]_0^{\frac{\pi}{6}}$$

> Integrate the standard forms.

$$= \left(-\frac{1}{4} \times \frac{1}{2} - \frac{1}{8} \times \left(-\frac{1}{2}\right)\right) - \left(-\frac{1}{4} - \frac{1}{8}\right)$$

$$= \frac{5}{16}$$

Example 20

a) Show that $(3 \sin x - \cos x)^2 = 5 - 3 \sin 2x - 4 \cos 2x$.

b) Hence calculate $\int_0^{\frac{\pi}{2}} (3 \sin x - \cos x)^2 \, dx$.

a) $(3 \sin x - \cos x)^2 = 9 \sin^2 x - 6 \sin x \cos x + \cos^2 x$ ← Expand the square.

$$= 9 \times \frac{1}{2}(1 - \cos 2x) - 3 \sin 2x + \frac{1}{2}(1 + \cos 2x)$$ ← Use the standard formulae.

$$= 5 - 3 \sin 2x - 4 \cos 2x$$ ← Simplify to the required form.

b) $\displaystyle\int_0^{\frac{\pi}{2}} (3 \sin x - \cos x)^2 \, dx = \int_0^{\frac{\pi}{2}} (5 - 3 \sin 2x - 4 \cos 2x) \, dx$ ← Use part **(a)**.

$$= \left[5x + \frac{3}{2} \cos 2x - 2 \sin 2x \right]_0^{\frac{\pi}{2}}$$ ← Integrate the standard forms.

$$= \left(\frac{5\pi}{2} - \frac{3}{2} - 0 \right) - \left(0 + \frac{3}{2} - 0 \right)$$ ← Be careful not to assume the limit at 0 is simply '0'.

$$= \frac{5\pi}{2} - 3$$ ← Simplify the answer.

Example 21

a) Find $\dfrac{d}{dx}(\cos x + x \sin x)$.

b) Hence calculate $\int_0^{\frac{\pi}{2}} (x \cos x) \, dx$.

Remember 'hence' means that you are expected to use a previous result.

a) $\dfrac{d}{dx}(\cos x + x \sin x) = -\sin x + \sin x + x \cos x = x \cos x$ ← Use the product rule.

b) $\displaystyle\int_0^{\frac{\pi}{2}} (x \cos x) \, dx = \int_0^{\frac{\pi}{2}} \left(\frac{d}{dx}(\cos x + x \sin x) \right) dx$ ← Use part **(a)**.

$$= \left[\cos x + x \sin x \right]_0^{\frac{\pi}{2}}$$ ← Integration is the reverse process to differentiation.

$$= \left(0 + \frac{\pi}{2} \right) - (1 + 0)$$ ← Substitute at upper and lower limits.

$$= \frac{\pi}{2} - 1$$

Example 22

a) Show that $\dfrac{\cot^2\theta}{1+\cot^2\theta} = \cos^2\theta.$ **b)** Hence calculate $\displaystyle\int_0^{\frac{\pi}{3}}\left(\dfrac{\cot^2\theta}{1+\cot^2\theta}\right)d\theta.$

---

a) $\dfrac{\cot^2\theta}{1+\cot^2\theta} = \dfrac{\cot^2\theta}{\csc^2\theta}$ ⟵ Use the standard identity.

$\qquad = \dfrac{\left(\dfrac{\cos^2\theta}{\sin^2\theta}\right)}{\left(\dfrac{1}{\sin^2\theta}\right)}$ ⟵ Express in terms of sin and cos only.

$\qquad = \cos^2\theta$

b) $\displaystyle\int_0^{\frac{\pi}{3}}\left(\dfrac{\cot^2\theta}{1+\cot^2\theta}\right)d\theta = \int_0^{\frac{\pi}{3}}\left(\cos^2\theta\right)d\theta$ ⟵ Use part (a).

$\qquad = \displaystyle\int_0^{\frac{\pi}{3}}\left(\dfrac{1+\cos2\theta}{2}\right)d\theta$ ⟵ Use the standard form to convert to an integrable function.

$\qquad = \left[\dfrac{\theta}{2}+\dfrac{\sin2\theta}{4}\right]_0^{\frac{\pi}{3}}$ ⟵ Integrate.

$\qquad = \left(\dfrac{\pi}{6}+\dfrac{\sqrt3}{8}\right)-(0+0)$ ⟵ Substitute at upper and lower limits.

$\qquad = \dfrac{\pi}{6}+\dfrac{\sqrt3}{8}$

Example 23

a) Show that $4\sin^4\theta = \dfrac{3}{2}-2\cos2\theta+\dfrac{\cos4\theta}{2}.$ **b)** Hence calculate $\displaystyle\int_0^{\frac{\pi}{2}}4\sin^4\theta\,d\theta.$

---

a) $4\sin^4\theta = (2\sin^2\theta)^2$

$\qquad = (1-\cos2\theta)^2$ ⟵ Using the double angle cosine formula twice to remove powers of $\cos\theta$ and $\sin\theta$ gives the required result. An alternative approach is to start with the right-hand side and work down to single angle expressions.

$\qquad = 1-2\cos2\theta+\cos^2 2\theta$

$\qquad = 1-2\cos2\theta+\left(\dfrac{1+\cos4\theta}{2}\right)$

$\qquad = \dfrac{3}{2}-2\cos2\theta+\dfrac{\cos4\theta}{2}$

b) $\displaystyle\int_0^{\frac{\pi}{2}}4\sin^4\theta\,d\theta = \int_0^{\frac{\pi}{2}}\left(\dfrac{3}{2}-2\cos2\theta+\dfrac{\cos4\theta}{2}\right)d\theta$ ⟵ Use part (a).

$\qquad = \left[\dfrac{3}{2}\theta-\sin2\theta+\dfrac{\sin4\theta}{8}\right]_0^{\frac{\pi}{2}}$ ⟵ Integration is the reverse process to differentiation.

$\qquad = \left(\dfrac{3\pi}{4}-0+0\right)-(0-0+0) = \dfrac{3\pi}{4}$ ⟵ Substitute at upper and lower limits.

Exercise 5.4

1. Find these integrals.

    a) $\int \cos^2 2x \, dx$ **b)** $\int \sin^2 3x \, dx$ **c)** $\int (1 + \tan^2 2x) \, dx$

    d) $\int (\sin x \cos x) \, dx$ **e)** $\int (\sin^2 x \cos^2 x) \, dx$ **f)** $\int (\sin x + \cos x)^2 \, dx$

2. Calculate the following definite integrals. Note that parts (**a**) to (**c**) use the same indefinite integral as you calculated in parts (**a**) to (**c**) of question **1**.

    a) $\int_0^{\frac{\pi}{4}} \cos^2 2x \, dx$ **b)** $\int_0^{\frac{\pi}{2}} \sin^2 3x \, dx$ **c)** $\int_0^{\frac{\pi}{12}} (1 + \tan^2 2x) \, dx$

    d) $\int_0^{\frac{\pi}{2}} (1 - \cos 2x)^2 \, dx$ **e)** $\int_0^{\pi} (1 + 2 \sin x)^2 \, dx$ **f)** $\int_0^{\frac{\pi}{2}} (\sin x - \cos x)^2 \, dx$

3. A curve passes through the origin and its gradient function is $\cos^2 x$. Find the equation of the curve.

4. Find the area bounded by the graph of $y = \sec^2 x$, the x-axis, and the lines $x = \frac{\pi}{6}$ and $x = \frac{\pi}{3}$.

5. Find the area bounded by the graphs of $y = \cos^2 x$, $y = \sin^2 x$, the y-axis, and the line $x = \frac{\pi}{4}$.

6. **a)** By writing $2x = 5x - 3x$ and $8x = 5x + 3x$, show that $\sin 3x \cos 5x = \frac{1}{2}(\sin 8x - \sin 2x)$.

    b) Hence calculate $\int_0^{\frac{\pi}{6}} \sin 3x \cos 5x \, dx$.

7. **a)** Show that $(\sin x + 5 \cos x)^2 = 13 + 5 \sin 2x + 12 \cos 2x$.

    b) Hence calculate $\int_0^{\frac{\pi}{2}} (\sin x + 5 \cos x)^2 \, dx$.

8. **a)** Show that $\frac{d}{dx}(\cos 2x + 2x \sin 2x - 2x^2 \cos 2x) = 4x^2 \sin 2x$.

    b) Hence calculate $\int_0^{\frac{\pi}{2}} (x^2 \sin 2x) \, dx$.

9. **a)** Show that $\dfrac{2 \tan \theta}{1 + \tan^2 \theta} = \sin 2\theta$.   **b)** Hence calculate $\int_0^{\frac{\pi}{6}} \left(\dfrac{2 \tan \theta}{1 + \tan^2 \theta} \right)^2 d\theta$.

10. **a)** Show that $\dfrac{1 - \tan^2 x}{1 + \tan^2 x} = \cos 2x$.

    Hence calculate

    b) $\int_0^{\frac{\pi}{6}} \left(\dfrac{1 - \tan^2 x}{1 + \tan^2 x} \right)^2 dx$ and   **c)** $\int_0^{\frac{\pi}{6}} \left(\dfrac{1 + \tan^2 x}{1 - \tan^2 x} \right)^2 dx$.

5.5 Numerical integration using the trapezium rule – Pure 2

So far in this chapter you have extended the number of types of function that you can integrate. If you continue to study mathematics you will continue to extend your range of integrable functions, but there will always be functions for which you need to calculate a definite integral, but you do not have a standard technique to use.

There are a number of numerical methods that can be used to approximate the area under the curve, using only the values of the function at particular points. In this course you only need one such method: the **trapezium rule**.

We will first consider using the trapezium rule with a simple function ($y = x^2$) which you can also integrate. In doing so, we will get some idea of how good the approximation is, and what factors affect the accuracy.

> Sometimes you will not be able to integrate a function because you have not yet met the technique, but at times you may be faced with a function for which no-one has yet discovered a way to integrate it – and it can be as simple as ($x \tan x$).

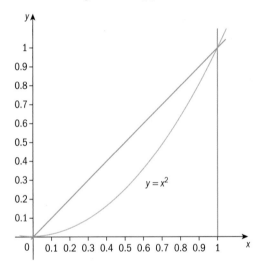

The area to the left of the green vertical line under the curve is given by $\int_0^1 x^2 \, dx = \left[\frac{1}{3}x^3\right]_0^1 = \frac{1}{3} - 0 = \frac{1}{3}$.

We will split the interval $(0, 1)$ up into an increasing number of strips and join the points on the curve by straight lines, creating trapeziums for which we have geometric formulae to calculate areas.

If there was just one strip, the approximation would actually be the area of a triangle because the left end starts on the x-axis – at the origin, and it would give an estimate of $\frac{1}{2}$ for the area.

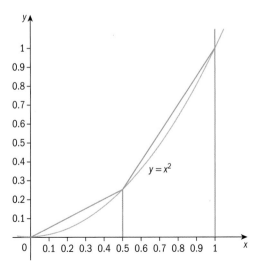

Even with two strips it is clear (visually) that the approximation will be reasonable. The two geometric shapes have areas $\frac{1}{16}$ and $\frac{5}{16}$, which when added together give an estimate of 0.375 for the area under the curve, compared to the true value of 0.333 (a 12.5% overestimate).

> Since the chords joining the points on the curve always lie above the curve itself, it is clear that every estimate with this method will be an overestimate. For most functions this will not be the case.

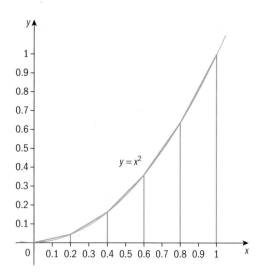

With five strips you can see that the chords joining the points on the curve follow the curve very closely, so it is no surprise that the area now is 0.34 which is overestimating the true area by only 2%. With 10 strips the overestimate is down to 0.5%.

Let us consider the simplest form of the trapezium rule: using a single interval to estimate the integral of a function f(x) over the interval (a, b). Then

$$\int_a^b f(x)\,dx \approx (b-a)\left(\frac{f(a)+f(b)}{2}\right)$$

(The area of a trapezium is the average length of the parallel sides multiplied by the distance between them.)

You have seen how the accuracy can improve if we divide the interval up into a number of strips, but we then have to calculate the area of each strip separately.

A less time-consuming way is the following:

If we choose equal strip widths, that will be a common factor in each area, and then consider the lengths which make the parallel sides of the trapeziums, apart from the first one (at a) and the last one (at b), each occurs as the end of one strip and again as the start of the next one. We can therefore express the estimate obtained by the trapezium rule in a formula rather than actually working out all the areas separately.

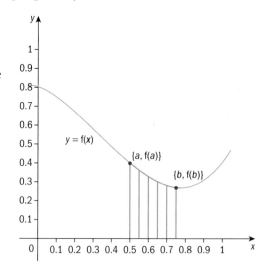

Some formal notation is needed to allow us to express this as a general rule.
If the interval (a, b) is divided into n intervals, each of width h, then let
$y_0 = f(a)$, $y_1 = f(a + h)$, $y_2 = f(a + 2h)$, ... , $y_n = f(a + nh) = f(b)$.

The trapezium rule can then be stated as

$$A = \frac{1}{2}h\{y_0 + y_n + 2(y_1 + y_2 + ... + y_{n-1})\} \approx \int_a^b f(x)\,dx, \text{where } h = \frac{b-a}{n}$$

The formula for the trapezium rule will not be provided in the examination, so you need to remember it one way or another.

Note that the first y value is labelled y_0 because with n strips there are actually $n + 1$ values being used. It is helpful to think about this formula in words relating to way it is derived: the bracket is 'the sum of the end values plus twice the sum of the other values'. The formula usually refers to n intervals, but sometimes it may refer to $n + 1$ ordinates. (The x-coordinate is known as the abscissa and the y-coordinate as the ordinate of a point in the plane.)

Example 24

a) Use the trapezium rule with five intervals to estimate $\int_0^{\frac{\pi}{2}} (x \cos x)\,dx$, giving your answer correct to 3 decimal places.

b) In Example 21, you saw that the exact value of this integral is $\frac{\pi}{2} - 1$. Calculate the percentage error in the estimate using the trapezium rule with five intervals.

a) In order to get the answer correct to 3 decimal places we need to use at least 4 decimal places in our working. The example below uses five decimal places.

n	0	1	2	3	4	5
x	0	$\frac{\pi}{10}$	$\frac{2\pi}{10} = \frac{\pi}{5}$	$\frac{3\pi}{10}$	$\frac{4\pi}{10} = \frac{2\pi}{5}$	$\frac{5\pi}{10} = \frac{\pi}{2}$
$y = f(x)$	0	0.298 78	0.508 32	0.553 97	0.388 32	0

You may find it helpful to underline the values that are to be added twice.

$$A = \frac{1}{2} \times \frac{\pi}{10} \times \{0 + 0 + 2 \times (0.298\,78 + 0.508\,32 + 0.553\,97 + 0.388\,32)\}$$
$$= 0.549\,58... - 0.550 \text{ (3 d.p.)}$$

Show the value obtained in your calculation to more accuracy than required and then give the rounded value.

▶ Continued on the next page

b) % error = $\dfrac{\left(\dfrac{\pi}{2} - 1\right) - 0.550}{\left(\dfrac{\pi}{2} - 1\right)} \times 100\% = 3.6\%$

This is the actual error divided by the true value, expressed as a percentage.

The diagram shows the function $y = x \cos x$ in the required interval, along with the five trapeziums whose areas have been used to calculate the estimate. You can see that multiplying $\cos x$ by x means the symmetry of the wave function has been lost.

It is also clear from the sketch that the trapezium rule will give an underestimate of the integral for any number of intervals.

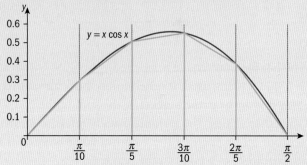

Example 25

a) Use the trapezium rule with five ordinates to estimate $\displaystyle\int_1^3 \dfrac{1}{x}\,dx$, giving your answer correct to 3 decimal places.

b) Show that the exact value of this integral is $\ln 3$.

c) Calculate the percentage error in the estimate found in part **(a)**.

a)

n	0	1	2	3	4
x	1	1.5	2	2.5	3
$y = f(x)$	1	$\dfrac{2}{3}$	$\dfrac{1}{2}$	$\dfrac{2}{5}$	$\dfrac{1}{3}$

$A = \dfrac{1}{2} \times \dfrac{1}{2} \times \left\{ 1 + \dfrac{1}{3} + 2 \times \left(\dfrac{2}{3} + \dfrac{1}{2} + \dfrac{2}{5} \right) \right\}$

$= \dfrac{67}{60} = 1.116\,66... \; 1.117 \; (3 \, d.p.)$

In this instance, the values of $f(x)$ can be left as fractions and the exact value of A can be calculated before rounding to the required accuracy.

b) $\displaystyle\int_1^3 \dfrac{1}{2}\,dx = [\ln x]_1^3 = \ln 3 - \ln 1 = \ln 3$

c) % error = $\dfrac{\left(\ln 3 - \dfrac{67}{60}\right)}{\ln 3} \times 100\% = -1.6\%$

so the percentage error is 1.6%.

This is the actual error divided by the true value, expressed as a percentage. The negative sign only shows that the estimate is larger than the true value, so the absolute value is given.

You may be given a sketch of the function being integrated and asked whether you can tell if the trapezium rule gives an overestimate or underestimate. However, you will not normally be asked about the percentage error in the estimate.

Example 26

a) Use the trapezium rule with two intervals to estimate
$\int_{0}^{\frac{1}{2}\pi} \sqrt{1 - \sin x}\, dx$, giving your answer correct to
2 decimal places.

b) The diagram shows the graph of $y = \sqrt{1 - \sin x}$. State,
with a reason, whether your answer to part (a) is an
overestimate or an underestimate of the integral.

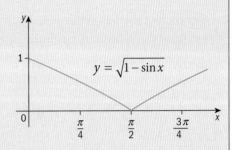

a)

n	0	1	2
x	0	$\dfrac{\pi}{4}$	$\dfrac{\pi}{2}$
y	1	0.5412	0

$A = \dfrac{1}{2} \times \dfrac{\pi}{4} \times \{1 + 0 + 2 \times 0.5412\}$

$\qquad = 0.8177... = 0.82\ (2\,\text{d.p.})$

> Two intervals only gives one value to be multiplied by 2 in the formula. It may feel different, but the process is exactly the same. Two intervals are used quite often in examinations to cut down on 'busy work'.

b) The graph is curving in such a way that no chord lies above the graph at any point in the interval $(0, \dfrac{\pi}{2})$, so the trapezium rule will give an underestimate for the integral.

Example 27

Use the trapezium rule with two intervals to estimate $\int_{0}^{1} e^{-x^2}\, dx$, giving your
answer correct to 2 decimal places.

n	0	1	2
x	0	0.5	1
y	1	0.7788	0.3679

> Sketch the graph of $y = e^{-x^2}$.

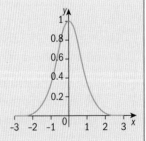

$A = \dfrac{1}{2} \times \dfrac{1}{2} \times \{1 + 0.3679 + 2 \times 0.7788\}$

$\qquad = 0.7313... = 0.73\ (2\,\text{d.p.})$

> If you study Probability and Statistics you may recognise the shape of this curve as that of the normal distribution.

Exercise 5.5

1. For each of the following, use the trapezium rule with two intervals to estimate the value of the definite integral to 3 significant figures.

   a) $\displaystyle\int_0^{\frac{\pi}{4}} \cos^3 2x \, dx$

   b) $\displaystyle\int_0^{\frac{\pi}{2}} \sqrt{1 + \sin 3x} \, dx$

   c) $\displaystyle\int_0^{\frac{\pi}{6}} \sqrt{1 + \tan^2 2x} \, dx$

   d) $\displaystyle\int_0^1 \sqrt{1 + \sqrt{4x}} \, dx$

   e) $\displaystyle\int_0^1 \frac{e^{2x}}{1 + e^x} \, dx$

   f) $\displaystyle\int_0^1 \frac{1}{1 + 9x^2} \, dx$

2. For each of the following, use the trapezium rule with the specified number of intervals to estimate the value of the definite integral to 3 significant figures.

   a) $\displaystyle\int_0^{\frac{\pi}{4}} \cos^3 2x \, dx$ [4 intervals]

   b) $\displaystyle\int_0^{\frac{\pi}{2}} \sqrt{1 + \sin 3x} \, dx$ [3 intervals]

   c) $\displaystyle\int_0^{\pi/6} \sqrt{1 + \tan^2 2x} \, dx$ [6 intervals]

   d) $\displaystyle\int_0^1 \sqrt{1 + \sqrt{4x}} \, dx$ [5 intervals]

   e) $\displaystyle\int_0^1 \frac{e^{2x}}{1 + e^x} \, dx$ [4 intervals]

   f) $\displaystyle\int_0^1 \frac{1}{1 + 9x^2} \, dx$ [5 intervals]

3. The diagram shows the graph of $y = \sqrt{1 + x^2}$. The region R is bounded by the curve, the x- and y-axes and the line $x = 2$.

   a) Use the trapezium rule with six ordinates to obtain an approximation for the area of R, giving your answer to a suitable degree of accuracy.

   b) Explain, with the aid of a sketch, whether the approximation is an overestimate or underestimate.

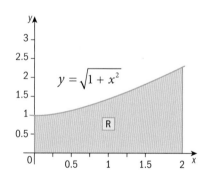

4. The diagram shows the graph of $y = xe^{-x}$.
   The region R is bounded by the curve, the x-axis
   and the line $x = 2$.

   a) Use the trapezium rule with 11 ordinates to obtain
      an approximation for the area of R, giving your
      answer to 3 significant figures.

   b) Explain, with the aid of a sketch, whether the
      approximation is an overestimate or
      underestimate.

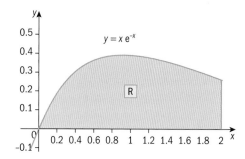

5. The diagram shows the graph of $y = \dfrac{1}{1+x}$.
   The region R is bounded by the curve, the x- and
   y-axes, and the line $x = 2$.

   a) Use the trapezium rule with
      i) two intervals and
      ii) 10 intervals
      to obtain approximations for
      the area of R, giving your answers to
      3 significant figures.

   b) Calculate the exact value of $\displaystyle\int_0^2 \dfrac{1}{1+x}\,dx$ and comment
      on the accuracy of your estimates.

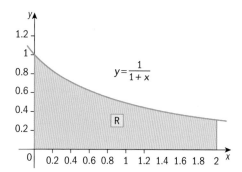

Summary exercise 5

1. Find the following indefinite integrals.

   a) $\displaystyle\int e^{2x-3}\,dx$
   b) $\displaystyle\int \dfrac{-3}{6x+2}\,dx$

   c) $\displaystyle\int \sin 4x\,dx$
   d) $\displaystyle\int \sin^2 6x\,dx$

   e) $\displaystyle\int \sqrt{e^{2-x}}\,dx$
   f) $\displaystyle\int \dfrac{12}{3x-4}\,dx$

   g) $\displaystyle\int (\sin x - 2\cos x)^2\,dx$

   h) $\displaystyle\int \dfrac{-6}{3x+5}\,dx$
   i) $\displaystyle\int -\cos\left(3x + \dfrac{\pi}{3}\right)\,dx$

2. Calculate the following definite integrals,
   leaving your answers as exact values.

   a) $\displaystyle\int_0^2 e^{5x}\,dx$
   b) $\displaystyle\int_0^1 \dfrac{4}{4x+2}\,dx$

   c) $\displaystyle\int_0^{\frac{1}{6}\pi} \sec^2\left(2x - \dfrac{\pi}{6}\right)\,dx$

   d) $\displaystyle\int_0^{\frac{1}{4}\pi} (1 - 3\sin 2x)^2\,dx$

   e) $\displaystyle\int_0^{\frac{1}{4}\pi} \sin\left(\dfrac{\pi}{4} - \dfrac{1}{2}x\right)\,dx$

   f) $\displaystyle\int_{\ln 2}^{\ln 3} e^{2x}\,dx$

g) $\int_0^{\frac{1}{6}\pi} (\sin^2 x - \cos^2 x)\,dx$

h) $\int_{-1}^{2} -e^{2x}\,dx$ **i)** $\int_{-1}^{1} \dfrac{5}{3x-5}\,dx$

EXAM-STYLE QUESTIONS

3. Find the area bounded by the graph of $y = e^{-2x+1}$, the x- and y-axes, and the line $x = 2$.

4. Find the area bounded by the graph of $y = \dfrac{1}{2+x}$, the x- and y-axes, and the line $x = 1$.

5. A curve passes through the point $\left(\dfrac{\pi}{4}, 0\right)$ and its gradient function is $\sin^2 x$. Find the equation of the curve.

6. Find the area bounded by the graphs of $y = \dfrac{1}{x-5}$, $y = \dfrac{1}{2x+1}$, the y-axis, and the line $x = 3$.

7. Find the area bounded by the graphs of $y = \dfrac{1}{2x-7}$, $y = 2 + e^{-x}$, the y-axis, and the line $x = 2$.

8. Calculate the following improper integrals, leaving your answers in terms of e where appropriate.

a) $\int_0^{\infty} e^{-2x}\,dx$

b) $\int_1^{\infty} e^{-0.5x}\,dx$

c) $\int_{-\infty}^{-1} \sqrt{e^x}\,dx$

EXAM-STYLE QUESTIONS

9. The line $y = 4\,e^{-2x}$ between $x = 1$ and $x = 2$ is rotated through 360° about the x-axis. Find the volume of the solid obtained.

10. a) By writing $x = 4x - 3x$ and $7x = 4x + 3x$, show that $\sin 3x \sin 4x = \dfrac{1}{2}(\cos x - \cos 7x)$.

b) Hence calculate $\int_0^{\frac{1}{3}\pi} \sin 3x \sin 4x \,dx$.

11. a) Show that
$(\sin 2x - 3\cos 2x)^2 = 5 - 3\sin 4x + 4\cos 4x.$

b) Hence calculate $\int_0^{\frac{1}{4}\pi} (\sin 2x - 3\cos 2x)^2\,dx$.

12. a) Show that $\dfrac{1 + \tan^2 \theta}{1 - \tan^2 \theta} = \sec 2\theta$.

b) Hence calculate $\int_0^{\frac{\pi}{6}} \left(\dfrac{1 + \tan^2 \theta}{1 - \tan^2 \theta}\right)^2 d\theta$.

EXAM-STYLE QUESTIONS – PURE 2

13. The diagram shows the graph of $y = \sqrt{1 - \sin x}$. The region R is bounded by the curve, the x- and y-axes and the line $x = \dfrac{3\pi}{2}$. (R is in two sections).

a) Use the trapezium rule with 7 ordinates to obtain an approximation for the total area of R, giving your answer correct to 2 decimal places.

b) Explain, with the aid of a sketch, whether the approximation is an overestimate or underestimate.

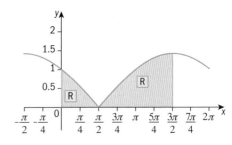

14. The diagram shows the graph of $y = x \ln x$. The region R is bounded by the curve, the x-axis, and the line $x = 2$.

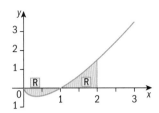

a) Explain why $\displaystyle\int_0^2 x \ln x \, dx$ (if you were able to calculate it) would not give the area of R.

b) Calculate an estimate for the area of R using the trapezium rule and intervals of width 0.5.

15. Use the trapezium rule with five intervals to find an approximation to

$$\int_{-2}^{3} |x^2 - 1| \, dx.$$

16. a) Find $\displaystyle\int 2\cos^2(3\theta) \, d\theta$.

b) Find the exact value of

$$\int_{-2}^{3} \frac{1}{2x + 5} \, dx.$$

Chapter summary

Integrating exponentials

- $\displaystyle\int e^{ax+b} = \frac{1}{a} e^{ax+b} + c$

Integrating functions of the form $\dfrac{1}{ax+b}$

- $\displaystyle\int \frac{1}{x} = dx = \ln|x| + c$

- $\displaystyle\int \frac{1}{ax+b} \, dx = \frac{1}{a} \ln|ax+b| + c$

Integrating standard trigonometric functions

- $\displaystyle\int \sin(ax + b) \, dx = -\frac{1}{a} \cos(ax + b) + c$

- $\displaystyle\int \cos(ax + b) \, dx = \frac{1}{a} \sin(ax + b) + c$

- $\displaystyle\int \sec^2(ax + b) \, dx = \frac{1}{a} \tan(ax + b) + c$

Integrating more trigonometric functions

- Standard trigonometrical relationships, such as identities and double angle formulae, can be used to rearrange functions so that they are integrable.

The trapezium rule – Pure 2 only

- The trapezium rule is a method to calculate a numerical approximation to a definite integral:

$$A = \frac{1}{2}h\{y_0 + y_n + 2(y_1 + y_2 + \ldots + y_{n-1})\} \approx \int_a^b f(x)\,dx, \text{ where } h = \frac{b-a}{n}$$

The formula for the trapezium rule will not be provided.

6 Numerical solution of equations

Computational mathematics is one of the fastest growing employment areas for mathematicians because of the ever-increasing ability of computers to work quickly through large quantities of calculations, and the increasingly sophisticated algorithms which are used. For example, numerical weather prediction uses mathematical models of the atmosphere and oceans to predict the weather based on current climate conditions. This chapter is an introduction to some of the basic techniques which are the foundation of that broad branch of mathematics.

Objectives

- Locate approximately a root of an equation, by means of graphical considerations and/or searching for a sign change.
- Understand the idea of, and use the notation for, a sequence of approximations which converges to a root of an equation.
- Understand how a given simple iterative formula of the form $x_{n+1} = F(x_n)$ relates to the equation being solved, and use a given iteration, or an iteration based on a given rearrangement of an equation, to determine a root to a prescribed degree of accuracy.

Before you start

You should know how to:

1. Substitute values into functions, and evaluate a function at a certain point.

   e.g. If $f(x) = \dfrac{\sqrt{(3x^2 + 4)}}{2}$,

   then $f(1) = \dfrac{\sqrt{(3 + 4)}}{2} = \dfrac{\sqrt{7}}{2}$,

   and $f(2) = \dfrac{\sqrt{(12 + 4)}}{2} = 2$.

Skills check:

Give your answers to **1** and **2** correct to 2 decimal places.

1. If $f(x) = \dfrac{\sqrt{(x^3 + 12)}}{2}$, find

   a) $f(1)$ b) $f(2)$ c) $f(-2)$.

2. Substitute values into exponential, logarithmic and trigonometric functions, and evaluate these functions.

   e.g. If $f(x) = e^{3x} + 2\ln(5x)$,

   then $f(1) = e^3 + 2\ln(5)$,

   and $f(2) = e^6 + 2\ln(10)$,

   Also, $f(-1) = e^{-3} + 2\ln(-5)$, which is undefined because $\ln(-5)$ is undefined.

3. Be able to rearrange equations.

   e.g. If $y = 3\tan^2 x + 1$, express x in terms of y.

   $$y = 3\tan^2 x + 1 \Rightarrow \tan^2 x = \left(\frac{y-1}{3}\right)$$

   $$\Rightarrow x = \tan^{-1}\left(\sqrt{\frac{y-1}{3}}\right)$$

2. If $f(x) = e^{-x} + \ln(5x + 2)$, find

   a) $f(1)$

   b) $f(2)$

   c) $f(-2)$.

3. If $y = e^{2x+1} - 3$, express x in terms of y.

6.1 Finding approximate roots by change of sign or graphical methods

In mathematics, the final step in solving a problem is often to solve an equation. As you learn more mathematics, you extend the range of equations to which you can produce an exact solution using algebra. You have previously learned to solve:

- linear equations, e.g. $3x + 7 = 5 \Rightarrow x = -\frac{2}{3}$
- quadratic equations of different types, e.g.

  - $x^2 - 3x + 2 = 0 \Rightarrow (x - 1)(x - 2) = 0 \Rightarrow x = 1$ or 2

  - $3x^2 + 2x - 7 = 0 \Rightarrow x = \frac{-2 \pm \sqrt{4 + 84}}{6} = \frac{-1 \pm \sqrt{22}}{3}$

- trigonometric equations, e.g. for $0 < \theta < 360°$, $2\sin\theta = 1 \Rightarrow \sin\theta = \frac{1}{2} \Rightarrow \theta = 30°, 150°$.

In the earlier chapters of this book, you have learned how to solve equations involving modulus, exponential, and logarithmic functions as well as harder trigonometric equations. If you plan to study P3, you will learn how to solve equations with complex numbers.

However, there are many types of equation that no-one has found an algebraic method of solving exactly, and many more for which there are methods of solving algebraically, but you have not yet studied them. Being able to find approximate values of solutions to these equations is a valuable skill.

You have probably used the simplest of these methods already (perhaps calling it 'trial and improvement'). We want to treat this method more formally now, and put a condition in.

Note: Trial and improvement is sometimes called 'trial and error' but that misses a key element of the process.

The **sign-change rule**: if f(x) is continuous in an interval $\alpha \le x \le \beta$ and if f(α) and f(β) have different signs, then f(x) = 0 has at least one root between α and β.

Note: The condition that f(x) is continuous is important because it means that the function cannot jump over the x-axis without creating a root. The wording 'at least one' is also important because there can be more than one root. Some illustrations are shown below.

f(1) is negative and f(5) is positive so there must be a root between 1 and 5 – in fact there are two – a (double) root where the graph touches the axis at 2 and another where it cuts at 4, which creates the change of sign.

f(1) is negative and f(3) is positive so there must be a root between 1 and 3.

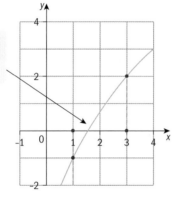

f(1) is negative and f(5) is positive so there must be a root between 1 and 5 – in fact there are three – so the sign actually changes three times in the interval

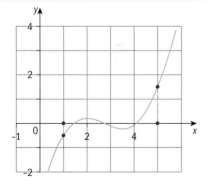

Note that in a case like this third illustration, the choice of the intermediate values will determine which root we obtain: if we consider f(2) instead of f(5) then we will find the root between 1 and 2, whereas if we considered f(3) or f(4) instead of f(1) we would find the root between 4 and 5.

If we have a pair of values giving an interval satisfying the sign-change rule, we can then conduct a systematic process of narrowing the interval until we identify a root correct to any degree of accuracy we want.

The simplest of these processes to narrow the interval is called a decimal search, where first we identify a pair of integers between which the root lies, followed by a pair of neighbouring values within 1 decimal place of each other between which the root lies, followed by a pair of neighbouring values within 2 decimal places of each other between which the root lies, etc. If we keep going with this process we will arrive at the value of the root to the required accuracy.

Example 1

Show that if $f(x) = x^3 - 4x + 5$, then $f(x) = 0$ has a root between $x = -3$ and $x = -2$.
Find the root correct to 1 decimal place.

$f(-3) = -27 + 12 + 5 = -10$

$f(-2) = -8 + 8 + 5 = 5$

By the sign-change rule, f has   ◄————————— State this explicitly each time.
a root between $x = -3$ and $x = -2$.

The next step is to trial a value between −3 and −2. If we take −2.5 then we will know whether the root lies in the interval (−3, −2.5) or in (−2.5, −2).

$f(-2.5) = -0.625 \Rightarrow$ root lies in $(-2.5, -2)$

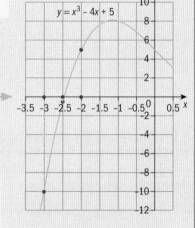

$f(-2.5) = -0.625$, which is a very small negative number, and tells us that the root is likely to be close to −2.5. The next trial should be −2.4.

$f(-2.4) = 0.776 \Rightarrow$ root lies in $(-2.5, -2.4)$

The sign of the function at the mid-interval point (i.e. at −2.45) will tell us whether −2.5 or −2.4 is closer to the root.

$f(-2.45) = 0.0938...$ So the root is −2.5 correct to 1 decimal place.

Just looking at the values of the function at −2.5 and −2.4 is not enough. We need to test the sign of the function at −2.45.

Finding approximate roots by change of sign or graphical methods

Example 2

Show that if $f(x) = x \cos x - \sin x + 1$, then $f(x) = 0$ has a root between $x = 4$ and $x = 5$, where x is in radians, and find the root correct to 2 decimal places.

$f(4) = -0.857\,77...$

$f(5) = 3.377\,23...$ The sign-change rule implies $f(x)$ has a root between $x = 4$ and $x = 5$.

> The next step is to trial a value between 4 and 5. Here the function values suggest the root is closer to 4 than to 5, so 4.2 might be sensible for the next trial value.

$f(4.2) = -0.187\,51... \Rightarrow$ root lies in $(4.2, 5)$

Try 4.3:

$f(4.3) = 0.192\,72... \Rightarrow$ root lies in $(4.2, 4.3)$

> The sign-change rule implies f has a root between $x = 4.2$ and $x = 4.3$.

Try 4.25:

$f(4.25) = -0.000\,88... \Rightarrow$ root lies in $(4.25, 4.3)$

> $f(4.25)$ is extremely small, which suggests 4.25 is close to the root. The next trial should be 4.26.

$f(4.26) = 0.037\,29... \Rightarrow$ root lies in $(4.25, 4.26)$

$f(4.255) = 0.018\,17...$ so the root is 4.25 correct to 2 decimal places.

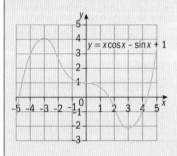

> The actual root is 4.250 390... Using a spreadsheet or programmable calculator allows roots of equations like this to be obtained to any degree of accuracy available in the program. There are other roots which are shown in the graph on the left.

Example 3

a) Verify by calculation that $\sqrt{(x^3 + 3)} = 2 - x$ has a root between $x = 0$ and $x = 1$.
b) Find the root correct to 1 decimal place.

a) We first need to write the equation in the form $f(x) = 0$ in order to be able to use the sign-change rule, so let $f(x) = x - 2 + \sqrt{(x^3 + 3)}$.

x	$f(x)$	
0	$-0.267\,95...$	The sign-change rule implies
1	1	there is a root between 0 and 1.

> Remember to state this.

b)

x	$f(x)$	
0.3	$0.039\,83...$	The root is between 0 and 0.3
0.2	$-0.065\,64...$	The root is between 0.2 and 0.3
0.25	$-0.013\,44...$	The root is between 0.25 and 0.3

> A table of values like this, showing the search process with the values, is often the simplest way of keeping track of where we know the root to be located.

The root is 0.3 correct to 1 decimal place.

Example 4

a) **Sketch** the graphs of $y = x^2 + 2$ and $y = \dfrac{1}{x}$ on the same set of axes.

b) Using your graphs, **explain** why there is only one root to the equation $x^3 + 2x - 1 = 0$.

a)

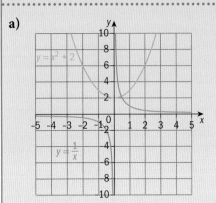

b) The intersection of the graphs $y = x^2 + 2$ and $y = \dfrac{1}{x}$ is the point at which $x^2 + 2 = \dfrac{1}{x}$.

Rearranging gives $x^3 + 2x - 1 = 0$.

Since there is only one point of intersection of the two graphs, there is only one root of $x^3 + 2x - 1 = 0$.

Exercise 6.1

1. Show that if $f(x) = x^3 - 5x^2 + 3x + 2$ then $f(x) = 0$ has a root between $x = 1$ and $x = 2$ and find the root correct to 1 decimal place.

2. Show that if $f(x) = 2^x - 3x - 3$ then $f(x) = 0$ has a root between $x = 3$ and $x = 4$ and find the root correct to 1 decimal place.

3. Show that if $f(x) = \dfrac{x}{3} - \cos^{-1} x$, where x is in radians, then $f(x) = 0$ has a root between $x = 0.5$ and $x = 1.0$ and find the root correct to 1 decimal place.

4. Show that if $f(x) = \dfrac{\pi}{3} + \dfrac{x}{5} - \tan x$, where x is in radians, then $f(x) = 0$ has a root between $x = 0$ and $x = 1$ and find the root correct to 1 decimal place.

5. a) Show that if $f(x) = x^3 - 3x + 1$ then $f(x) = 0$ has a root between $x = 0$ and $x = 1$ and find the root correct to 1 decimal place.

   b) Show that there are two other roots of $f(x) = 0$ which lie in the intervals $(-2, -1)$ and $(1, 2)$, and find those roots correct to 1 decimal place.

6. Sketch the graphs of $y = e^{2x}$ and $y = 7 - 3x$ on the same axes for values of x between -2 and 3. Show that $e^{2x} = 7 - 3x$ has a root between $x = 0$ and $x = 1$ and find the root correct to 2 decimal places.

7. Show that if $f(x) = 5 - x\,e^{2x}$ then $f(x) = 0$ has a root between $x = 0$ and $x = 1$ and find the root correct to 2 decimal places.

8. Show that if $f(x) = \ln(x^2 + 3) - 2x + 5$ then $f(x) = 0$ has a root between $x = 3$ and $x = 4$ and find the root correct to 2 decimal places.

Finding approximate roots by change of sign or graphical methods

6.2 Finding roots using iterative relationships

Consider the sequence with nth term given by $x_n = 1-(0.1)^{n-1}$. The sequence is 0, 0.9, 0.99, 0.999, 0.9999, 0.99999, ... Every term is getting bigger in the sequence, but we are always subtracting a positive number from 1 so we know there is an upper limit of 1 on the terms in the sequence.

As n increases we can see that the sequence will get as close to 1 as we require (and will stay close to it), and we say that the sequence converges to 1.

> We can put a formal condition on convergence by saying that L is the limit of a sequence $\{x_n\}$ if for any positive value ε, we can find an integer N so that $|x_n - L| < \varepsilon$ for every $n > N$.

The iterative relation $x_{n+1} = F(x_n)$, together with an initial value x_1, defines a sequence. If that sequence converges to a limit, then the limit will be a solution to the equation $x = F(x)$.

Example 5

a) Show that the equation $\csc x = 4 - \frac{1}{2}x^2$ can be rearranged to $x = \sin^{-1}\left(\frac{2}{8-x^2}\right)$.

b) Use the iterative relationship $x_{n+1} = \sin^{-1}\left(\frac{2}{8-x_n^2}\right)$, with $x_1 = 0$, to find four further approximations x_2, x_3, x_4, x_5, writing down 6 decimal places from your calculator.

c) Determine the root of the equation correct to 3 s.f.

....

a) $\csc x = 4 - \frac{1}{2}x^2$

$\Rightarrow \dfrac{1}{\sin x} = \dfrac{8-x^2}{2}$

$\Rightarrow x = \sin^{-1}\left(\dfrac{2}{8-x^2}\right)$

b) $x_1 = 0$

$x_2 = 0.252680 \ldots$

$x_3 = 0.254758 \ldots$

$x_4 = 0.254792 \ldots$

$x_5 = 0.254793 \ldots$

c) The root is 0.255 (to 3 s.f.). ←

> Because the sequence appears to be converging fast, the root is probably 0.25479 to 5 s.f., but you will not be asked to make judgements like this.

 A standard scientific calculator allows you to produce each successive iterated value by a single key press once the function has been set up:

- Enter the initial value and press '='.
- Enter the expression $F(x_n)$ using 'Ans' for each occurrence of x_n.
- Press '=' to generate x_2, and again to generate x_3, and so on.

You need to record a number of digits (often specified in the question) to show your working but the calculator is using all the available accuracy when generating the next approximation.

If you have an equation $f(x) = 0$, you can rearrange into the form $x = F(x)$ for some function $F(x)$. From here, you can define an iterative equation $x_{x+1} = F(x_n)$ which will often converge to the root of $f(x) = 0$.

Example 6

The graph of $y = x^3 - 4x - 1$ is shown.

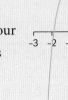

a) Show that the equation $x^3 - 4x - 1 = 0$ can be rearranged to
$$x = \frac{1}{4}(x^3 - 1).$$

b) Use the iterative relationship $x_{n+1} = \frac{1}{4}(x_n^3 - 1)$, with $x_1 = 0$, to find four further approximations x_2, x_3, x_4, x_5, writing down 6 decimal places (where appropriate) from your calculator.

c) Determine a root correct to 3 s.f.

d) Use the same iterative relationship with $x_1 = -2$ to find values for x_2, x_3, x_4, x_5.

e) Comment on the behaviour of your sequence.

a) $x^3 - 4x - 1 = 0 \Rightarrow 4x = x^3 - 1 \Rightarrow x = \frac{1}{4}(x^3 - 1)$

b) $x_1 = 0$

$x_2 = -0.25$

$x_3 = -0.253\,906 \ldots$

$x_4 = -0.254\,092 \ldots$

$x_5 = -0.254\,101 \ldots$

c) The root is -0.254, correct to 3 s.f.

d) $x_1 = -2$

$x_2 = -2.25$

$x_3 = -3.097\,656 \ldots$

$x_4 = -7.680\,870 \ldots$

$x_5 = -113.5347 \ldots$

e) The sequence does not appear to be converging – we see from the graph that there is a root just above -2, but the sequence is taking us further and further away from that root.

Note: You need to know that not all iterative relationships derived by rearranging an equation into the form $x = F(x)$ will converge. In Example 6, the three roots are -1.861, -0.2541, and 2.115. Any initial value taken between -1.861 and 2.115 and used in the iterative relationship we derived will converge to the middle root (-0.2541). Any initial value below -1.861 or above 2.115 will diverge.

Finding roots using iterative relationships

We can use an iterative method to find a root to a specified degree of accuracy. The number of *iterations* required for this degree of accuracy will differ from equation to equation, and will depend on the initial value we choose.

Example 7

a) Show that the equation $xe^x = 1$ has a root between 0 and 1.

b) Show that the equation $xe^x = 1$ can be rearranged into the forms **i)** $x = e^{-x}$ and **ii)** $x = -\ln x$.

c) Use the iterative relationships **i)** $x_{n+1} = e^{-x_n}$ and **ii)** $x_{n+1} = -\ln x_n$, with $x_1 = 0.6$ in both cases, to find eight further approximations $x_2, x_3, x_4, \ldots, x_9$, writing down 6 decimal places from your calculator.

a) Writing $xe^x = 1$ as $xe^x - 1 = 0$, let $f(x) = xe^x - 1$.
   Then $f(0) = -1$, $f(1) = e - 1 = 1.718 \ldots$, so by the sign-change rule there is a root between 0 and 1.

b) **i)** $xe^x = 1 \Rightarrow x = \dfrac{1}{e^x} = e^{-x}$ **ii)** $xe^x = 1 \Rightarrow \ln x + x = 0 \Rightarrow x = -\ln x$

c) **i)** $x_{n+1} = e^{-x_n}$

   $x_1 = 0.6$

   $x_2 = 0.548\,811 \ldots$

   $x_3 = 0.577\,635 \ldots$

   $x_4 = 0.561\,223 \ldots$

   $x_5 = 0.570\,510 \ldots$

   $x_6 = 0.565\,236 \ldots$

   $x_7 = 0.568\,225 \ldots$

   $x_8 = 0.566\,529 \ldots$

   $x_9 = 0.567\,491 \ldots$

   ii) $x_{n+1} = -\ln x_n$

   $x_1 = 0.6$

   $x_2 = 0.510\,825$

   $x_3 = 0.671\,726 \ldots$

   $x_4 = 0.397\,903 \ldots$

   $x_5 = 0.921\,546 \ldots$

   $x_6 = 0.081\,702 \ldots$

   $x_7 = 2.504\,673 \ldots$

   $x_8 = -0.918\,158 \ldots$

   x_9 gives an error message
   since we cannot take logs of a negative number.

1 In Example 5, the iteration $x_{n+1} = \sin^{-1}\left(\dfrac{2}{8 - x_n^2}\right)$ converged very quickly, allowing the root to be confirmed to 3 decimal places after only 4 iterations, despite the first iteration ($x_1 = 0$) being some distance away from the root. In Example 7, the iterative relationship $x_{n+1} = e^{-x_n}$ was started much closer to the root than in Example 5, but after 8 iterations we are still not entirely sure of the root to 3 d.p.

2 Consecutive iterations for $x_{n+1} = \sin^{-1}\left(\dfrac{2}{8 - x_n^2}\right)$ are alternately above and below the root (which is 0.567 1432 ...). As such, it is easier to conclude what the root will be than where the sequence produced by the iterative relationship approaches the root from one side only (as in Examples 5 and 6). We will look in the next section at how you can tell whether an iterative relationship will alternate or approach from one side.

Example 7 also illustrates that the iterative relationship $x_{n+1} = -\ln x_n$ initially generated a diverging sequence, but after 9 iterations it required an impossible calculation and thus stopped.

In all the examples so far, we have been given the equation to be solved numerically. Questions can be set where you are asked to show that an equation has to be satisfied in a given context.

Example 8

In the diagram on the right, O is the centre of a circle, and AB is a chord with angle $AOB = \theta$ radians.

The smaller segment (shaded) created by the chord is one half of the area of the larger segment.

a) Show that θ satisfies the equation $3\theta - 3\sin\theta = 2\pi$.

b) Show that this equation can be rearranged into the

    form $\theta = \dfrac{2}{3}\pi + \sin\theta$.

c) Using the iterative relation $\theta_{n+1} = \dfrac{2}{3}\pi + \sin\theta_n$, with initial

    value $\theta_1 = 2.6$, determine the angle θ correct to 2 decimal places.

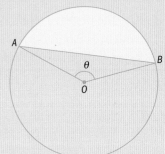

a) The minor sector created by radii OA and OB has area $\dfrac{1}{2}\theta r^2$ and the triangle OAB has

    area $\dfrac{1}{2}r^2\sin\theta$. If the area of the minor segment is one half of the area of the larger segment,

    then it is one third of the area of the full circle, i.e. $\dfrac{1}{3}\pi r^2$.

$$\frac{1}{2}\theta r^2 - \frac{1}{2}r^2\sin\theta = \frac{1}{3}\pi r^2$$

$$\left(\times\frac{6}{r^2}\right) \Rightarrow 3\theta - 3\sin\theta = 2\pi$$

b) $3\theta - 3\sin\theta = 2\pi \Rightarrow \theta = \dfrac{2}{3}\pi + \sin\theta$

c)

$x_1 = 2.6$	$x_2 = 2.60989\ldots$
$x_3 = 2.60139\ldots$	$x_4 = 2.60870\ldots$
$x_5 = 2.60241\ldots$	$x_6 = 2.60782\ldots$
$x_7 = 2.60317\ldots$	$x_8 = 2.60717\ldots$
$x_9 = 2.60373\ldots$	$x_{10} = 2.60668\ldots$
$x_{11} = 2.60415\ldots$	$x_{12} = 2.60333\ldots$
$x_{13} = 2.60445\ldots$	$x_{14} = 2.60607\ldots$
$x_{15} = 2.60468\ldots$	$x_{16} = 2.60587\ldots$
$x_{17} = 2.60485\ldots$	$x_{18} = 2.60573\ldots$
$x_{19} = 2.60497\ldots$	$x_{20} = 2.60562\ldots$
$x_{21} = 2.60506\ldots$	

Exam questions will not require this number of iterations – the convergence is slow and the root lies close to the border between 2.60 and 2.61 radians.

Because the iterative sequence is alternating as it moves towards the root, as soon as the lower sequence exceeded 2.605, it is safe to conclude that the root will be 2.61 correct to 2 d.p.

The angle is 2.61 radians correct to 2 decimal places.

Example 9

The relationship for the golden ratio, x, can be described by the equation $\dfrac{1}{x} = \dfrac{x-1}{1}$.

a) Show that this equation can be rearranged into the form $x = \sqrt[3]{x(x+1)}$.

b) Use the iterative relationship $x_{n+1} = \sqrt[3]{x_n(x_n + 1)}$ with initial value $x_1 = 1.6$ to find four further approximations to the golden ratio, showing at least 4 decimal places for each approximation.

c) Give an approximation to the golden ratio correct to 2 decimal places.

• •

a) $\dfrac{1}{x} = \dfrac{x-1}{1}$ $(\times x^2) \Rightarrow x = x^3 - x^2 \Rightarrow x^3 = x^2 + x$

$\Rightarrow x^3 = x(x+1) \Rightarrow x = \sqrt[3]{x(x+1)}$

b) $x_1 = 1.6$

$x_2 = 1.6082 \ldots$

$x_3 = 1.6127 \ldots$

$x_4 = 1.6151 \ldots$

$x_5 = 1.6165.$

c) The golden ratio is 1.62 correct to 2 decimal places.

Exercise 6.2

1. Find, and simplify, the equations that the following iterative relationships provide roots for, assuming that the iterative sequences converge. (You do not need to check whether or not the iterative sequences converge.)

a) $x_{n+1} = \dfrac{1}{3 + x_n}$ b) $x_{n+1} = \dfrac{\sqrt{x_n^3 + 2}}{3 + x_n}$ c) $x_{n+1} = e^{-2x_n}$

d) $x_{n+1} = 3 - \ln(2x_n)$ e) $x_{n+1} = \cos x_n + \dfrac{\pi}{4}$ f) $x_{n+1} = \sqrt[10]{x_n^6 + 3}$

2. Find at least three possible rearrangements for each of the following equations into a form $x = F(x)$.

a) $x^4 - 3x + 2 = 0$ b) $x^7 - 4x^3 + 1 = 0$ c) $x^5 - x^2 - 1 = 0$

d) $x^2 - 3x - e^{-x} = 0$ e) $e^{x^2} - 3x = 0$ f) $\cot x + 3 - x^2 = 0$

3. For each iterative formula and given initial value, find values of $x_2, x_3, x_4,$ and x_5 (giving the first 5 decimal places of each approximation).

a) $x_{n+1} = \dfrac{1}{3 + x_n};\ x_1 = 0.5$

b) $x_{n+1} = \dfrac{\sqrt{x_n^3 + 2}}{3 + x_n};\ x_1 = 0.5$

c) $x_{n+1} = e^{-2x_n};\ x_1 = 0.5$

d) $x_{n+1} = 3 - \ln 2x_n;\ x_1 = 1.75$

e) $x_{n+1} = \cos x_n + \dfrac{\pi}{4};\ x_1 = 1.1$

f) $x_{n+1} = \sqrt[10]{x_n^6 + 3};\ x_1 = 1.2$

4. **a)** Show that the equation $3x^2 - x - 5 = 0$ has a root between $x = 1$ and $x = 2$.

   b) Show that the equation $3x^2 - x - 5 = 0$ can be rearranged to $x = \sqrt{\dfrac{x+5}{3}}$.

   c) Using an initial value of $x_1 = 1.5$ and the iterative relationship $x_{n+1} = \sqrt{\dfrac{x_n + 5}{3}}$, find the root of $3x^2 - x - 5 = 0$ that lies between 1 and 2, correct to 2 decimal places.

5. **a)** Show that the equation $e^x - x^2 - 5 = 0$ has a root between $x = 2$ and $x = 3$.

   b) Show that the equation $e^x - x^2 - 5 = 0$ can be rearranged to $x = \ln(x^2 + 5)$.

   c) Using an initial value of $x_1 = 2.5$ and the iterative relationship $x_{n+1} = \ln(x_n^2 + 5)$, find values of x_2, x_3, x_4, and x_5 (giving the first 5 decimal places of each approximation).

6. **a)** Show that the equation $\theta = \dfrac{3\pi}{5} + \cos\theta$ has a root between $\theta = 1.7$ and $\theta = 1.8$.

   b) Using an initial value of $\theta_1 = 1.75$ and the iterative relationship $\theta_{n+1} = \dfrac{3\pi}{5} + \cos\theta_n$, find values of θ_2, θ_3, θ_4, and θ_5 (giving the first 5 decimal places of each approximation).

   c) Explain why you know that this iterative relationship will (eventually) converge.

7. The shaded segment of the circle has an area of $7\,\text{cm}^2$.

   a) Show that angle $AOB = \theta$ (where θ is in radians) satisfies the equation $18\theta - 18\sin\theta = 7$.

   b) Use the iterative relationship $\theta_{n+1} = \sin\theta_n + \dfrac{7}{18}$ with initial value $\theta_1 = 1.5$ to find the angle θ in radians correct to 2 decimal places.

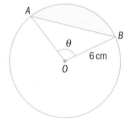

8. **a)** Show that both $x_{n+1} = \cos x_n$ and $x_{n+1} = \cos^{-1} x_n$ are iterative relationships that provide a root to the equation $x - \cos x = 0$ if the sequence converges.

   b) Using $x_1 = 0.75$ as an initial value, show that one sequence diverges and the other converges, and find the root correct to 2 decimal places.

9. **a)** Show that the equation $x^4 - x^2 - 5 = 0$ can be rearranged to each of the following forms.

   i) $x = \sqrt{\dfrac{5}{x^2 - 1}}$ **ii)** $x = \dfrac{x^2 + 5}{x^3}$ **iii)** $x = \sqrt[4]{x^2 + 5}$

   b) Using an initial value of $x_1 = 1.7$, show that the iterative relationships based on forms **(i)** and **(ii)** do not converge, and use $x_{n+1} = \sqrt[4]{x_n^2 + 5}$ to find the root correct to 2 decimal places.

6.3 Convergence behaviour of iterative functions

Although you are not required to know conditions for convergence of an iterative relationship, there are some simple principles about conditions for convergence which may help you to feel confident that you know what is going on as a sequence progresses, giving you a feeling for how quickly an iterative process may converge, and whether the terms will approach the root from one side or will be alternately above and below the root.

We know that the equation $x^2 - 3x - 5 = 0$ can be solved using the quadratic formula, with solutions $x = \dfrac{3 \pm \sqrt{9+20}}{2} = -1.192\,58\ldots$ or $4.192\,58\ldots$

> The Cambridge International 9709 syllabus does not require you to know conditions for convergence of an iterative relationship, although you do need to know that an iteration may fail to converge.

The equation $x^2 - 3x - 5 = 0$ can also be rearranged into various forms of $x = F(x)$, which are not normally helpful because to work out x on the left-hand side we need a numerical value for x on the right-hand side. However, rearranging equations into the form $x = F(x)$ and creating an iterative relationship $x_{n+1} = F(x_n)$ will often produce a sequence of approximations which converge to a root of $x = F(x)$.

Example 10

Verify that the equation can be rearranged to $x^2 - 3x - 5 = 0$ in each case.

a) $x = \sqrt{(3x+5)}$ b) $x = \dfrac{3x+5}{x}$ c) $x = \dfrac{5}{x-3}$

......

a) $x = \sqrt{(3x+5)}$

$\Rightarrow x^2 = 3x + 5$

$\Rightarrow x^2 - 3x - 5 = 0$

b) $x = \dfrac{3x+5}{x}$

$\Rightarrow x^2 = 3x + 5$

$\Rightarrow x^2 - 3x - 5 = 0$

c) $x = \dfrac{5}{x-3}$

$\Rightarrow x(x-3) = 5$

$\Rightarrow x^2 - 3x - 5 = 0$

Starting close to a particular root does not mean that the iterative relationship will converge to that root (or indeed that it will converge at all). Consider what happens if we form iterative relationships from the three equations in Example 10 and take initial values of -1 and 4:

a) For iterative relationship $x_{n+1} = \sqrt{(3x_n + 5)}$, if we choose $x_1 = 4$ then x_n converges to 4.19. If we choose $x_1 = -1$, we see that x_n still converges to 4.19.

b) Iterative relationship $x_{n+1} = \dfrac{3x_n + 5}{x_n}$ produces the same behaviour as in part (a) above.

c) For iterative relationship $x_{n+1} = \dfrac{5}{(x_n - 3)}$, choosing either $x_1 = 4$ or $x_1 = -1$ results in x_n converging to -1.19.

In Example 6, we saw that initial values between the largest and smallest roots gave sequences which converged to the middle root, while any other initial value which was not an exact root itself generated a divergent sequence. So we could start very close to a root and either move towards a different root or not converge to any root at all.

In Example 7, there was an iterative relationship which produced an undefined calculation.

Sketching the graphs of $y = x$ and $y = F(x)$ on the same axes helps us to have some idea of what will happen in any iterative relationship.

- If the gradient of $F(x)$ close to the root is positive, then any convergent sequence will approach the root from one side, and if it is negative the sequence will alternate values above and below the root.

- Generally, the smaller the modulus of the gradient, the faster the convergence will be. For example, for small gradients it will take fewer iterations to find the root to a given accuracy.

We will illustrate this by some examples.

Drawing the graphs of $y = x$ and $y = \frac{1}{4}(x^3 - 1)$ (which we studied in Example 6) shows that the gradient close to the middle root is small and positive so we would expect the convergence to be from one side (a 'staircase diagram') and fast.

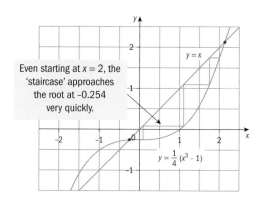

Even starting at $x = 2$, the 'staircase' approaches the root at -0.254 very quickly.

You will not be asked to produce staircase or cobweb diagrams in examinations, but seeing them here may help illuminate why the iterative process works in these cases.

The iterative process in Example 8 was very slow and alternated values above and the root. The graphs show the functions $y = x$ and $y = \frac{2}{3}\pi + \sin x$, on a very expanded scale in the neighbourhood of the root ($x = 2.61$). In this neighbourhood, the gradient of $y = \frac{2}{3}\pi + \sin x$ is negative and not numerically small (approximately –0.9). Thus, convergence happens very slowly, and the graph of the iterations produces what is known as a 'cobweb diagram'.

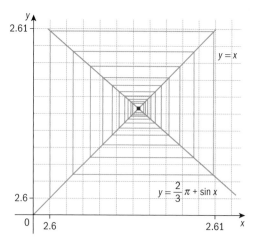

Summary exercise 6

1. Show that if $f(x) = x^3 - 3x^2 + 4x - 1$ then $f(x) = 0$ has a root between $x = 0$ and $x = 1$ and find the root correct to 1 decimal place.

2. Show that if $f(x) = 3x - 5x - 4$ then $f(x) = 0$ has a root between $x = -1$ and $x = 0$ and another between $x = 2$ and $x = 3$. Find the roots correct to 1 decimal place.

3. i) Show that if $f(x) = x^3 + 4x^2 - 2x - 7$ then $f(x) = 0$ has a root between $x = 1$ and $x = 2$ and find the root correct to 1 decimal place.

   ii) Show that there are two other roots in the intervals $(-5, -4)$ and $(-2, -1)$ and find the roots correct to 1 decimal place.

4. Show that if $f(x) = 6 - x^2 e^x$ then $f(x) = 0$ has a root between $x = 1$ and $x = 2$ and find the root correct to 1 decimal place.

5. Find the equation that each iterative relationship provides a root for, if the sequence converges. (You do not need to check whether it does converge.)

   i) $x_{n+1} = \dfrac{1}{4 + 2x_n}$

   ii) $x_{n+1} = \sqrt{\dfrac{x_n^3 - 1}{5}}$

6. Find at least three possible rearrangements of each equation into the form $x = F(x)$.

   i) $x^4 - 4x^2 + 3 = 0$ ii) $x^5 - 3x^2 + 2 = 0$

7. For each iterative formula and given initial value, find values of $x_2, x_3, x_4,$ and x_5 (giving the first 5 decimal places of each approximation).

   i) $x_{n+1} = \dfrac{1}{5 - x_n}$; $x_1 = 0.5$

   ii) $x_{n+1} = \dfrac{\sqrt{2x_n^3 + 3}}{5 + x_n}$; $x_1 = 0.5$

8. i) Show that the equation $e^x - 2x^2 - 1 = 0$ has a root between $x = 2$ and $x = 3$.

   ii) Show that the equation $e^x - 2x^2 - 1 = 0$ can be rearranged to $x = \ln(2x^2 + 1)$.

   iii) Using an initial value of 2.85, and the iterative relationship $x_{n+1} = \ln(2x_n^2 + 1)$, find the root correct to 2 decimal places.

9. i) Show that both $x_{n+1} = \tan x_n - 1$ and $x_{n+1} = \tan^{-1}(x_n + 1)$ are iterative relationships that provide a root to the equation $x - \tan x + 1 = 0$ if the sequence converges.

   ii) Using $x_1 = 1$ as an initial value, show that one sequence diverges and the other converges and find the root correct to 2 decimal places.

10.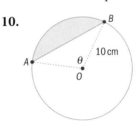

The shaded segment of the circle has an area of $19\,\text{cm}^2$.

   i) Show that angle $AOB = \theta$ radians satisfies the equation $50\theta - 50 \sin \theta = 19$.

   ii) Use the iterative relationship $\theta_{n+1} = \sin \theta_n + \dfrac{19}{50}$ with an initial value of 1.5 to find the angle in radians correct to 2 decimal places.

11. i) By sketching a suitable pair of graphs, show that the equation

$$\operatorname{cosec} x = 3 - x^2,$$

where x is in radians, has a root in the interval $0 < x < \frac{1}{4}\pi$.

ii) Verify by calculation that this root lies between 0.3 and 0.5.

iii) Show that this root also satisfies the equation

$$x = \sin^{-1}\left(\frac{1}{3-x^2}\right)$$

iv) Use an iterative formula based on the equation in part **(iii)** to determine the root correct to 2 decimal places. Give the result of each iteration to 4 decimal places.

12. The sequence of values given by the iterative formula

$$x_{n+1} = \frac{2x_n}{3} + \frac{10}{x_n^2}$$

with initial value $x_1 = 3$, converges to α.

i) Use this iterative formula to find α correct to 2 decimal places, giving the result of each iteration to 4 decimal places.

ii) State an equation satisfied by α and hence find the exact value of α.

13.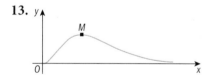

The diagram shows part of the curve $y = \dfrac{x^2}{1+e^{2x}}$ and its maximum point M.

The x-coordinate of M is denoted by m.

i) Find $\dfrac{dy}{dx}$ and hence show that m satisfies the equation $x = 1 + e^{-2x}$.

ii) Show by calculation that m lies between 0.7 and 0.8.

ii) Use an iterative formula based on the equation in part **(i)** to find m correct to 3 decimal places. Give the result of each iteration to 5 decimal places.

14.

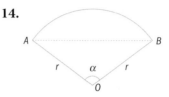

The diagram shows a sector AOB of a circle with centre O and radius r. The angle AOB is α radians, where $0 < \alpha < \pi$. The area of triangle AOB is one third the area of the sector.

i) Show that α satisfies the equation

$$x = 3 \sin x.$$

ii) Verify by calculation that α lies between $\frac{2}{3}\pi$ and $\frac{3}{4}\pi$.

iii) Show that, if a sequence of values given by the iterative formula

$$x_{n+1} = \frac{1}{2}(x_n + 3\sin x_n)$$

converges, then it converges to a root of the equation in part **(i)**.

iv) Use this iterative formula, with initial value $x_1 = 2.3$, to find α correct to 2 decimal places. Give the result of each iteration to 4 decimal places.

Chapter summary

Locating a root of an equation approximately using the sign-change rule or a graphical method

- The sign-change rule:

  If f(x) is continuous in an interval $\alpha \leq x \leq \beta$ and if f(α) and f(β) have different signs, then f(x) = 0 has at least one root between α and β.

- The simplest process to narrow down an interval in which the root of an equation lies is a decimal search. First we identify a pair of integers between which the root lies, followed by a pair of neighbouring values within 1 decimal place, then within 2 decimal places, etc. If we keep going with this process we will arrive at the value of the root to the required accuracy.

Finding roots using iterative relationships

- The iterative relation $x_{n+1} = F(x_n)$, together with an initial value x_1, defines a sequence. If that sequence converges to a limit, then the limit will be a solution to the equation $x = F(x)$.

- If you have an equation f(x) = 0, you can rearrange it into the form $x = F(x)$ for some function F(x). From here, you can define an iterative equation $x_{n+1} = F(x_n)$ which will often converge to a root of f(x) = 0. Use an iterative relationship and an initial value to produce further approximations to a root.

- We can use an iterative method to find a root to a specified degree of accuracy. The number of iterations required for this degree of accuracy will differ from equation to equation, and will depend on the value of the first iteration we choose.

Convergence behaviour of iterative functions

- If the gradient of F(x) close to the root is positive, then any convergent sequence will approach the root from one side, and if it is negative the sequence will alternate values above and below the root.

- Generally, the smaller the modulus of the gradient, the faster the convergence will be. For example, for small gradients it will take fewer iterations to find the root to a given accuracy.

Maths in real-life

Nature of mathematics

The Fibonacci sequence is extremely simple to define: it starts 1, 1 and then the next term is generated by the sum of the previous two terms – so the third is 2 = 1 + 1 and the fourth is 3 = 1 + 2. It continues infinitely:

1, 1, 2, 3, 5, 8, 13, 21, 34, 55, 89, 144, …

Perhaps surprisingly, these numbers occur frequently in nature. Exploring some geometry will help you see why this happens, but here are some examples.

Flower heads with seeds often have an arrangement that seems to optimise the packing of seeds which are the same size – and it looks as though there are spirals curving both right and left from the centre.
The numbers of these spirals in each direction are usually neighbouring Fibonacci numbers.

As a nautilus shell grows, the size of the chambers follows the pattern of the Fibonacci numbers. The shape of the nautilus shell suggests that this is also an extremely efficient way for growth to occur.

This satellite picture of a hurricane looks very like the shape of the nautilus shell. The shape is built up by adding squares on the new longest side, generating the Fibonacci numbers. This movement seems to be the way that hurricanes limit the energy losses from the system.

Similar patterns appear on the grandest scale imaginable – in the way that stars group together in spiral galaxies in our universe.

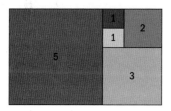

Look at how the geometry grows: starting with a square of side 1, add another above it of side 1. Then to the right add a square of side 2, below a square of side 3, left a square of side 5 and so on. Moving round, getting larger each time, creates the spiral shape seen in the nautilus shell, the hurricane and the galaxies.

As the Fibonacci sequence continues, the ratio of successive terms approaches the 'golden ratio' – which was felt by the Ancient Greeks to be the most pleasing on the eye and they used it a lot in architecture – for example in the Parthenon (shown here).

The golden ratio occurs when the short side (AB) of a rectangle (ABCD) has unit length, and the long side (AD) has length x, such that when you remove a square (ABEF) whose sides have length 1, you are left with a rectangle (CDFE) whose sides (lengths 1 and $x - 1$) are in the same ratio as the sides in the original rectangle (ABCD) were. The quadratic equation satisfied by the golden ratio comes from solving

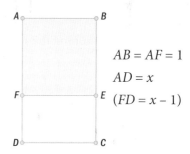

$AB = AF = 1$

$AD = x$

$(FD = x - 1)$

$$\frac{1}{x} = \frac{x-1}{1}$$

$$\Rightarrow 1 = x(x - 1)$$

$$\Rightarrow x^2 - x - 1 = 0$$

using the quadratic formula gives:

$$x = \frac{1 \pm \sqrt{1+4}}{2} = \frac{1+\sqrt{5}}{2} = 1.618 \ (3 \text{ d.p.})$$

Note that the second solution is discarded because it is negative. Also, some sources reference the golden ratio as 0.61803 ... = 1 ÷ 1.618 ...

7 Further algebra

Real-world applications of algebra include the use of binomial expansions in computing, for example, the automatic distribution of internet provider addresses.

Economists use binomial expansions to help them make realistic predictions about the way economies will behave in the next few years. Architects and engineers also use them in designing infrastructure and in calculating the costs and time associated with bringing their plans to realisation while ensuring that projects are profitable.

Objectives

- Recall an appropriate form for expressing rational functions in partial fractions, and carry out the decomposition, in cases where the denominator is no more complicated than $(ax + b)(cx + d)(ex + f)$ or $(ax + b)(cx + d)^2$ or $(ax + b)(cx^2 + d)$, and where the degree of the numerator does not exceed that of the denominator.
- Use the expansion of $(1 + x)^n$, where n is a rational number and $|x| < 1$. (Finding a general term is not included, but adapting the standard series to expand e.g. $\left(2 - \frac{1}{2}x\right)^{-1}$ is included, and so is determining the set of values of x for which the expansion is valid in such cases.)

Before you start

You should know how to:

1. Equate coefficients,

    e.g. Given $3x^2 - Bx + 5 \equiv Ax^2 + 2x - C$, find the values of A, B, and C.

    Coefficient of x^2: $A = 3$

    Coefficient of x: $B = -2$

    Constant: $C = -5$

    e.g. Given $2Ax^2 - 9x + C - 8 \equiv 6x^2 + 3Bx - 7$ find the values of A, B, and C.

    Coefficient of x^2: $2A = 6 \Rightarrow A = 3$

    Coefficient of x: $-9 = 3B \Rightarrow B = -3$

    Constant: $C - 8 = -7 \Rightarrow C = 1$

Skills check:

For each of the following identities, find the values of A, B, and C.

1. $Ax^2 - 4x + 9 \equiv x^2 + Bx - C$

2. $3Ax^2 - 2Bx - 2 \equiv 6x^2 + 8x - 2 + C$

3. $(1 - A)x^2 - 2x + 5C \equiv 8x^2 - 3Bx - 10$

4. $12x^2 + 2Bx \equiv 4Ax^2 + x - 7C$

5. $2Ax^2 - 3Bx + 6 \equiv 6x^2 + 4 - 2C$

7.1 Partial fractions

We use the method of partial fractions to split a rational function into simpler component parts. We choose which method to use by looking at the denominator of the function.

Partial fractions type 1: Algebraic fractions with linear factors in the denominator can be split into partial fractions in the following way:

$$\frac{1}{(x+p)(x+q)} \equiv \frac{A}{(x+p)} + \frac{B}{(x+q)}$$

Example 1

Express $\dfrac{x+3}{(x-2)(x+1)}$ in partial fractions.

$$\frac{x+3}{(x-2)(x+1)} \equiv \frac{A}{x-2} + \frac{B}{x+1}$$

The denominator has two linear factors so we use two fractions.

$$x + 3 \equiv A(x+1) + B(x-2)$$

Multiply each term by $(x-2)(x+1)$.

Method 1: Choose appropriate values for x.

We use the symbol $\equiv$ to mean it is true for all values of x.

Let $x = 2$

Choose a value of x that makes one bracket $= 0$.

$$5 = A(3) + 0 \implies A = \frac{5}{3}$$

Let $x = -1$

Choose a value of x that makes the other bracket $= 0$.

$$2 = 0 + B(-3) \implies B = -\frac{2}{3}$$

Hence $\dfrac{x+3}{(x-2)(x+1)} \equiv \dfrac{5}{3(x-2)} - \dfrac{2}{3(x+1)}$

Substitute the values of A and B into the fractions.

Method 2: Expand the brackets and equate coefficients.

$$x + 3 \equiv A(x+1) + B(x-2)$$

$$x + 3 \equiv Ax + A + Bx - 2B$$

Expand the brackets.

$$(1)x + (3) \equiv (A+B)x + (A-2B)$$

Hence $A + B = 1$ and $A - 2B = 3$

Equate the coefficients of x and equate the constants.

Solving simultaneously gives $A = \dfrac{5}{3}$ and $B = -\dfrac{2}{3}$

Put the minus sign (−) before the second fraction.

Hence $\dfrac{x+3}{(x-2)(x+1)} \equiv \dfrac{5}{3(x-2)} - \dfrac{2}{3(x+1)}$

Example 2

Express $\dfrac{x}{1-x-2x^2}$ in partial fractions.

$\dfrac{x}{(1+x)(1-2x)} \equiv \dfrac{A}{1+x} + \dfrac{B}{1-2x}$ ⟵ Factorise the denominator into two linear factors.

$x \equiv A(1-2x) + B(1+x)$ ⟵ Multiply each term by $(1+x)(1-2x)$.

Let $x = -1$

$\quad -1 = A(3) + 0 \implies A = -\dfrac{1}{3}$

Let $x = \dfrac{1}{2}$ ⟵ When $1-2x = 0, x = \dfrac{1}{2}$.

$\quad \dfrac{1}{2} = 0 + B\left(\dfrac{3}{2}\right) \implies B = \dfrac{1}{3}$

Hence $\dfrac{x}{(1+x)(1-2x)} \equiv -\dfrac{1}{3(1+x)} + \dfrac{1}{3(1-2x)}$ ⟵ Substitute the values of A and B into the fractions.

Note: You could also use the method of equating coefficients.

Exercise 7.1A

Express the following in partial fractions.

1. $\dfrac{x-4}{(x-1)(x-2)}$

2. $\dfrac{x}{(x+1)(x-4)}$

3. $\dfrac{3x+5}{x^2+2x-3}$

4. $\dfrac{2x-1}{x(2x+1)}$

5. $\dfrac{2x^2+17x+21}{(x+2)(x+3)(x-3)}$

Partial fractions type 2: Algebraic fractions with a repeated factor in the denominator can be split into partial fractions in the following way:

$$\frac{1}{(x + p)^2} \equiv \frac{A}{(x + p)} + \frac{B}{(x + p)^2}$$

Example 3

Express $\dfrac{3x^2 + 2}{x(x - 1)^2}$ in partial fractions.

$\dfrac{3x^2 + 2}{x(x - 1)^2} \equiv \dfrac{A}{x} + \dfrac{B}{x - 1} + \dfrac{C}{(x - 1)^2}$ Separate the repeated factor $(x - 1)^2$ into two denominators.

$3x^2 + 2 \equiv A(x - 1)^2 + Bx(x - 1) + Cx$

Let $x = 0$ This means B and C are eliminated.

$\quad 2 = A(-1)^2 + 0 + 0 \Rightarrow A = 2$

Let $x = 1$

$\quad 5 = 0 + 0 + C \Rightarrow C = 5$ This means A and B are eliminated.

Equating coefficients of x^2:

$3 = A + B \Rightarrow 3 = 2 + B \Rightarrow B = 1$ Substitute for $A = 2$ in $3 = A + B$.

$\dfrac{3x^2 + 2}{x(x - 1)^2} \equiv \dfrac{2}{x} + \dfrac{1}{x - 1} + \dfrac{5}{(x - 1)^2}$ **Note:** You could also equate coefficients of x or equate the constant terms.

Example 4

Express $\dfrac{2x}{(x + 3)^2}$ in partial fractions.

$\dfrac{2x}{(x + 3)^2} \equiv \dfrac{A}{x + 3} + \dfrac{B}{(x + 3)^2}$ Separate the repeated factor $(x + 3)^2$ into two denominators.

$2x \equiv A(x + 3) + B$

Let $x = -3$ This means A is eliminated.

$\quad -6 = B \Rightarrow B = -6$

Equating coefficients of x: Instead, you could equate the constants, i.e. $0 = 3A + B$.

$\quad 2 = A \Rightarrow A = 2$

$\dfrac{2x}{(x + 3)^2} \equiv \dfrac{2}{x + 3} - \dfrac{6}{(x + 3)^2}$

Exercise 7.1B

Express the following in partial fractions.

1. $\dfrac{x}{(x-2)^2}$

2. $\dfrac{3}{x(3x-1)^2}$

3. $\dfrac{x}{(x-1)(x-2)^2}$

4. $\dfrac{1}{(x+1)(x-1)^2}$

5. $\dfrac{1}{x(x^2-2x+1)}$

Partial fractions type 3: Algebraic fractions with a quadratic factor in the denominator can be split into partial fractions in the following way:

$$\frac{1}{(x+p)(x^2+q)} \equiv \frac{A}{x+p} + \frac{Bx+C}{x^2+q}$$

Example 5

Express $\dfrac{2x^2-x+6}{(x+1)(x^2+2)}$ in partial fractions.

$\dfrac{2x^2-x+6}{(x+1)(x^2+2)} \equiv \dfrac{A}{x+1} + \dfrac{Bx+C}{x^2+2}$ ← Put $Bx + C$ as the numerator of $x^2 + 2$.

$2x^2 - x + 6 \equiv A(x^2+2) + (Bx+C)(x+1)$

Let $x = -1$ ← This means B and C are eliminated.

$\quad 9 = A(3) + 0 \quad \Rightarrow \quad A = 3$

Equating coefficients of x^2:

$2 = A + B \quad \Rightarrow \quad 2 = 3 + B$ ← Substitute for $A = 3$ in $2 = A + B$.

$\qquad\qquad \Rightarrow \quad B = -1$

Equating the constant terms:

$6 = 2A + C \quad \Rightarrow \quad 6 = 6 + C$ ← Substitute for $A = 3$ in $6 = 2A + C$.

$\qquad\qquad \Rightarrow \quad C = 0$

$\dfrac{2x^2-x+6}{(x+1)(x^2+2)} \equiv \dfrac{3}{x+1} - \dfrac{x}{x^2+2}$

Note: You could also equate coefficients of x, but it is simpler in this case to equate coefficients of x^2 and equate the constant terms.

Example 6

Express $\dfrac{5x^2 - 5x + 4}{(2x - 1)(3 - x^2)}$ in partial fractions.

$\dfrac{5x^2 - 5x + 4}{(2x - 1)(3 - x^2)} \equiv \dfrac{A}{2x - 1} + \dfrac{Bx + C}{3 - x^2}$ ← Put $Bx + C$ as the numerator of $3 - x^2$.

$5x^2 - 5x + 4 \equiv A(3 - x^2) + (Bx + C)(2x - 1)$

Let $x = \dfrac{1}{2}$ ← This means B and C are eliminated.

$\dfrac{11}{4} = A\left(3 - \dfrac{1}{4}\right) + 0 \quad \Rightarrow \quad A = 1$

Equating coefficients of x^2:

$5 = -A + 2B \Rightarrow 5 = -1 + 2B$ ← Substitute for $A = 1$ in $5 = -A + 2B$.

$\Rightarrow B = 3$

Equating coefficients of the constant terms:

$4 = 3A - C \quad \Rightarrow \quad 4 = 3 - C$ ← Substitute for $A = 3$ in $4 = 3A - C$.

$\Rightarrow C = -1$

$\dfrac{5x^2 - 5x + 4}{(2x - 1)(3 - x^2)} \equiv \dfrac{1}{2x - 1} + \dfrac{3x - 1}{3 - x^2}$ **Note:** You could also equate coefficients of x.

Exercise 7.1C

Express the following in partial fractions.

1. $\dfrac{1 - 5x - 2x^2}{(1 - x)(x^2 + 2)}$

2. $\dfrac{2(1 - x^2)}{x(2 + x^2)}$

3. $\dfrac{13x^2 + 2x - 13}{(x + 2)(2x^2 - 1)}$

4. $\dfrac{5}{(x^2 + 4)(x + 1)}$

5. $\dfrac{16x^2 - 13x - 44}{(5 - x)(4x^2 - 3)}$

6. $\dfrac{3 - x}{(1 - 2x)(1 + x^2)}$

7. $\dfrac{7x - 4x^2}{(3 - x^2)(x - 2)}$

8. $\dfrac{-22x^2 + x - 5}{(2x + 1)(2 + 3x^2)}$

Partial fractions type 4: Improper algebraic fractions occur when a polynomial of degree greater than or equal to n is divided by another polynomial of degree n.

They can be split into partial fractions by first doing long division, and then splitting the remainder into partial fractions using one of the techniques discussed in types 1, 2, and 3:

$$\frac{x^2}{(x+p)(x+q)} \equiv A + \frac{Bx+C}{(x+p)(x+q)}$$

Example 7

Express $\dfrac{x^2 - x + 5}{(x-1)(x+2)}$ in partial fractions.

Method 1: Assigning values to x and equating coefficients.

We have a polynomial of degree 2 divided by another polynomial of degree 2, so the leading term in the quotient will be a constant. Let this constant be A.

$$\frac{x^2 - x + 5}{(x-1)(x+2)} \equiv A + \frac{B}{x-1} + \frac{C}{x+2} \qquad \longleftarrow \quad \text{There are now no improper fractions.}$$

$$x^2 - x + 5 \equiv A(x-1)(x+2) + B(x+2) + C(x-1) \qquad \longleftarrow \quad \text{Multiply each term by } (x-1)(x+2).$$

Let $x = 1$: $\quad 5 = 0 + B(3) + 0 \Rightarrow B = \dfrac{5}{3}$

Let $x = -2$: $\ 11 = 0 + 0 + C(-3) \Rightarrow C = -\dfrac{11}{3}$

Equating coefficients of x^2: $1 = A \qquad \longleftarrow \quad$ It is easiest to equate coefficients of x^2 first.

$$\frac{x^2 - x + 5}{(x-1)(x+2)} \equiv 1 + \frac{5}{3(x-1)} - \frac{11}{3(x+2)}$$

Method 2: Using long division, $(x^2 - x + 5) \div (x^2 + x - 2) = 1 \text{ remainder } -2x + 7$

Thus $\dfrac{x^2 - x + 5}{(x-1)(x+2)} = 1 + \dfrac{-2x+7}{(x-1)(x+2)} \qquad \longleftarrow \quad$ This is a **type 1** partial fraction.

Consider $\dfrac{-2x+7}{(x-1)(x+2)} \equiv \dfrac{A}{x-1} + \dfrac{B}{x+2}$

$-2x + 7 \equiv A(x+2) + B(x-1)$

Let $x = 1$: $\qquad 5 = A(3) + 0 \quad \Rightarrow \quad A = \dfrac{5}{3}$ $\qquad$ Do not forget to put the constant term in the answer.

Let $x = -2$: $\qquad 11 = 0 + B(-3) \Rightarrow B = -\dfrac{11}{3}$

$$\frac{x^2 - x + 5}{(x-1)(x+2)} \equiv 1 + \frac{5}{3(x-1)} - \frac{11}{3(x+2)}$$

Example 8

Find the values of the constants A, B, C, and D such that

$$\frac{3x^3 + 2x^2 + 6x + 4}{x^2(x + 1)} = A + \frac{B}{x} + \frac{C}{x^2} + \frac{D}{x + 1}$$

We have a polynomial of degree 3 which is divided by another polynomial of degree 3.

Using long division, $(3x^3 + 2x^2 + 6x + 4) \div (x^3 + x^2) = 3$ remainder $-x^2 + 6x + 4$

Hence $\dfrac{3x^3 + 2x^2 + 6x + 4}{x^2(x + 1)} = 3 + \dfrac{-x^2 + 6x + 4}{x^2(x + 1)}$

Thus $A = 3$.

Consider $\dfrac{-x^2 + 6x + 4}{x^2(x + 1)} \equiv \dfrac{B}{x} + \dfrac{C}{x^2} + \dfrac{D}{x + 1}$ ← Repeated factor and linear factor

$-x^2 + 6x + 4 \equiv Bx(x + 1) + C(x + 1) + Dx^2$

Let $x = -1$: $\quad -3 = D(1) \Rightarrow D = -3$

Let $x = 0$: $\quad 4 = C(1) \Rightarrow C = 4$

Coeff x^2: $\quad -1 = B + D \Rightarrow B = 2$

$A = 3$, $B = 2$, $C = 4$, $D = -3$ ← Remember to state the values of A, B, C, and D at the end of your working.

Exercise 7.1D

1. Express $\dfrac{x^2 + 5}{(x - 2)(x - 3)}$ in partial fractions.

2. Express $\dfrac{x^2}{x^2 - 1}$ in partial fractions.

3. Express $\dfrac{3x^3 - 5x^2 - 3x - 40}{(x^2 + 4)(x - 3)}$ in partial fractions.

4. Express $\dfrac{-x^2 - 8x - 11}{(x + 3)^2}$ in partial fractions.

5. Find the values of the constants A, B, and C such that $\dfrac{2x - 2x^2}{x^2 - 4} \equiv A + \dfrac{B}{x + 2} + \dfrac{C}{x - 2}$.

Exercise 7.1E

This exercise covers all four types of partial fractions that you have studied in this chapter so far.
Express the following in partial fractions.

1. $\dfrac{x - 1}{(x + 1)(x + 3)}$

2. $\dfrac{x^2 - 4x + 3}{(x + 1)(x + 2)}$

3. $\dfrac{2x - 7}{(x - 5)^2}$

4. $\dfrac{11x + 3}{(1 - 5x)(1 + x^2)}$

5. $\dfrac{x - 8}{(x + 1)(x - 2)}$

6. $\dfrac{3x^2 + 5x + 3}{x(x - 3)}$

7. $\dfrac{1 + 3x^2}{x(x - 1)^2}$

8. $\dfrac{2x^2 - 3}{x(x^2 + 2)}$

9. $\dfrac{x - 1}{x^2 + x}$

10. $\dfrac{x(x + 2)}{(x + 3)(x - 3)}$

11. $\dfrac{8x^2 + 5x + 3}{(x - 2)(x + 1)^2}$

12. $\dfrac{x - 3}{(x - 1)(x^2 + 1)}$

13. $\dfrac{x}{(x - 1)(x - 2)^2}$

14. $\dfrac{x^2 + 1}{x^2 - 1}$

15. $\dfrac{2x^2 + 2x + 3}{(x + 2)(x^2 + 3)}.$

7.2 Binomial expansions of the form $(1 + x)^n$ when n is not a positive integer

In P1, you met the binomial expansion $(1 + x)^n = 1 + nx + \dfrac{n(n - 1)}{2!}x^2 + \dfrac{n(n - 1)(n - 2)}{3!}x^3 + \dots + x^n,$
where n is a positive integer.

We now consider the expansion when n is not necessarily a positive integer.

$$(1 + x)^n = 1 + nx + \dfrac{n(n - 1)}{2!}x^2 + \dfrac{n(n - 1)(n - 2)}{3!}x^3 + \dots$$

where n is any integer or fraction and $|x| < 1$

In the case where n is **not** a positive integer, the binomial expansion of $(1 + x)^n$
does not terminate (it forms an infinite series). In this case, we must restrict the
domain of the expansion to $|x| < 1$ in order for the infinite series to converge. Most
often we use this expansion to find an approximation to $(1 + x)^n$ up to a certain order of n.

Example 9

Expand $(1 - 2x)^{-2}$ in ascending powers of x, up to and including the term in x^3,
simplifying the coefficients.

$(1 - 2x)^{-2} = 1 + (-2)(-2x) + \dfrac{(-2)(-3)}{2 \times 1}(-2x)^2 + \dfrac{(-2)(-3)(-4)}{3 \times 2 \times 1}(-2x)^3 + \dots$ ← $n = -2$, and we use $-2x$ instead of x.

$= 1 + 4x + 12x^2 + 32x^3 + \dots$

valid for $|-2x| < 1$ ← Instead of $|x| < 1$ we write $|-2x| < 1$.

i.e. valid for $-\dfrac{1}{2} < x < \dfrac{1}{2}$ ← Write this condition in the form shown.

Example 10

a) Expand $\left(1 - \frac{x}{2}\right)^{\frac{1}{3}}$ in ascending powers of x, up to and including the term in x^3.

b) Use your expansion to evaluate $\sqrt[3]{0.999}$, giving your answer to 6 decimal places.

a) $\left(1 - \frac{x}{2}\right)^{\frac{1}{3}} = 1 + \left(\frac{1}{3}\right)\left(-\frac{x}{2}\right) + \frac{\left(\frac{1}{3}\right)\left(-\frac{2}{3}\right)}{2 \times 1}\left(-\frac{x}{2}\right)^2 + \frac{\left(\frac{1}{3}\right)\left(-\frac{2}{3}\right)\left(-\frac{5}{3}\right)}{3 \times 2 \times 1}\left(-\frac{x}{2}\right)^3 + \ldots$

$$= 1 - \frac{1}{6}x - \frac{1}{36}x^2 - \frac{5}{648}x^3 + \ldots \text{ provided } \left|-\frac{x}{2}\right| < 1, \text{ i.e. } -2 < x < 2$$

b) Let $\left(1 - \frac{x}{2}\right)^{\frac{1}{3}} = \sqrt[3]{0.999} = (0.999)^{\frac{1}{3}}$

> Do not forget to write this condition.

$1 - \frac{x}{2} = 0.999$

$1 - 0.999 = \frac{x}{2}$

$x = 0.002$

> We use this value of x to find $\sqrt[3]{0.999}$.

$\sqrt[3]{0.999} = 1 - \frac{1}{6}(0.002) - \frac{1}{36}(0.002)^2 - \frac{5}{648}(0.002)^3 + \ldots$

> Substitute $x = 0.002$ into your expansion in (a).

$$= 1 - 0.000\,333\,333 - 0.000\,000\,111 - 0.000\,000\,02666$$

$$= 0.999\,666\,5293 = 0.999\,667 \text{ to 6 d.p.}$$

Example 11

Find the coefficient of x^3 in the expansion of $\dfrac{3 + x}{1 + 3x}$.

$\dfrac{3 + x}{1 + 3x} = (3 + x)(1 + 3x)^{-1}$

> The denominator is the same as $(1 + 3x)^{-1}$.

$(1 + 3x)^{-1} = 1 + (-1)(3x) + \dfrac{(-1)(-2)}{2 \times 1}(3x)^2 + \dfrac{(-1)(-2)(-3)}{3 \times 2 \times 1}(3x)^3 + \ldots$

$$= 1 - 3x + 9x^2 - 27x^3 + \ldots$$

> Expand the denominator.

$(3 + x)(1 + 3x)^{-1} = (3 + x)(1 - 3x + 9x^2 - 27x^3 + \ldots)$

Term in x^3 is $(3)(-27x^3) + (x)(9x^2) = -72x^3$

> You do not need to fully expand the brackets.

The coefficient of x^3 is -72.

> Do not put x^3 in your answer.

Example 12

Find the first four terms in the expansion of $(1 + x^2)^{-3}$.

$(1 + x^2)^{-3} = 1 + (-3)\,x^2 + \dfrac{(-3)(-4)}{2 \times 1}(x^2)^2 + \dfrac{(-3)(-4)(-5)}{3 \times 2 \times 1}(x^2)^3 + \ldots$ ←   Use x^2 instead of x in the expansion.

$= 1 - 3x^2 + 6x^4 - 10x^6 + \ldots$ for $-1 < x < 1$

Exercise 7.2

1. Expand the following in ascending powers of x, up to and including the term in x^3, and simplify the coefficients. State the range of values of x for which each expansion is valid.

   a) $(1 - x)^{-4}$

   b) $(1 + 3x)^{-5}$

   c) $\sqrt[3]{(1 + x)}$

   d) $\left(1 + \dfrac{x}{2}\right)^{-3}$

   e) $\dfrac{x}{1 - x}$

   f) $\dfrac{1}{(1 - 5x)^2}$

2. Find the first three terms in the expansion of $\dfrac{1}{\sqrt{1 + x^2}}$ in ascending powers of x.

3. a) Find the first four terms in the expansion of $\dfrac{1}{(1 - x)^2}$ in ascending powers of x.

   b) Hence find the 50th term of this expansion.

4. a) Expand $\sqrt{(1 - 2x)}$ in ascending powers of x, up to and including the term in x^4.

   b) Use your expansion to estimate $\sqrt{0.8}$, giving your answer to 4 decimal places.

5. Find the first three terms in the expansion of $\dfrac{1 - x}{\sqrt{1 + x}}$ in ascending powers of x.

6. Find the first four terms in the expansion of $(1 + 4x^2)^{-2}$.

7. Expand $\dfrac{1 + 2x}{(1 - x)^2}$ in ascending powers of x, up to and including the term in x^3.

8. a) Expand $(1 + x)^{\frac{1}{2}}$ in ascending powers of x, up to and including the term in x^3.

   b) Use your expansion to estimate $\sqrt{1.08}$, giving your answer to 4 decimal places.

9. The first three terms in the expansion of $(1 + ax)^n$ are $1 + 4x + 10x^2$.

   a) Find the value of a and the value of n.

   b) Hence find the coefficient of the term in x^3.

10. Expand $\dfrac{1 - 3x}{1 + x^2}$ in ascending powers of x, up to and including the term in x^4.
    State the values of x for which the expansion is valid.

Binomial expansions of the form $(1 + x)^n$ when n is not a positive integer

11. The first three terms in the expansion of $(1 + kx)^n$ are $1 + 12x + 81x^2$.

    a) Find the value of k and the value of n.   **b)** Hence find the term in x^3.

12. Find the first five terms in the expansion of $\dfrac{1 - 2x}{1 + 2x^2}$ in ascending powers of x.

7.3 Binomial expansions of the form $(a + x)^n$ where n is not a positive integer

We now consider the expansion when it is of the form $(a + x)^n$ where $a \neq 1$.

We want to use the binomial expansion:

$$(1 + x)^n = 1 + nx + \frac{n(n-1)}{2!}x^2 + \frac{n(n-1)(n-2)}{3!}x^3 + \dots$$

where n is any integer or fraction and $|x| < 1$.

> However, when the first term in the bracket is not 1, we must first factorise the expression into the form $a^n \left(1 + \dfrac{x}{a}\right)^n$ and then use the binomial expansion of $(1 + x)^n$, so
>
> $$(a + x)^n = a^n \left(1 + \frac{x}{a}\right)^n = a^n \left(1 + n\left(\frac{x}{a}\right) + \frac{n(n-1)}{2!}\left(\frac{x}{a}\right)^2 + \dots \right)$$

Example 13

Express the following in the form $a^n(1 + x)^n$.

a) $(5 - x)^{-3}$ **b)** $\sqrt{(2 - 6x)}$ **c)** $\dfrac{32}{(4 + 2x)^2}$

. . . .

a) $(5 - x)^{-3} = \left[5\left(1 - \dfrac{x}{5}\right)\right]^{-3}$ ⟵ Take out 5 and divide each term in the bracket by 5.

$= 5^{-3}\left(1 - \dfrac{x}{5}\right)^{-3}$ ⟵ Remember 5 is also to the power of −3.

$= \dfrac{1}{125}\left(1 - \dfrac{x}{5}\right)^{-3}$

b) $\sqrt{(2 - 6x)} = [2(1 - 3x)]^{\frac{1}{2}}$ ⟵ It is helpful to use square brackets and round brackets.

$= 2^{\frac{1}{2}}(1 - 3x)^{\frac{1}{2}}$

c) $\dfrac{32}{(4 + 2x)^2} = 32(4 + 2x)^{-2}$ ⟵ First get rid of the fraction.

$= 32\left[4\left(1 + \dfrac{1}{2}x\right)\right]^{-2}$

$= 32\left[4^{-2}\left(1 + \dfrac{1}{2}x\right)^{-2}\right]$

$= 2\left(1 + \dfrac{1}{2}x\right)^{-2}$ ⟵ $4^{-2} = \dfrac{1}{16}$

Example 14

Expand $\dfrac{1}{\sqrt{(4+x)}}$ in ascending powers of x, up to and including the term in x^2, simplifying the coefficients.

$\dfrac{1}{\sqrt{(4+x)}} = (4+x)^{-\frac{1}{2}}$ ← Rearrange to the form $(a+b)^n$.

$= \left[4\left(1+\dfrac{x}{4}\right)\right]^{-\frac{1}{2}} = 4^{-\frac{1}{2}}\left(1+\dfrac{x}{4}\right)^{-\frac{1}{2}}$ ← We want it to be in the form $p(1+q)^n$.

$= \dfrac{1}{2}\left(1+\dfrac{x}{4}\right)^{-\frac{1}{2}}$ ← $4^{-\frac{1}{2}} = \dfrac{1}{\sqrt{4}} = \dfrac{1}{2}$

Now $\left(1+\dfrac{x}{4}\right)^{-\frac{1}{2}} = 1 + \left(-\dfrac{1}{2}\right)\left(\dfrac{x}{4}\right) + \dfrac{\left(-\dfrac{1}{2}\right)\left(-\dfrac{3}{2}\right)}{2 \times 1}\left(\dfrac{x}{4}\right)^2$ ← $n = -\dfrac{1}{2}$

$= 1 - \dfrac{1}{8}x + \dfrac{3}{128}x^2 + \ldots$

Thus $\dfrac{1}{\sqrt{(4+x)}} = \dfrac{1}{2}\left(1 - \dfrac{1}{8}x + \dfrac{3}{128}x^2 + \ldots\right)$

$= \dfrac{1}{2} - \dfrac{1}{16}x + \dfrac{3}{256}x^2 + \ldots$

for $-4 < x < 4$ ← Do not forget to write this condition.

Exercise 7.3

1. Express the following in the form $a(1+x)^n$.

    a) $(25 - 50x)^{-\frac{1}{2}}$ b) $\dfrac{9}{(3-x)^2}$ c) $\sqrt[3]{(27+4x)}$

2. Expand $(2-x)^{-5}$ in ascending powers of x, up to and including the term in x^3, simplifying the coefficients.

3. Find the first three terms in the expansion of $\sqrt{(4-x)}$ in ascending powers of x.

4. Find the coefficient of the term in x^2 in the expansion of $\dfrac{1}{3+x}$.

5. Find the first four terms in the expansion of $\dfrac{1}{(2+x^2)}$ in ascending powers of x.

6. a) Expand $\dfrac{1}{(4-x)^2}$ in ascending powers of x, up to and including the term in x^3, simplifying the coefficients.

    b) Hence find the coefficient of x^3 in the expansion of $\dfrac{2-x}{(4-x)^2}$.

7. Expand $(1+3x)\sqrt{(9-x)}$ in ascending powers of x, up to and including the term in x^2, simplifying the coefficients.

8. Expand $\dfrac{1}{\sqrt{(4+2x^2)}}$ in ascending powers of x, up to and including the term in x^6, simplifying the coefficients.

9. The first three terms in the expansion of $(2+px)^{-2}$ are $\dfrac{1}{4}+6x+qx^2$.

   a) Find the value of p and the value of q.   **b)** Hence find the term in x^3.

10. In the expansion of $(4-kx)^{\frac{1}{2}}$ where $k\neq 0$, the coefficient of the term in x is 8 times the coefficient of the term in x^2. Find the value of k.

7.4 Binomial expansions and partial fractions

We can use partial fractions to help us simplify the expansions of more complex expressions.

Example 15

a) Express $\dfrac{1}{(x+1)(x-1)^2}$ in partial fractions.

b) Hence obtain the expansion of $\dfrac{1}{(x+1)(x-1)^2}$ in ascending powers of x, up to and including the term in x^2.

a) In Exercise 7.1B question 4, you found that

$$\dfrac{1}{(x+1)(x-1)^2}\equiv\dfrac{1}{4(x+1)}-\dfrac{1}{4(x-1)}+\dfrac{1}{2(x-1)^2}$$

b) $\dfrac{1}{4(x+1)}=\dfrac{1}{4}(1+x)^{-1}=\dfrac{1}{4}[1+(-1)x+\dfrac{(-1)(-2)}{2\times1}x^2+...]$

    $=\dfrac{1}{4}(1-x+x^2+...)=\dfrac{1}{4}-\dfrac{1}{4}x+\dfrac{1}{4}x^2+...$

    for $|x|<1$, i.e. for $-1<x<1$

> Expand each of the three terms separately.

$\dfrac{1}{4(x-1)}=\dfrac{1}{4}(-1)^{-1}(1-x)^{-1}=-\dfrac{1}{4}[1+(-1)(-x)+\dfrac{(-1)(-2)}{2\times1}(-x)^2+...]$

    $=-\dfrac{1}{4}(1+x+x^2+...)=-\dfrac{1}{4}-\dfrac{1}{4}x-\dfrac{1}{4}x^2+...$

    for $|-x|<1$, i.e. for $-1<x<1$

> We want to expand $(1-x)^{-1}$ not $(x-1)^{-1}$.

$\dfrac{1}{2(x-1)^2}=\dfrac{1}{2}(-1)^{-2}(1-x)^{-2}=\dfrac{1}{2}[1+(-2)(-x)+\dfrac{(-2)(-3)}{2\times1}(-x)^2+...]$

    $=\dfrac{1}{2}[1+2x+3x^2+...]=\dfrac{1}{2}+x+\dfrac{3}{2}x^2+...$

    for $|-x|<1$, i.e. for $-1<x<1$

> Add the first three terms of each expansion.

Hence

$$\dfrac{1}{(x+1)(x-1)^2}=\dfrac{1}{4}-\dfrac{1}{4}x+\dfrac{1}{4}x^2-\left(-\dfrac{1}{4}-\dfrac{1}{4}x-\dfrac{1}{4}x^2\right)+\dfrac{1}{2}+x+\dfrac{3}{2}x^2=1+x+2x^2+...$$

    for $-1<x<1$

> This is true for all the separate expansions.

Example 16

a) Express $\dfrac{2x^2 - x + 6}{(x + 1)(x^2 + 2)}$ in partial fractions.

b) Hence obtain the expansion of $\dfrac{2x^2 - x + 6}{(x + 1)(x^2 + 2)}$ in ascending powers of x, up to and including the term in x^3.

· ·

a) See Example 5, where we showed that

$$\frac{2x^2 - x + 6}{(x + 1)(x^2 + 2)} \equiv \frac{3}{x+1} - \frac{x}{x^2 + 2}$$

b) $\dfrac{3}{x+1} = 3(1 + x)^{-1} = 3\left[1 + (-1)x + \dfrac{(-1)(-2)}{2 \times 1}x^2 + \dfrac{(-1)(-2)(-3)}{3 \times 2 \times 1}x^3 + \ldots\right]$

$\qquad = 3(1 - x + x^2 - x^3 + \ldots) = 3 - 3x + 3x^2 - 3x^3 + \ldots$

$\qquad$ for $|x| < 1$, i.e. for $-1 < x < 1$

$\dfrac{x}{x^2 + 2} = x(2 + x^2)^{-1} = x(2)^{-1}\left(1 + \dfrac{x^2}{2}\right)^{-1} = \dfrac{x}{2}\left(1 + \dfrac{x^2}{2}\right)^{-1}$

> We want to expand a bracket in the form $(1 + x)^n$.

$\qquad = \dfrac{x}{2}\left[1 + (-1)\left(\dfrac{x^2}{2}\right) + \dfrac{(-1)(-2)}{2 \times 1}\left(\dfrac{x^2}{2}\right)^2 + \dfrac{(-1)(-2)(-3)}{3 \times 2 \times 1}\left(\dfrac{x^2}{2}\right)^3 + \ldots\right]$

> Use $\dfrac{x^2}{2}$ instead of x in the expansion.

$\qquad = \dfrac{x}{2}\left[1 - \dfrac{x^2}{2} + \dfrac{x^4}{4} + \ldots\right] = \dfrac{x}{2} - \dfrac{x^3}{4} + \ldots$

$\qquad$ for $\left|\dfrac{x^2}{2}\right| < 1$, i.e. for $0 < x^2 < 2$, or $-\sqrt{2} < x < \sqrt{2}$

Hence $\dfrac{2x^2 - x + 6}{(x + 1)(x^2 + 2)} = (3 - 3x + 3x^2 - 3x^3 + \ldots) - \left(\dfrac{x}{2} - \dfrac{x^3}{4} + \ldots\right)$

> The overlap of $-1 < x < 1$ and $-\sqrt{2} < x < \sqrt{2}$ is $-1 < x < 1$.

$\qquad = 3 - \dfrac{7}{2}x + 3x^2 - \dfrac{11}{4}x^3 + \ldots$, for $-1 < x < 1$

Exercise 7.4

1. a) Express $\dfrac{3x + 5}{(1 - x)(1 + 3x)}$ in partial fractions.

b) Hence obtain the expansion of $\dfrac{3x + 5}{(1 - x)(1 + 3x)}$ in ascending powers of x, up to and including the term in x^2.

2. a) Express $\dfrac{x^2}{(1 + 2x)(1 + x)^2}$ in partial fractions.

b) Hence find the coefficient of x^3 in the expansion of $\dfrac{x^2}{(1 + 2x)(1 + x)^2}$.

3. Let $f(x) = \dfrac{x + 2}{x^2 - 1}$.

    a) Express $f(x)$ in partial fractions.

    b) Hence obtain the expansion of $f(x)$ in ascending powers of x, up to and including the term in x^3.

4. a) Express $\dfrac{8 - 11x + 4x^2}{(1 - x)^2(2 - x)}$ in partial fractions.

    b) Hence find the first three terms in the expansion of $\dfrac{8 - 11x + 4x^2}{(1 - x)^2(2 - x)}$ in ascending powers of x, simplifying your answer.

5. a) Express $\dfrac{1}{(1 - x)(x - 2)}$ in partial fractions.

    b) Hence find the coefficient of x^2 in the expansion of $\dfrac{1}{(1 - x)(x - 2)}$.

6. a) Express $\dfrac{1}{(1 - x)(1 + x^2)}$ in partial fractions.

    b) Hence, by working out all expansions to the term in x^4, find the first three terms in the expansion of $\dfrac{1}{(1 - x)(1 + x^2)}$ in ascending powers of x.

7. Let $f(x) = \dfrac{3 - x}{(1 - 2x)(1 + x^2)}$.

    a) Express $f(x)$ in partial fractions.

    b) Hence obtain the expansion of $f(x)$ in ascending powers of x, up to and including the term in x^2.

Summary exercise 7

1. Express in partial fractions $\dfrac{3x + 11}{x^2 - x - 6}$.

2. Express in partial fractions $\dfrac{2x - 1}{(x + 1)(x^2 + 1)}$.

3. Express in partial fractions $\dfrac{x^2 + 3x - 1}{(x - 2)(x + 1)}$.

4. Express in partial fractions $\dfrac{3x + 1}{(x - 1)^2(x + 2)}$.

5. Express in partial fractions

$$\dfrac{32 + 45x - 8x^2}{10(x + 3)(x - 2)(2x - 1)}$$

6. Express in partial fractions $\dfrac{96 + 12x - 5x^2}{(x + 4)^2(x - 4)}$.

7. Express in partial fractions $\dfrac{7x^2 + 8x - 11}{(x^2 - 2)(x + 3)}$.

8. Find the values of the constants A, B, C, and D such that

$$\frac{5x^3 - 11x^2 + 7x - 3}{x(x - 1)^2} \equiv A + \frac{B}{x} + \frac{C}{x-1} + \frac{D}{(x - 1)^2}$$

9. A function $f(x)$ is defined as $f(x) = (2 - 5x)^{-2}$. Obtain the expansion of $f(x)$ in ascending powers of x, as far as the term in x^3, and simplify your answer.

10. a) Expand $(1 - x)^{\frac{1}{2}}$ in ascending powers of x, up to and including the term in x^3.

    b) Use your expansion to estimate $\sqrt{0.9}$, giving your answer to 4 decimal places.

11. Find the binomial expansion of $\dfrac{1}{\left(4 + 3x\right)^3}$ in ascending powers of x, up to and including the term in x^2. Simplify your answer.

12. a) Expand $\dfrac{1}{\sqrt{4 + 2x}}$ in ascending powers of x, up to and including the term in x^3.

    b) Use your expansion to estimate the value of $4.02^{-\frac{1}{2}}$.

13. Expand $(2 - x)(\sqrt{1 - 2x})$ in ascending powers of x, up to and including the term in x^3, simplifying the coefficients.

14. a) Express $\dfrac{2x + 5}{(1 + x)(2 + x)}$ in partial fractions.

    b) Hence obtain the expansion of $\dfrac{2x + 5}{(1 + x)(2 + x)}$ in ascending powers of x, up to and including the term in x^3, simplifying your answer.

15. $f(x) = \dfrac{3x^2 - x + 4}{(1 + x)(1 - x)^2}$

    a) Express $f(x)$ in partial fractions.

    b) Hence obtain the expansion of $f(x)$ in ascending powers of x, up to and including the term in x^2.

Chapter summary

Partial fractions

- **Type 1:** Algebraic fractions with **linear factors in the denominator** can be split into partial fractions in the following way: $\dfrac{1}{(x + p)(x + q)} \equiv \dfrac{A}{x + p} + \dfrac{B}{x + q}$

- **Type 2:** Algebraic fractions with a **repeated factor in the denominator** can be split into partial fractions in the following way: $\dfrac{1}{(x + p)^2} \equiv \dfrac{A}{x + p} + \dfrac{B}{(x + p)^2}$

- **Type 3:** Algebraic fractions with a **quadratic factor in the denominator** can be split into partial fractions in the following way: $\dfrac{1}{(x + p)(x^2 + q)} \equiv \dfrac{A}{x + p} + \dfrac{Bx + C}{x^2 + q}$

- **Type 4: Improper algebraic fractions** occur when a polynomial of degree n or higher is divided by another polynomial of degree n. They can be split into partial fractions by first doing long division, and then splitting the remainder into partial fractions using one of the techniques discussed in Types 1, 2, or 3:

$$\frac{x^2}{(x+1)(x+2)} = 1 + \frac{-3x-2}{(x+1)(x+2)}$$

Binomial expansions of $(1 + x)^n$ where n is not a positive integer

- $(1 + x)^n = 1 + nx + \dfrac{n(n-1)}{2!}x^2 + \dfrac{n(n-1)(n-2)}{3!}x^3 + \dots$

  where n is any integer or fraction and $|x| < 1$.

Binomial expansions of $(a + x)^n$ where n is not a positive integer

- First put it in the form $a^n\left(1 + \dfrac{x}{a}\right)^n$.

Binomial expansion and partial fractions

- We can use partial fractions to help us simplify the expansions of more complex expressions.

8 Further integration

In recent years, many parts of the world have been badly hit by severe flooding caused by changing weather patterns. It often happens when more rain falls on already saturated ground, or rain falls more quickly than it can be absorbed into the ground and runs off to lower levels. Differential equations and integration techniques are important mathematical tools employed in the search for ways to predict when and where flooding might occur and ways to lessen the effects of it.

Objectives

- Integrate rational functions by means of decomposition into partial fractions (restricted to the types of partial fractions specified in Chapter 7).

- Extend the idea of 'reverse differentiation' to include the integration of $\dfrac{1}{x^2 + a^2}$.

- Recognise an integrand of the form $\dfrac{kf'(x)}{f(x)}$, and integrate, for example, $\dfrac{x}{x^2 + 1}$ or $\tan x$.

- Recognise when an integrand can usefully be regarded as a product, and use integration by parts to integrate, for example, $x \sin 2x$, $x^2 e^{-x}$, $\ln x$ or $x \tan^{-1} x$.

- Use a given substitution to simplify and evaluate either a definite or an indefinite integral.

Before you start

You should know how to:

1. Decompose rational functions into partial fractions,

   e.g. a) $\dfrac{x+3}{(x-2)(x+1)} \equiv \dfrac{5}{3(x-2)} - \dfrac{2}{3(x+1)}$

   b) $\dfrac{3x^2 + 2}{x(x-1)^2} \equiv \dfrac{2}{x} + \dfrac{1}{x-1} + \dfrac{5}{(x-1)^2}$

2. Differentiate the logarithm of a function,

   e.g. $\dfrac{d}{dx}[\ln(e^x - 3x)] = \dfrac{e^x - 3}{e^x - 3x}$

Skills check:

1. Express the following as partial fractions.

   a) $\dfrac{x^2 + 3x - 1}{(x-2)(x+1)}$
   b) $\dfrac{3x+1}{(x-1)^2(x+2)}$

   c) $\dfrac{7x^2 + 8x - 11}{(x^2 - 2)(x+3)}$

2. Find $\dfrac{dy}{dx}$ when $y = \ln\sqrt{(2x+1)}$.

In Chapter 5 you extended integration to include a number of different types of function which you had learned how to differentiate in Chapter 4, including logarithms, exponentials, and trigonometric functions. In Chapter 4 you also learned how to differentiate products and quotients, and techniques of implicit and parametric differentiation. This chapter will extend the range of integrals you can tackle.

8.1 Integration using partial fractions

In section 7.1 you learned how to express four different types of function in partial fractions. One of the applications of this is that almost all the partial fraction forms are the sums of integrable functions where the initial rational function is not directly integrable. The first four examples here use functions which were split into partial fractions in Chapter 7, so we are looking at the extra steps to be done when an integral is required.

> To integrate difficult algebraic fractions, first look to see if you can split the fraction into partial fractions, where each partial fraction is integrable.

Example 1 deals with fractions which have linear factors in the denominator.

Example 1

Find $\int \dfrac{x+3}{(x-2)(x+1)} \, dx$.

$\dfrac{x+3}{(x-2)(x+1)} \equiv \dfrac{5}{3(x-2)} - \dfrac{2}{3(x+1)}$ ← From Example 1 in Chapter 7

So $\int \dfrac{x+3}{(x-2)(x+1)} \, dx = \int \left(\dfrac{5}{3(x-2)} - \dfrac{2}{3(x+1)} \right) dx$ ← Two terms which are each integrable as logarithmic functions

$\qquad = \dfrac{5}{3}\ln(x-2) - \dfrac{2}{3}\ln(x+1) + c$

Example 2 deals with fractions which have repeated linear factors in the denominator.

Example 2

Find $\int \dfrac{3x^2+2}{x(x-1)^2} \, dx$.

$\dfrac{3x^2+2}{x(x-1)^2} \equiv \dfrac{2}{x} + \dfrac{1}{x-1} + \dfrac{5}{(x-1)^2}$ ← From Example 3 in Chapter 7

$\int \dfrac{3x^2+2}{x(x-1)^2} \, dx = \int \left(\dfrac{2}{x} + \dfrac{1}{x-1} + \dfrac{5}{(x-1)^2} \right) dx$ ← Three terms which are each integrable

$\qquad = 2\ln x + \ln(x-1) - \dfrac{5}{(x-1)} + c$ ← Integrate each term carefully, and include the constant of integration.

Example 3 deals with fractions which have a quadratic factor in the denominator.

Example 3

Find $\displaystyle\int \frac{2x^2 - x + 6}{(x+1)(x^2+2)}\,dx$.

$$\frac{2x^2 - x + 6}{(x+1)(x^2+2)} \equiv \frac{3}{x+1} - \frac{x}{x^2+2}$$

From Example 5 in Chapter 7

$$\int \frac{2x^2 - x + 6}{(x+1)(x^2+2)}\,dx = \int\left(\frac{3}{x+1}\right)dx - \int\left(\frac{x}{x^2+2}\right)dx$$

The first of these you already know how to integrate and the second you will meet in section 8.3.

$$= 3\ln(x+1) - \int\left(\frac{x}{x^2+2}\right)dx$$

Example 4 deals with improper algebraic fractions.

Example 4

Find $\displaystyle\int \frac{x^2 - x + 5}{(x-1)(x+2)}\,dx$.

$$\frac{x^2 - x + 5}{(x-1)(x+2)} \equiv 1 + \frac{5}{3(x-1)} - \frac{11}{3(x+2)}$$

From Example 7 in Chapter 7

$$\int \frac{x^2 - x + 5}{(x-1)(x+2)}\,dx = \int\left(1 + \frac{5}{3(x-1)} - \frac{11}{3(x+2)}\right)dx$$

Three terms which are each integrable

$$= x + \frac{5}{3}\ln(x-1) - \frac{11}{3}\ln(x+2) + c$$

Example 5

By first expressing $\displaystyle\frac{1}{(x-2)(x-3)}$ in partial fractions, find $\displaystyle\int\left(\frac{1}{(x-2)(x-3)}\right)dx$.

$$\frac{1}{(x-2)(x-3)} \equiv \frac{A}{x-2} + \frac{B}{x-3}$$

Or any standard technique to put into partial fractions

$$1 = A(x-3) + B(x-2); \quad x = 2 \Rightarrow A = -1; \quad x = 3 \Rightarrow B = 1$$

$$\int\left(\frac{1}{(x-2)(x-3)}\right)dx = \int\left(\frac{-1}{x-2} + \frac{1}{x-3}\right)dx$$

$$= -\ln(x-2) + \ln(x-3) + c$$

Integrate each term.

$$= \ln\left(k\left(\frac{x-3}{x-2}\right)\right)$$

Combine into a single logarithm.

Example 6

By first expressing $\dfrac{4}{x^2(x-2)}$ in partial fractions, find $\displaystyle\int\left(\dfrac{4}{x^2(x-2)}\right)dx$.

$$\dfrac{4}{x^2(x-2)} \equiv \dfrac{A}{x-2} + \dfrac{B}{x} + \dfrac{C}{x^2}$$

Separate the repeated factor x^2 into two denominators.

$$4 \equiv Ax^2 + Bx(x-2) + C(x-2)$$

$$x = 2 \Rightarrow 4 = A(4) + 0 \Rightarrow A = 1$$

This means B and C are eliminated.

$$4 = x^2 + Bx^2 - 2Bx + Cx - 2C \Rightarrow B = -1, C = -2$$

Substituting for A means B and C can be found by equating coefficients.

$$\Rightarrow \dfrac{4}{x^2(x-2)} = \dfrac{1}{x-2} - \dfrac{1}{x} - \dfrac{2}{x^2}$$

$$\int\left(\dfrac{4}{x^2(x-2)}\right)dx = \int\left(\dfrac{1}{x-2} - \dfrac{1}{x} - \dfrac{2}{x^2}\right)dx$$

$$= \ln(x-2) - \ln x + \dfrac{2}{x} + k$$

Integrate each term; avoid using 'c' as constant of integration, having used 'C' before.

$$= \ln\left(\dfrac{x-2}{x}\right) + \dfrac{2}{x} + k$$

Simplify the logarithmic term.

Example 7

By first expressing $\dfrac{2x-4}{x(x^2+4)}$ in partial fractions, find $\displaystyle\int\left(\dfrac{2x-4}{x(x^2+4)}\right)dx$.

$$\dfrac{2x-4}{x(x^2+4)} \equiv \dfrac{A}{x} + \dfrac{Bx+C}{(x^2+4)}$$

Put $Bx + C$ as the numerator of $x^2 + 4$.

$$2x - 4 \equiv A(x^2+4) + (Bx+C)x$$

$$x = 0 \Rightarrow -4 = A(4) + 0 \Rightarrow A = -1$$

This means B and C are eliminated.

$$2x - 4 = -x^2 - 4 + Bx^2 + Cx \Rightarrow B = 1, C = 2$$

Substituting for A means B and C can be found by equating coefficients.

$$\Rightarrow \dfrac{2x-4}{x(x^2+4)} = \dfrac{-1}{x} + \dfrac{x+2}{(x^2+4)}$$

$$\int\left(\dfrac{x+2}{x(x^2+4)}\right)dx = \int\left(-\dfrac{1}{x} + \dfrac{x+2}{(x^2+4)}\right)dx$$

You will learn how to integrate the second term later in this chapter.

$$= -\ln x + \int\left(\dfrac{x+2}{(x^2+4)}\right)dx$$

Example 8

Find $\int\left(\dfrac{3x^3 - 4x^2 - 2}{x^2(x-1)}\right)dx$.

$\dfrac{3x^3 - 4x^2 - 2}{x^2(x-1)} \equiv A + \dfrac{B}{x} + \dfrac{C}{x^2} + \dfrac{D}{(x-1)}$ ⟵ There are now no improper fractions.

$3x^3 - 4x^2 - 2 \equiv Ax^2(x-1) + Bx(x-1) + C(x-1) + Dx^2$ ⟵ Multiply each term by $x^2(x-1)$.

$x = 0 \Rightarrow -2 = -C \Rightarrow C = 2$

$x = 1 \Rightarrow 3 - 4 - 2 = -3 = D \Rightarrow D = -3$

$3x^3 - 4x^2 - 2 = Ax^2(x-1) + Bx(x-1) + 2(x-1) - 3x^2$ ⟵ Substitute values for C and D.

$\qquad = Ax^3 - Ax^2 + Bx^2 - Bx + 2x - 2 - 3x^2$

$(x^3): A = 3; \ (x^2): -4 = -3 + B - 3 \Rightarrow B = 2$ ⟵ Equate coefficients to find the last two constants; use the extra term to check.

Check coefficient of $x: 0 = -B + 2 \Rightarrow B = 2$

$\dfrac{3x^3 - 4x^2 - 2}{x^2(x-1)} = 3 + \dfrac{2}{x} + \dfrac{2}{x^2} - \dfrac{3}{(x-1)}$

$\int\left(\dfrac{3x^3 - 4x^2 - 2}{x^2(x-1)}\right)dx = \int\left(3 + \dfrac{2}{x} + \dfrac{2}{x^2} - \dfrac{3}{(x-1)}\right)dx$

$\qquad = 3x + 2\ln x - \dfrac{2}{x} - 3\ln(x-1) + c$ ⟵ Integrate each term.

$\qquad = 3x - \dfrac{2}{x} + \ln\left(\dfrac{x^2}{(x-1)^3}\right) + c$ ⟵ Combine the logarithmic terms.

Example 9

Evaluate $\displaystyle\int_4^6 \dfrac{1}{(x-2)(x-3)}dx$.

$\displaystyle\int_4^6 \dfrac{1}{(x-2)(x-3)}dx = \left[-\ln(x-2) + \ln(x-3)\right]_4^6$ ⟵ Using Example 5

$\qquad = (-\ln 4 + \ln 3) - (-\ln 2 + \ln 1) = \ln\dfrac{3}{2}$ ⟵ Combine into a single logarithm.

Exercise 8.1

1. By first expressing each rational expression using partial fractions, find

a) $\int\left(\dfrac{7}{(x+3)(2x-1)}\right)dx$ b) $\int\left(\dfrac{2}{(x+5)(x+7)}\right)dx$ c) $\int\left(\dfrac{7}{(x+5)(x-2)}\right)dx$

d) $\int\left(\dfrac{2-x}{(2x+3)(x+5)}\right)dx$ e) $\int\left(\dfrac{3-x}{(2+3x)(1-4x)}\right)dx$ f) $\int\left(\dfrac{5-3x}{(x-2)(3-2x)}\right)dx$

g) $\int_1^2\left(\dfrac{4}{(x+3)(x+5)}\right)dx$ h) $\int_{-0.5}^{0.5}\left(\dfrac{5}{(3+2x)(1-x)}\right)dx$ i) $\int_3^4\left(\dfrac{5x+5}{(x-2)(x+3)}\right)dx.$

2. By first expressing each rational expression using partial fractions, find

a) $\int\left(\dfrac{1}{x(x-1)^2}\right)dx$ b) $\int\left(\dfrac{1}{x(x+1)^2}\right)dx$ c) $\int\left(\dfrac{8x+24}{(x+1)(x-3)^2}\right)dx$

d) $\int_7^{12}\left(\dfrac{4-x}{(x-2)^2}\right)dx$ e) $\int_4^5\left(\dfrac{25}{(2x-3)^2(1+x)}\right)dx$ f) $\int_3^4\left(\dfrac{4}{x^3-4x^2+4x}\right)dx.$

3. By first expressing each improper fraction using partial fractions, find

a) $\int\left(\dfrac{x^2-2x+6}{(x+1)(x-2)}\right)dx$ b) $\int\left(\dfrac{2x^2-x+1}{(x+1)(x-3)}\right)dx$

c) $\int_4^5\left(\dfrac{x^2-3x+10}{x^2-x-6}\right)dx$ d) $\int_3^4\left(\dfrac{10x^2-26x+10}{(2x^2-5x)}\right)dx.$

4. By first expressing $\dfrac{2\sqrt3}{x^2-3}$ in partial fractions, show that $\int_{2\sqrt3}^{3\sqrt3}\left(\dfrac{2\sqrt3}{x^2-3}\right)dx=\ln\left(\dfrac32\right).$

5. By first expressing $\dfrac{4x^2+9x+4}{2x^2+5x+3}$ in partial fractions, show that $\int_0^1\left(\dfrac{4x^2+9x+4}{2x^2+5x+3}\right)dx=2+\dfrac12\ln\left(\dfrac{5}{12}\right).$

6. Find the area bounded by the graph of $y=\dfrac{1}{(x-3)(x-2)}$, the x-axis, the y-axis, and the line $x=1$.

7. Find the area bounded by the graph of $y=\dfrac{x^3+3x^2-7x}{(x+2)(x-1)^2}$, the x-axis, and the lines $x=2$ and $x=3$.

8. Show that the area bounded by the graph of $y=\dfrac{2x^2-4x}{(1-2x)(1-x^2)}$, the x-axis, and the lines $x=-3$ and $x=-2$ is $\ln(2.1)$.

8.2 Integration of $\dfrac{1}{x^2+a^2}$

In section 4.6 you saw that if $y = \tan^{-1} x$, then $\dfrac{dy}{dx} = \dfrac{1}{1+x^2}$. This means that

$$\int \left(\frac{1}{1+x^2}\right) dx = \tan^{-1} x + c$$

Example 10

$y = \tan^{-1}\left(\dfrac{x}{a}\right)$. Find an expression for $\dfrac{dy}{dx}$.

. . . .

$y = \tan^{-1}\left(\dfrac{x}{a}\right) \Rightarrow x = a \tan y$

$1 = a \sec^2 y \dfrac{dy}{dx}$ ⟵   Differentiate with respect to x implicitly.

$1 = a(1 + \tan^2 y) \dfrac{dy}{dx}$ ⟵   Use $1 + \tan^2 y = \sec^2 y$.

Rearranging gives

$\dfrac{1}{(1 + \tan^2 y)} = a \dfrac{dy}{dx}$

But $\dfrac{x}{a} = \tan y$, so

$\dfrac{1}{\left(1 + \left(\dfrac{x}{a}\right)^2\right)} = a \dfrac{dy}{dx}$ ⟵   Substitute $\dfrac{x}{a}$ for $\tan y$ in $\dfrac{1}{1 + \tan^2 y}$.

Rearranging and simplifying:

$\dfrac{a^2}{a^2 + x^2} = a \dfrac{dy}{dx}$

$\dfrac{dy}{dx} = \dfrac{a}{a^2 + x^2}$

Example 11

$y = \tan^{-1}(bx)$. Find an expression for $\dfrac{dy}{dx}$.

$y = \tan^{-1}(bx) \Rightarrow x = \dfrac{1}{b}\tan y$

$1 = \dfrac{1}{b}\sec^2 y \dfrac{dy}{dx}$

Differentiate with respect to x implicitly.

$1 = \dfrac{1}{b}(1 + \tan^2 y)\dfrac{dy}{dx}$

Use $1 + \tan^2 y = \sec^2 y$.

Rearranging:

$\dfrac{b}{(1 + \tan^2 y)} = \dfrac{dy}{dx}$

But $bx = \tan y$, so

$\dfrac{b}{(1 + (bx)^2)} = \dfrac{dy}{dx}$

Substitute bx for $\tan y$ in $\dfrac{1}{1 + \tan^2 y}$.

Rearranging and simplifying:

$\dfrac{dy}{dx} = \dfrac{b}{1 + b^2 x^2}$

$$\int\left(\dfrac{1}{a^2 + x^2}\right)dx = \dfrac{1}{a}\tan^{-1}\left(\dfrac{x}{a}\right) + c$$

You are given this formula in the formula book.

$$\int\left(\dfrac{1}{1 + b^2 x^2}\right)dx = \dfrac{1}{b}\tan^{-1}(bx) + c$$

You are **not** given this formula in the formula book. It is useful if you can remember it accurately.

Example 12

Find $\displaystyle\int\left(\dfrac{1}{x^2 + 25}\right)dx$.

$\displaystyle\int\left(\dfrac{1}{x^2 + 25}\right)dx = \dfrac{1}{5}\tan^{-1}\left(\dfrac{x}{5}\right) + c$

Use $a = 5$ in the standard result.

Example 13

Find $\displaystyle\int \frac{3}{x^2 - 8x + 17}\,dx$.

$$\int \frac{3}{x^2 - 8x + 17}\,dx = \int \left(\frac{3}{(x-4)^2 + 1}\right)dx$$

Complete the square on the bottom.

$$= 3\tan^{-1}(x - 4) + c$$

Use the standard result.

Example 14

The diagram shows the graph of $y = \dfrac{1}{16 + x^2}$.

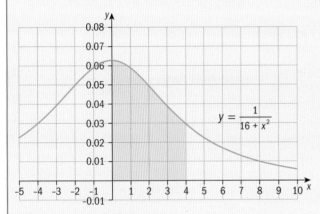

$$y = \frac{1}{16 + x^2}$$

Calculate the exact value of the shaded area.

The shaded area is the area under the curve between $x = 0$ and $x = 4$.

So area $= \displaystyle\int_0^4 \left(\frac{1}{16 + x^2}\right)dx$

$$\int_0^4 \left(\frac{1}{16 + x^2}\right)dx = \left[\frac{1}{4}\tan^{-1}\left(\frac{x}{4}\right)\right]_0^4 = \frac{1}{4}(\tan^{-1}1 - \tan^{-1}0) = \frac{\pi}{16}$$

Note that the value of y goes from 0.0625 to around 0.03, so an 'eyeball' estimate of the shaded area is around 0.2.

Example 15

Find $\int \dfrac{1}{1+9x^2}\,dx$.

....

$1+9x^2 = 1+(3x)^2$

$\int \dfrac{1}{1+9x^2}\,dx = \dfrac{1}{3}\tan^{-1}(3x)+c$ ⟵ Use the standard result.

Example 16

Find $\int \dfrac{1}{16+25x^2}\,dx$.

....

$16+25x^2 = 16\left(1+\left(\dfrac{5}{4}x\right)^2\right)$ ⟵ It is helpful to get the integrand into a standard form.

$\int \dfrac{1}{16+25x^2}\,dx = \dfrac{1}{16}\int \dfrac{1}{1+\left(\dfrac{5}{4}x\right)^2}\,dx = \dfrac{1}{16}\times\dfrac{4}{5}\tan^{-1}\left(\dfrac{5}{4}x\right)+c$ ⟵ Use the standard result.

$= \dfrac{1}{20}\tan^{-1}\left(\dfrac{5}{4}x\right)+c$ ⟵ Simplify the expression. Remember to include $+c$ except where there are limits.

Exercise 8.2

1. Find the following integrals.

   a) $\int \dfrac{1}{4+x^2}\,dx$

   b) $\int \dfrac{5}{25+x^2}\,dx$

   c) $\int \dfrac{9}{x^2+81}\,dx$

2. Find the following integrals.

   a) $\int \dfrac{1}{1+9x^2}\,dx$

   b) $\int \dfrac{1}{1+16x^2}\,dx$

   c) $\int \dfrac{1}{49x^2+1}\,dx$

3. Find the following integrals.

   a) $\int \dfrac{1}{1+3x^2}\,dx$

   b) $\int \dfrac{1}{1+8x^2}\,dx$

   c) $\int \dfrac{1}{6x^2+1}\,dx$

4. Find the following integrals.

   a) $\int \dfrac{1}{4+9x^2}\,dx$

   b) $\int \dfrac{12}{9+16x^2}\,dx$

   c) $\int \dfrac{1}{49x^2+81}\,dx$

5. Calculate the exact value of the following definite integrals.

a) $\displaystyle\int_{-3}^{3} \frac{1}{9+x^2}\,dx$

b) $\displaystyle\int_{\sqrt{3}}^{3} \frac{\sqrt{3}}{3+x^2}\,dx$

c) $\displaystyle\int_{0}^{5} \frac{20}{25+x^2}\,dx$

6. Calculate the exact value of the following definite integrals.

a) $\displaystyle\int_{0}^{\frac{\sqrt{3}}{2}} \frac{1}{1+4x^2}\,dx$

b) $\displaystyle\int_{-0.2}^{0.2} \frac{1}{1+25x^2}\,dx$

c) $\displaystyle\int_{\frac{1}{4\sqrt{3}}}^{\frac{\sqrt{3}}{4}} \left(\frac{8}{1+16x^2}\right)dx$

8.3 Integration of $\dfrac{kf'(x)}{f(x)}$

In section 4.2 you saw that if $y = \ln[f(x)]$ then $\dfrac{dy}{dx} = \dfrac{f'(x)}{f(x)}$. This means that

$$\int \left(\frac{kf'(x)}{f(x)}\right)dx = k\ln(f(x)) + c$$

Example 17

Find $\displaystyle\int \left(\frac{2x}{x^2+2}\right)dx$.

$\displaystyle\int \left(\frac{2x}{x^2+2}\right)dx = \ln(x^2+2) + c$ ⟵ Since $\dfrac{d}{dx}(x^2+2) = 2x$

Example 18

Find $\displaystyle\int \frac{2x^2-x+6}{(x+1)(x^2+2)}\,dx$.

$\displaystyle\int \frac{2x^2-x+6}{(x+1)(x^2+2)}\,dx = 3\ln(x+1) - \int\left(\frac{x}{x^2+2}\right)dx$ ⟵ From Example 3

$\displaystyle = 3\ln(x+1) - \frac{1}{2}\ln(x^2+2) + c$ ⟵ Using the result from Example 17

Example 19

Find $\displaystyle\int \left(\frac{e^{2x}}{e^{2x}+3}\right)dx$.

$\displaystyle\int \left(\frac{e^{2x}}{e^{2x}+3}\right)dx = \frac{1}{2}\ln(e^{2x}+3) + c$ ⟵ Since $\dfrac{d}{dx}(e^{2x}+3) = 2e^{2x}$

Example 20

Find $\displaystyle\int\left(\frac{1+2xe^{x^2}}{x+e^{x^2}}\right)dx$.

$\displaystyle\int\left(\frac{1+2xe^{x^2}}{x+e^{x^2}}\right)dx = \ln(x+e^{x^2}) + c$ ← Since $\dfrac{d}{dx}(x+e^{x^2}) = 1+2xe^{x^2}$

Example 21

Find $\displaystyle\int \tan 2x\,dx$.

$\displaystyle\int \tan 2x\,dx = \int\left(\frac{\sin 2x}{\cos 2x}\right)dx$ ← Writing tan in this form allows you to view it as a logarithmic integral, since $\dfrac{d}{dx}(\cos 2x) = -2\sin 2x$.

$\displaystyle = -\frac{1}{2}\ln(\cos 2x) + c$

$\displaystyle = \frac{1}{2}\ln(\sec 2x) + c$ ← This is an optional simplification.

Example 22

Find $\displaystyle\int_0^1\left(\frac{e^x - e^{-x}}{e^x + e^{-x}}\right)dx$.

$\displaystyle\int_0^1\left(\frac{e^x - e^{-x}}{e^x + e^{-x}}\right)dx = \left[\ln(e^x + e^{-x})\right]_0^1$ ← Since $\dfrac{d}{dx}(e^x + e^{-x}) = e^x - e^{-x}$

$= (\ln(e + e^{-1})) - (\ln(1+1))$ ← Substitute the upper and lower limits.

$= \ln(e + e^{-1}) - \ln 2$

$= \ln\left(\dfrac{e^2 + 1}{2e}\right)$ ← Simplify the logarithmic function.

Exercise 8.3

1. Find the following integrals.

a) $\displaystyle\int \frac{6x^2}{x^3 + 1}\,dx$

b) $\displaystyle\int \frac{\cos 2x}{1 + \sin 2x}\,dx$

c) $\displaystyle\int \frac{x-2}{x^2 - 4x + 3}\,dx$

d) $\displaystyle\int \frac{e^{3x}}{4 + 2e^{3x}}\, dx$

e) $\displaystyle\int \frac{x}{x^2 + 4}\, dx$

f) $\displaystyle\int \cot x\, dx$

g) $\displaystyle\int_0^{\frac{1}{6}\pi} \left(\frac{\cos 2x}{1 + \sin 2x} \right) dx$

h) $\displaystyle\int_0^1 \left(\frac{e^{2x}}{1 + e^{2x}} \right) dx$

i) $\displaystyle\int_0^1 \left(\frac{2x + 5}{(x + 3)(x + 2)} \right) dx$

2. Find $\displaystyle\int_0^2 \frac{x - 1}{x^2 - 2x + 2}\, dx.$

3. a) Show that $\dfrac{d}{dx}(\sin^2 x) = \sin 2x.$

  b) Hence find $\displaystyle\int \frac{1 + \sin 2x}{x + \sin^2 x}\, dx.$

4. a) Show that $\dfrac{d}{dx}\left(\tan x + \dfrac{1}{3}\tan^3 x \right) = \sec^4 x.$

  b) Hence find $\displaystyle\int_{\frac{\pi}{6}}^{\frac{\pi}{3}} \sec^4 x\, dx.$

5. a) Find $\dfrac{d}{dx}(\tan x - x).$

  b) Hence find $\displaystyle\int_{\frac{\pi}{6}}^{\frac{\pi}{4}} \frac{\tan^2 x}{\tan x - x}\, dx.$

6. Find $\displaystyle\int \frac{e^{2x} - e^{-2x}}{e^{2x} + e^{-2x}}\, dx.$

7. a) Show that $\dfrac{1 - \tan x}{1 + \tan x} = \dfrac{\cos x - \sin x}{\cos x + \sin x}.$

  b) Hence show that $\displaystyle\int_0^{\frac{\pi}{4}} \frac{1 - \tan x}{1 + \tan x}\, dx = \frac{1}{2}\ln 2.$

8. a) Show that $\dfrac{1}{1 + e^x} = \dfrac{e^{-x}}{e^{-x} + 1}.$

  b) Hence find $\displaystyle\int_1^2 \frac{1}{1 + e^x}\, dx,$ giving your answer correct to 3 decimal places.

Integration of $\dfrac{kf'(x)}{f(x)}$

8.4 Integration by parts

In section 4.3 you met the product rule for differentiation,

$$\frac{d}{dx}(uv) = u\frac{dv}{dx} + v\frac{du}{dx}$$

Since integration is the reverse process to differentiation, it follows that

(integrating both sides) $uv = \int\left(u\frac{dv}{dx}\right)dx + \int\left(v\frac{du}{dx}\right)dx.$

This can be rewritten to give the following rule.

> **Integration by parts**
> $$\int\left(u\frac{dv}{dx}\right)dx = uv - \int\left(v\frac{du}{dx}\right)dx$$

However, when you are faced with an integral that is not presented in the form of the left-hand side, but is expressed as a product – something like $\int(x\,e^{2x})dx$ – then you have to decide which to treat as u and which as $\frac{dv}{dx}$. The key to making this choice lies in the form of the second term on the right: $-\int\left(v\frac{du}{dx}\right)dx$. Choose u which makes $\int\left(v\frac{du}{dx}\right)dx$ integrable directly, or at least simpler. So in the example choose u to be x and then v is whatever satisfies $\frac{dv}{dx} = e^{2x}$, so $v = \frac{1}{2}e^{2x}$.

Using integration by parts then gives:

$$\int(x\,e^{2x})dx = x\left(\frac{1}{2}e^{2x}\right) - \int 1\times\frac{1}{2}e^{2x}\,dx$$

$$= \frac{1}{2}x\,e^{2x} - \frac{1}{4}e^{2x} + c$$

> Differentiating x gives 1, so the second term now is recognisable as a standard integral form.

Generally, if the integral is in the form $\int x^n f(x)\,dx$ and $f(x)$ is an exponential or trigonometric function (sin or cos), then take u as x^n because it will become simpler when it is differentiated (you may need to do integration by parts more than once) and the function $f(x)$ does not get any more complicated when it is integrated repeatedly.

Example 23

Find $\int x\cos 3x\, dx$.

$$\int x\cos 3x\, dx = x\frac{1}{3}\sin 3x - \int \left(\frac{1}{3}\sin 3x\right)dx$$

$u = x; \quad \dfrac{dv}{dx} = \cos 3x \Rightarrow v = \dfrac{1}{3}\sin 3x$

$$= \frac{1}{3}x\sin 3x + \frac{1}{9}\cos 3x + c$$

Don't forget the '+ c' at the end here and be careful with negatives with the sin and cos integrals.

Example 24

Find $\int x^2 e^{2x}\, dx$.

$$\int x^2 e^{2x}\, dx = x^2 \frac{1}{2}e^{2x} - \int \left(2x\frac{1}{2}e^{2x}\right)dx$$

$u = x^2; \quad \dfrac{dv}{dx} = e^{2x} \Rightarrow v = \dfrac{1}{2}e^{2x}$

$$= \frac{1}{2}x^2 e^{2x} - \int (xe^{2x})\, dx$$

This needs integration by parts again.

$$= \frac{1}{2}x^2 e^{2x} - x\frac{1}{2}e^{2x} + \int \left(\frac{1}{2}e^{2x}\right)dx$$

Be careful about the negative signs when you use parts more than once.

$$= \frac{1}{2}x^2 e^{2x} - \frac{1}{2}xe^{2x} + \frac{1}{4}e^{2x} + c$$

Example 25

Find $\int x^2 \cos 2x\, dx$.

$$\int x^2 \cos 2x\, dx = x^2 \frac{1}{2}\sin 2x - \int \left(2x\frac{1}{2}\sin 2x\right)dx$$

$u = x^2; \quad \dfrac{dv}{dx} = \cos 2x \Rightarrow v = \dfrac{1}{2}\sin 2x$

$$= \frac{1}{2}x^2 \sin 2x - \int (x\sin 2x)\, dx$$

This needs integration by parts again.

$$= \frac{1}{2}x^2 \sin 2x + x\frac{1}{2}\cos 2x - \int \left(\frac{1}{2}\cos 2x\right)dx$$

When integrating by parts more than once involving sin and cos, you need to be very careful with the negatives.

$$= \frac{1}{2}x^2 \sin 2x + \frac{1}{2}x\cos 2x - \frac{1}{4}\sin 2x + c$$

All the above examples used x or x^2 as u, the function to be differentiated to provide a simpler integral on the right-hand side of the integration by parts rule. There is one very important exception you need to look at now (and remember). This is where $\ln x$ appears in the integral; because $\frac{d}{dx}(\ln x) = \frac{1}{x}$ but you do not know any function whose derivative is $\ln x$, you cannot use $\frac{dv}{dx} = \ln x$ but should instead take $\ln x$ to be u.

Example 26

Find $\int (x \ln x)\, dx$.

$$\int (x\ln x)\,dx = \left(\frac{1}{2}x^2\right)\ln x - \int\left(\frac{1}{2}x^2\right)\left(\frac{1}{x}\right)dx$$

$u = \ln x; \quad \frac{dv}{dx} = x \Rightarrow v = \frac{1}{2}x^2$

$$= \left(\frac{1}{2}x^2\right)\ln x - \int\left(\frac{1}{2}x\right)dx$$

Now the second term on the right is recognisable as a standard integral.

$$= \left(\frac{1}{2}x^2\right)\ln x - \frac{1}{4}x^2 + c$$

Using the same process we can now find $\int (\ln x)\,dx$ by thinking of $\ln x$ as the product $1 \times \ln x$.

It is not immediately obvious that writing in the form of a product will be helpful, but often in mathematics these little devices prove the key step in a useful result.

Example 27

Find $\int (\ln x)\, dx$.

$$\int (\ln x)\,dx = (x)\ln x - \int (x)\left(\frac{1}{x}\right)dx$$

$u = \ln x; \quad \frac{dv}{dx} = 1 \Rightarrow v = x$

$$= x\ln x - \int (1)\,dx$$

Again the second term on the right is recognisable as a standard integral.

$$= x\ln x - x + c$$

Definite integration which involves integration by parts requires attention to detail in notation about the use of the limits. You can do the substitution of limits into the integrated expression at the first opportunity, but you would then be switching back and forth between evaluation at limits and more integration.

The example below leaves all the evaluation at limits until after all the integration is complete and then it is done all in one go.

Example 28

Find $\displaystyle\int_1^2 (x^3 \ln x)\,dx$.

$$\int_1^2 (x^3 \ln x)\,dx = \left[\left(\frac{1}{4}x^4\right)\ln x\right]_1^2 - \int_1^2 \left(\frac{1}{4}x^4\right)\left(\frac{1}{x}\right)dx$$ $u = \ln x; \quad \dfrac{dv}{dx} = x^3 \Rightarrow v = \dfrac{1}{4}x^4$

$$= \left[\left(\frac{1}{4}x^4\right)\ln x\right]_1^2 - \int_1^2 \left(\frac{1}{4}x^3\right)dx$$ Now the second term on the right is recognisable as a standard integral.

$$= \left[\left(\frac{1}{4}x^4\right)\ln x\right]_1^2 - \left[\frac{1}{16}x^4\right]_1^2$$ Finish the integration processes.

$$= (4\ln 2 - 0) - \left(1 - \frac{1}{16}\right)$$ Now substitute the limits into both terms at the same time.

$$= 4\ln 2 - \frac{15}{16}$$

When doing integration, it is important to remember results you have learned in other areas of mathematics. These will often help you to integrate functions you might not think are integrable at first glance.

Example 29

Find $\displaystyle\int (\ln x^3)\,dx$.

$$\int (\ln x^3)\,dx = \int (3\ln x)\,dx$$ Simplify the logarithm.

$$= 3(x\ln x - x) + c$$ Using the result in Example 27

Example 30

Find $\int (2x^3 e^{x^2})\,dx$.

$$\int (2x^3 e^{x^2})\,dx = \int (x^2\, 2xe^{x^2})\,dx$$
$$= (x^2 e^{x^2}) - \int (2xe^{x^2})\,dx$$
$$= x^2 e^{x^2} - e^{x^2} + c$$

$u = x^2;\quad \dfrac{dv}{dx} = 2xe^{x^2} \Rightarrow v = e^{x^2}$

$\dfrac{du}{dx} = 2x$

Now the second term is an exact integral.

The next example illustrates a surprising extension to the range of functions which can be integrated by parts. In all the examples above, we ended up with an integrable function on the right-hand side after using integration by parts once or twice. However, if we try to integrate $e^x \sin x$ or similar then we will never find an integrable function, no matter how many times we integrate by parts. Fortunately, there is a way around this. Integrating by parts twice (whilst keeping the exponential and trigonometric functions in the same role each time), we will end up with $-k$ times the expression on the left as the integral appearing on the right-hand side, so by collecting like terms we can then evaluate the integral. Example 31 shows how this is done.

Remember: With an indefinite integral it is important to remember to introduce the constant of integration at the end.

Example 31

Find $\int (e^{2x}\sin x)\,dx$

$$\int (e^{2x}\sin x)\,dx = \frac{1}{2}e^{2x}\sin x - \int \left(\frac{1}{2}e^{2x}\cos x\right)dx$$

$$= \frac{1}{2}e^{2x}\sin x - \frac{1}{4}e^{2x}\cos x + \int \left(\frac{1}{4}e^{2x}(-\sin x)\right)dx$$

$$\Rightarrow \frac{5}{4}\int \left(e^{2x}\sin x\right)dx = \frac{1}{2}e^{2x}\sin x - \frac{1}{4}e^{2x}\cos x$$

$$\Rightarrow \int \left(e^{2x}\sin x\right)dx = \frac{4}{5}\left(\frac{1}{2}e^{2x}\sin x - \frac{1}{4}e^{2x}\cos x\right) + c$$

$$= \frac{1}{5}\left(2\,e^{2x}\sin x - e^{2x}\cos x\right) + c$$

$u = \sin x;\quad \dfrac{dv}{dx} = e^{2x} \Rightarrow v = \frac{1}{2}e^{2x}$
For the second integral use
$u = \cos x;\quad \dfrac{dv}{dx} = \frac{1}{2}e^{2x}$

After integrating by parts twice, the last term is the same function as the left-hand side.

Collect like terms.

Remember the constant.

Exercise 8.4

1. Find the following integrals using integration by parts.

   a) $\displaystyle\int 4x\,e^{2x}\,dx$

   b) $\displaystyle\int 3x\sin 2x\,dx$

   c) $\displaystyle\int (2x+1)e^x\,dx$

   d) $\displaystyle\int (x+3)\cos 3x\,dx$

   e) $\displaystyle\int x\,e^{3x-1}\,dx$

   f) $\displaystyle\int (x\ln 5x)\,dx$

   g) $\displaystyle\int x\tan^{-1}x\,dx$

2. Evaluate the following definite integrals. Note that the first three are
   integrating the same function as parts **(a)** to **(c)** of question 1.

   a) $\displaystyle\int_0^1 4x\,e^{2x}\,dx$

   b) $\displaystyle\int_0^{\frac{\pi}{2}} 3x\sin 2x\,dx$

   c) $\displaystyle\int_0^2 (2x+1)e^x\,dx$

   d) $\displaystyle\int_0^{\frac{\pi}{2}} x\cos\!\left(\frac{1}{2}x\right)dx$

   e) $\displaystyle\int_1^e x^4\ln x\,dx$

   f) $\displaystyle\int_0^{\frac{\pi}{2}} x\sin x\,dx$

3. Find the following integrals using integration by parts twice.

   a) $\displaystyle\int 18x^2\,e^{3x}\,dx$

   b) $\displaystyle\int 3x^2\cos 2x\,dx$

   c) $\displaystyle\int_0^2 (2x^2+1)e^x\,dx$

   d) $\displaystyle\int_1^e x^2\,e^x\,dx$

   e) $\displaystyle\int_0^{\frac{\pi}{2}} x^2\sin x\,dx$

   f) $\displaystyle\int_{\frac{\pi}{6}}^{\frac{\pi}{3}} x^2\sin 2x\,dx$

   g) $\displaystyle\int_{-\infty}^0 x^2\,e^{3x}\,dx$

   h) $\displaystyle\int_0^{\frac{\pi}{4}} e^{3x}\cos 2x\,dx$

4. Find the area bounded by the graph of $y=x\sin x$, the x-axis,
   and the line $x=\frac{\pi}{2}$ (as shown in the diagram).

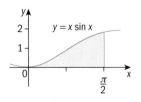

5. Find the area bounded by the graphs of $y = x\,e^{-2x}$, the x-axis, and the line $x = 2$ (as shown in the diagram).

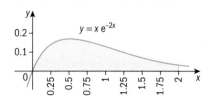

6. **a)** Find the area bounded by the graph of $y = x\cos 2x$, the x-axis, and the lines $x = 0$ and $x = \dfrac{\pi}{6}$.

   b) Find the volume of the solid of revolution obtained by rotating this region about the x-axis.

8.5 Integration using substitution

In P1 you met the chain rule for differentiation, $\dfrac{dy}{dx} = \dfrac{dy}{du} \times \dfrac{du}{dx}$.

The technique of integration by substitution is essentially the reverse of the chain rule for differentiation.

$\displaystyle\int \dfrac{3x}{\sqrt{x+1}}\,dx$ is not an integral that you can deal with

directly at the moment, but using the substitution $u^2 = x + 1$ transforms it into an integral that you can do.

In order to evaluate this we need to substitute for all the x terms and also for dx (and, for definite integrals, we must also change the limits to limits in the new variable).

> It is sensible, at least at first, to write down everything in the integral in terms of the new variable u (including the dx term), and then rewrite the integral completely in one go.

You could write this integral as

$3\displaystyle\int \dfrac{(x+1)-1}{\sqrt{x+1}}\,dx$ and create

two integrable terms.

Don't ever write an integral which is a mix of x and u.

$u^2 = x + 1$

$\Rightarrow \sqrt{x+1} = u; \quad 3x = 3(u^2 - 1);$

$2u\dfrac{du}{dx} = 1 \Rightarrow 2u\dfrac{du}{dx}\,dx = dx$

While this looks strange at the moment, when the substitutions are made you will see why the technique works.

Substituting gives

$\displaystyle\int \dfrac{3x}{\sqrt{x+1}}\,dx = \int \dfrac{3(u^2-1)}{u}2u\dfrac{du}{dx}\,dx$

This is where the chain rule is used to replace $\dfrac{du}{dx}\,dx$ by du.

$\qquad = \displaystyle\int 6(u^2 - 1)\,du$

$\qquad = 2u^3 - 6u + c$

Your original integral was in terms of x, so you need to express your answer in terms of x:

$$2u^3 - 6u + c = 2(\sqrt{x+1})^3 - 6\sqrt{x+1} + c$$

So $\displaystyle\int \frac{3x}{\sqrt{x+1}}\,dx = 2(\sqrt{x+1})^3 - 6\sqrt{x+1} + c$ ◄——— There is some simplification which could be done by using $(\sqrt{x+1})^3 = (x+1)\sqrt{x+1}$ but unless you are asked to show a particular form for the answer, this is perfectly acceptable.

In this course, you are not expected to identify appropriate substitutions for yourself, but you will probably be able to spot relationships between the integral form and the substitution you are told to use that would allow you to do without that help later on.

Example 32

Find $\displaystyle\int \sqrt{4-x^2}\,dx$ using the substitution $x = 2\sin\theta$.

· ·

$x = 2\sin\theta \Rightarrow \sqrt{4-x^2} = \sqrt{4-4\sin^2\theta} = 2\cos\theta$ ◄——— Only one expression here

$1 = 2\cos\theta \dfrac{d\theta}{dx} \Rightarrow dx = 2\cos\theta \dfrac{d\theta}{dx}\,dx$ ◄——— You need this to be able to complete the transformation of the integral to the new variable.

Substituting gives

$\displaystyle\int \sqrt{4-x^2}\,dx = \int 2\cos\theta \times 2\cos\theta \dfrac{d\theta}{dx}\,dx$

$= \displaystyle\int 4\cos^2\theta\,d\theta = \int 2(1+\cos 2\theta)\,d\theta$ ◄——— Use the standard identity.

$= 2\theta + \sin 2\theta + c = 2\theta + 2\sin\theta\cos\theta + c$ ◄——— Use the double angle relation.

$= 2\sin^{-1}\left(\dfrac{1}{2}x\right) + \dfrac{1}{2}x\sqrt{4-x^2} + c$ ◄——— Express in terms of the original variable.

When dealing with definite integrals, make sure you express the limits in terms of the new variable, u.

Example 33

Find $\int_1^2 x\sqrt{5x-2}\,dx$ using the substitution $u = 5x - 2$. Give your answer to 3 significant figures.

$u = 5x - 2 \Rightarrow x = \dfrac{u+2}{5}; \sqrt{5x-2} = u^{\frac{1}{2}}$ — The two terms in the integral

$x = 1 \Rightarrow u = 3; \quad x = 2 \Rightarrow u = 8$ — Change the limits.

$5 = \dfrac{du}{dx} \Rightarrow dx = \dfrac{1}{5}\dfrac{du}{dx}dx$ — So we can complete the transformation of the integral to the new variable

$\int_1^2 x\sqrt{5x-2}\,dx = \int_3^8 \left(\dfrac{u+2}{5}\right)u^{\frac{1}{2}}\dfrac{1}{5}\dfrac{du}{dx}\,dx$ — Substitute for the limits at the same time.

$= \dfrac{1}{25}\int_3^8 \left(u^{\frac{3}{2}} + 2u^{\frac{1}{2}}\right)du$ — Taking the common factor outside the integral simplifies things.

$= \dfrac{1}{25}\left[\dfrac{2}{5}u^{\frac{5}{2}} + \dfrac{4}{3}u^{\frac{3}{2}}\right]_3^8$ — These are standard integral forms now.

$= \dfrac{1}{25}\left(\dfrac{256}{5}\sqrt{2} + \dfrac{64}{3}\sqrt{2}\right) - \dfrac{1}{25}\left(\dfrac{18}{5}\sqrt{3} + 4\sqrt{3}\right)$ — Because the limits are changed at the start, the definite integral does not need to be expressed in terms of x again.

$= \dfrac{1}{25}\left(\dfrac{1088}{15}\sqrt{2} - \dfrac{38}{5}\sqrt{3}\right) = 3.5765... = 3.58\,(3\,\text{s.f.})$

Example 34

Find $\int_1^e \left(\dfrac{1}{x}\ln x\right)dx$ using the substitution $x = e^u$.

$x = e^u \Rightarrow \dfrac{1}{x} = e^{-u}; \quad \ln x = u$ — Express both parts in terms of u.

$x = 1 \Rightarrow u = 0; \quad x = e \Rightarrow u = 1$ — So we can complete the transformation of the integral to the new variable

$1 = e^u \dfrac{du}{dx} \Rightarrow dx = e^u \dfrac{du}{dx}dx$

Substituting gives — Substitute for all terms and the limits.

$\int_1^e \left(\dfrac{1}{x}\ln x\right)dx = \int_0^1 e^{-u}ue^u\dfrac{du}{dx}dx$ — Simplify the function and use the chain rule.

$= \int_0^1 u\,du$

$= \left[\dfrac{1}{2}u^2\right]_0^1$ — Integrate as a function of u.

$= \dfrac{1}{2} - 0 = \dfrac{1}{2}$ — Evaluate the function at the limits.

Example 35

Find $\int \dfrac{1}{9+x^2}\,dx$ using the substitution $x = 3\tan\theta$.

$x = 3\tan\theta \;\Rightarrow\; 9 + x^2 = 9 + 9\tan^2\theta = 9\sec^2\theta$ — Express in its simplest form.

$1 = 3\sec^2\theta\,\dfrac{d\theta}{dx} \;\Rightarrow\; dx = 3\sec^2\theta\,\dfrac{d\theta}{dx}\,dx$ — So we can complete the transformation of the integral to the new variable

$\displaystyle\int\dfrac{1}{9+x^2}\,dx = \int\left(\dfrac{1}{9\sec^2\theta}\right)3\sec^2\theta\,\dfrac{d\theta}{dx}\,dx$ — Substitute for all terms.

$\displaystyle = \int\dfrac{1}{3}\,d\theta$ — Simplify the function and use the chain rule.

$= \dfrac{1}{3}\theta + c$ — Integrate as a function of θ.

$= \dfrac{1}{3}\tan^{-1}\left(\dfrac{x}{3}\right) + c$ — Express this in terms of the original variable.

Example 36

Find $\int \dfrac{1}{\sqrt{5-4x^2}}\,dx$ using the substitution $x = \dfrac{\sqrt{5}}{2}\sin\theta$.

$x = \dfrac{\sqrt{5}}{2}\sin\theta \;\Rightarrow\; \sqrt{5-4x^2} = \sqrt{5-5\sin^2\theta} = \sqrt{5}\cos\theta$ — Only the one expression here

$1 = \dfrac{\sqrt{5}}{2}\cos\theta\,\dfrac{d\theta}{dx} \;\Rightarrow\; dx = \dfrac{\sqrt{5}}{2}\cos\theta\,\dfrac{d\theta}{dx}\,dx$ — So we can complete the transformation of the integral to the new variable

$\displaystyle\int\dfrac{1}{\sqrt{5-4x^2}}\,dx = \int\dfrac{1}{\sqrt{5}\cos\theta}\dfrac{\sqrt{5}}{2}\cos\theta\,\dfrac{d\theta}{dx}\,dx$ — Substitute for all terms.

Substituting gives

$\displaystyle = \int\dfrac{1}{2}\,d\theta$ — Simplify the function and use the chain rule.

$= \dfrac{1}{2}\theta + c$ — Integrate as a function of θ.

$= \dfrac{1}{2}\sin^{-1}\left(\dfrac{2}{\sqrt{5}}x\right) + c$ — Express in terms of the original variable.

Examples 32–36 are all functions that you could not integrate without using a substitution or manipulating into a standard form. Examples 37–39 show how substitutions can also help you see exactly what you are doing with some integrals which you should recognise as exact derivatives of other functions.

Apart from functions which are integrable directly (like Examples 37–39) you will always be given the substitution to use in this course and in the examination.

Example 37

Find $\displaystyle\int \frac{1}{(5x-2)^2}\, dx$ using the substitution $u = 5x - 2$.

$u = 5x - 2 \implies \dfrac{1}{(5x-2)^2} = \dfrac{1}{u^2}$

As normal – express the function and dx in terms of u.

$5 = \dfrac{du}{dx} \implies dx = \dfrac{1}{5}\dfrac{du}{dx}\, dx$

$\displaystyle\int \frac{1}{(5x-2)^2}\, dx = \int \left(\frac{1}{u^2}\right)\frac{1}{5}\frac{du}{dx}\, dx$

$= \dfrac{1}{5}\displaystyle\int \left(\frac{1}{u^2}\right) du$

This is a standard integral form now.

$= -\dfrac{1}{5u} + c$

$= -\dfrac{1}{5(5x-2)} + c$

Express in terms of the original variable.

Writing $\dfrac{1}{(5x-2)^2}$ as $(5x-2)^{-2}$ allows this integral to be done without a formal substitution.

Example 38

Find $\displaystyle\int_0^{\frac{\pi}{4}} \cos 2x \sin^3 2x\, dx$ using the substitution $u = \sin 2x$.

$u = \sin 2x \implies \sin^3 2x = u^3$

Only the one expression here

$x = 0 \implies u = 0; \quad x = \dfrac{\pi}{4} \implies u = 1$

Change the limits.

$\dfrac{du}{dx} = 2\cos 2x \implies \dfrac{1}{2}\dfrac{du}{dx}\, dx = \cos 2x\, dx$

This is the second part of the function to be integrated.

$\displaystyle\int_0^{\frac{\pi}{4}} \cos 2x \sin^3 2x\, dx = \int_0^1 \frac{1}{2}u^3\, du$

Substitution makes the roles of the multiplying constants easier to see.

$= \left[\dfrac{1}{8}u^4\right]_0^1$

Now the integration and evaluation are almost trivial.

$= \dfrac{1}{8} - 0 = \dfrac{1}{8}$

For integrals like that in Example 38, you would be expected to notice that $\cos 2x$ is the derivative of $\sin 2x$ and integrate this without being given a formal substitution. If you find difficulty in getting the constant part of this type of integral consistently right, then you might try using the substitution method yourself.

Example 39

Find $\displaystyle\int_0^{\frac{1}{12}\pi}\left(\frac{\cos 2x}{1+\sin 2x}\right)dx$ using the substitution $u = 1 + \sin 2x$.

$u = 1 + \sin 2x$ ← Nothing more to do here

$x = 0 \Rightarrow u = 1; \quad x = \dfrac{\pi}{12} \Rightarrow u = 1.5$ ← Calculate the limits for the new variable.

$\dfrac{du}{dx} = 2\cos 2x \Rightarrow \cos 2x\, dx = \dfrac{1}{2}\dfrac{du}{dx}\, dx$ ← So we can complete the transformation of the integral to the new variable

$\displaystyle\int_0^{\frac{1}{12}\pi}\left(\frac{\cos 2x}{1+\sin 2x}\right)dx = \int_1^{1.5}\left(\frac{1}{u}\right)\frac{1}{2}\frac{du}{dx}\,dx$ ← Substituting throughout

$\displaystyle = \int_1^{1.5}\left(\frac{1}{2u}\right)du$ ← This is the standard logarithmic integral form.

$\displaystyle = \left[\frac{1}{2}\ln u\right]_1^{1.5}$ ← Integrate and evaluate at the limits.

$\displaystyle = \frac{1}{2}\ln 1.5 - 0 = \frac{1}{2}\ln 1.5$

This integral was in the form $\dfrac{k f'(x)}{f(x)}$ and could have been integrated without the formal substitution.

Exercise 8.5

1. Find these indefinite integrals using the substitutions given.

a) $\displaystyle\int \frac{6x}{\sqrt{2x+1}}\,dx; \quad u^2 = 2x + 1$

b) $\displaystyle\int x^2\sqrt{1+x^3}\,dx; \quad u = 1 + x^3$

c) $\displaystyle\int \frac{1}{\sqrt{9-x^2}}\,dx; \quad x = 3\sin\theta$

d) $\displaystyle\int \frac{x}{\sqrt{x+2}}\,dx; \quad u = x + 2$

e) $\displaystyle\int \frac{1}{25+4x^2}\,dx; \quad x = \frac{5}{2}\tan\theta$

f) $\displaystyle\int \frac{2e^{2x}+1}{e^{2x}+x}\,dx; \quad u = e^{2x} + x$

2. Evaluate the following definite integrals using the substitutions given.
   Note that the first three are integrating the same function as
   parts (a) to (c) of question 1.

a) $\displaystyle\int_0^1 \frac{6x}{\sqrt{2x+1}}\,dx;\quad u^2 = 2x+1$

b) $\displaystyle\int_0^2 x^2\sqrt{1+x^3}\,dx;\ u = 1+x^3$

c) $\displaystyle\int_0^{\frac{3}{2}} \frac{1}{\sqrt{9-x^2}}\,dx;\quad x = 3\sin\theta$

d) $\displaystyle\int_0^{\frac{\pi}{6}} \frac{\cos x}{\sqrt{1+2\sin x}}\,dx;\quad u = 1+2\sin x$

e) $\displaystyle\int_0^1 (1+x)\sqrt{2x+x^2}\,dx;\quad u = 2x+x^2$

f) $\displaystyle\int_0^1 \frac{2e^{2x}+1}{e^{2x}+x}\,dx;\quad u = e^{2x}+x$

g) $\displaystyle\int_0^{\frac{\pi}{3}} \sec^2 x\,\tan^5 x\,dx;\quad u = \tan x$

h) $\displaystyle\int_0^{\sqrt{\frac{\pi}{3}}} 4x\sin(x^2)\,dx;\ u = \cos x^2$

3. Find $\displaystyle\int_2^3 \frac{1}{x-\sqrt{x}}\,dx$ using the substitution $x = u^2$.

4. Using the substitution $u = \ln x$, find the area bounded by the graph of
   $y = \dfrac{(\ln x)^4}{x}$, the x-axis, and the lines $x = 1$ and $x = e$.

5. Using the substitution $u = 2x + 1$, find the
   area bounded by the graph of $y = \dfrac{x}{\sqrt{2x+1}}$,
   the x-axis, and the line $x = 2$ (as shown in the diagram).

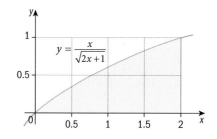

6. Find the area bounded by the graph of
   $y = 2x\cos\left(x^2 + \dfrac{\pi}{6}\right)$, the x-axis, and the lines
   $x = 0$ and $x = \sqrt{\dfrac{\pi}{6}}$ (as shown in the diagram).

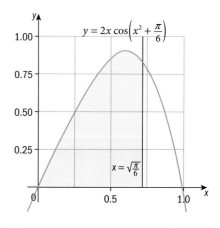

1. By first expressing each rational expression using partial fractions, find

   a) $\int \left(\dfrac{5}{(x-2)(x+3)} \right) dx$

   b) $\int \left(\dfrac{5x+2}{(2x-1)(x+4)} \right) dx$

   c) $\int \left(\dfrac{x-2}{x(x+1)^2} \right) dx$

   d) $\int \left(\dfrac{x^2-3x-1}{(2x-1)(x+1)^2} \right) dx$

   e) $\int \left(\dfrac{3x^2-5x}{x^2-1} \right) dx$.

2. Show that $\dfrac{4}{x(x^2+4)} = \dfrac{1}{x} - \dfrac{x}{x^2+4}$. Using the substitution $u = x^2 + 4$, find $\int \dfrac{4}{x(x^2+4)}\, dx$.

3. Find the following integrals.

   a) $\int \dfrac{4x^3}{x^4+6}\, dx$

   b) $\int \dfrac{\cos 3x}{3-2\sin 3x}\, dx$

   c) $\int \dfrac{3x-2}{3x^2-4x+7}\, dx$

   d) $\int \dfrac{e^{5x}}{6-5e^{5x}}\, dx$

   e) $\int \dfrac{2x}{x^2+9}\, dx$

   f) $\int \dfrac{x^2}{x^3+10}\, dx$

   g) $\int_0^{\frac{1}{6}\pi} \left(\dfrac{\cos 3x}{2+\sin 3x} \right) dx$

   h) $\int_0^1 \left(\dfrac{e^{3x}}{2+e^{3x}} \right) dx$

   i) $\int_0^1 \left(\dfrac{2x+3}{(x+1)(x+2)} \right) dx$

4. Find these integrals using integration by parts.

   a) $\int x e^{3x}\, dx$

   b) $\int 4x \cos 2x\, dx$

   c) $\int (2x-3)e^x\, dx$

   d) $\int (x^2 - x + 3)\cos 3x\, dx$

   e) $\int_0^1 x^2 e^{x-1}\, dx$

   f) $\int_1^e (x^2 \ln 4x)\, dx$

5. Find these integrals using the substitutions given.

   a) $\int \dfrac{6x}{\sqrt{3x-1}}\, dx; \quad u^2 = 3x - 1$

   b) $\int x^3 \sqrt{4+x^4}\, dx; \quad u = 4 + x^4$

   c) $\int \dfrac{1}{\sqrt{25-16x^2}}\, dx; \quad x = \dfrac{5}{4}\sin\theta$

   d) $\int \dfrac{x}{\sqrt{2x+7}}\, dx; \quad u = 2x + 7$

   e) $\int_0^1 (1+6x)\sqrt{x+3x^2}\, dx; \quad u = x + 3x^2$

   f) $\int_0^1 \dfrac{e^{2x}+x}{e^{2x}+x^2}\, dx; \quad u = e^{2x} + x^2$

   g) $\int_0^{\frac{\pi}{4}} \sec^2 x \tan^6 x\, dx; \quad u = \tan x$

   h) $\int_0^{\sqrt{\frac{\pi}{6}}} 4x \cos\left(x^2 + \dfrac{\pi}{6}\right) dx; u = \sin\left(x^2 + \dfrac{\pi}{6}\right)$

6. The diagram shows the area trapped between the graphs of $y = \dfrac{x}{(x-4)^2}$ and $y = x$. Show that the graphs intersect at $(0, 0)$ and $(3, 3)$ and that the area is $\dfrac{3}{2} + \ln 4$.

(Use the substitution $u = x - 4$ when dealing with the curve.)

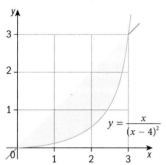

$$y = \frac{x}{(x-4)^2}$$

7. Calculate the exact value of $\displaystyle\int_0^{\frac{\sqrt{3}}{2}} \frac{1}{9+4x^2}\,dx$.

8. Calculate the exact value of

$$\int_{-3}^{-1} \frac{3}{x^2+6x+13}\,dx.$$

9. Use the substitution $u = 1 + 8\tan x$ to find the exact value of

$$\int_0^{\frac{1}{4}\pi} \frac{\sqrt{(1+8\tan x)}}{\cos^2 x}\,dx$$

Chapter summary

Integration of rational functions

To integrate difficult algebraic fractions, first look to see if you can split the fraction into partial fractions, where each partial fraction is integrable.

Integration of $\dfrac{1}{x^2 + a^2}$

- $\displaystyle\int \left(\frac{1}{1+x^2}\right) dx = \tan^{-1} x + c$

Integration of $\dfrac{k f'(x)}{f(x)}$

- $\displaystyle\int \left(\frac{k f'(x)}{f(x)}\right) dx = k\ln(f(x)) + c$

Integration by parts

- $\displaystyle\int \left(u\frac{dv}{dx}\right) dx = uv - \int \left(v\frac{du}{dx}\right) dx$

Integration using substitution

- Before making a variable substitution, first write down everything in the integral in terms of the new variable, u, (including the dx term) and then rewrite the integral completely in one go.

- When dealing with definite integrals, make sure you express the limits in terms of the new variable, u.

- For indefinite integrals if your original integral was in terms of x, you need to express your answer also in terms of x.

The main navigation mechanism used by most insects is called path integration. This involves a zig-zag path which can be modelled as a series of vectors. For example, biologists believe that ants can measure direction and distance, which helps them represent each step in a foraging path by a vector. Neural processing enables them to add vectors as they go along, so they always know how to get back home quickly when they have found food, or if they are attacked by a predator.

Objectives

- Use standard notations for vectors, i.e. $\begin{pmatrix} x \\ y \end{pmatrix}$, $x\mathbf{i} + y\mathbf{j}$, $\begin{pmatrix} x \\ y \\ z \end{pmatrix}$, $x\mathbf{i} + y\mathbf{j} + z\mathbf{k}$, $\overrightarrow{AB}$, $\mathbf{a}$.

- Carry out addition and subtraction of vectors and multiplication of a vector by a scalar, and interpret these operations in geometrical terms.

- Calculate the magnitude of a vector, and use unit vectors, displacement vectors, and position vectors.

- Understand the significance of all the symbols used when the equation of a straight line is expressed in the form $\mathbf{r} = \mathbf{a} + t\mathbf{b}$, and find the equation of a line, given sufficient information.

- Determine whether two lines are parallel, intersect or are skew, and find the point of intersection of two lines when it exists.

- Use formulae to calculate the scalar product of two vectors, and use scalar products in problems involving lines and points.

Before you start

You should know how to:

1. Find the vector $\overrightarrow{AB}$ which describes the translation from a point A to a point B,

   e.g. if A is the point $(2, 1)$ and B is the point $(5, 0)$, then $\overrightarrow{AB} = \begin{pmatrix} 3 \\ -1 \end{pmatrix}$ and $\overrightarrow{BA} = \begin{pmatrix} -3 \\ 1 \end{pmatrix}$.

Skills check:

1. Find the vectors $\overrightarrow{AB}$ and $\overrightarrow{BA}$ in each case.

   a) $A(0, 2)$ $B(3, 6)$

   b) $A(-3, 1)$ $B(0, 0)$

   c) $A(-2, 6)$ $B(-3, 4)$

2. Add and subtract column vectors and multiply a vector by a scalar,

e.g. $\begin{pmatrix} 3 \\ 4 \end{pmatrix} + \begin{pmatrix} -1 \\ 5 \end{pmatrix} = \begin{pmatrix} 2 \\ 9 \end{pmatrix}$,

$\begin{pmatrix} 3 \\ 4 \end{pmatrix} - \begin{pmatrix} -1 \\ 5 \end{pmatrix} = \begin{pmatrix} 4 \\ -1 \end{pmatrix}$,

$2\begin{pmatrix} 1 \\ -3 \end{pmatrix} = \begin{pmatrix} 2 \\ -6 \end{pmatrix}$.

2. Calculate the following.

a) $\begin{pmatrix} -3 \\ 4 \end{pmatrix} + \begin{pmatrix} 2 \\ -2 \end{pmatrix}$

b) $\begin{pmatrix} -2 \\ 1 \end{pmatrix} - \begin{pmatrix} 1 \\ 4 \end{pmatrix}$

c) $3\begin{pmatrix} 2 \\ 1 \end{pmatrix} + 2\begin{pmatrix} 0 \\ 4 \end{pmatrix}$

9.1 Vector notation

A **vector** is a quantity which has both magnitude (size) and direction.

The vector $\begin{pmatrix} a \\ b \end{pmatrix}$ can be used to describe a displacement of a units in the x-direction and b units in the y-direction.

A **scalar** is a quantity which has magnitude but no associated direction.

Distance is an example of a scalar.

In two dimensions a displacement can be represented as $\begin{pmatrix} x \\ y \end{pmatrix}$.

Positive and negative are used to denote directions as in the standard two-dimensional plane.

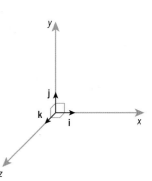

The **column vector** $\begin{pmatrix} x \\ y \end{pmatrix}$ can be represented in the form $x\mathbf{i} + y\mathbf{j}$,

where $\mathbf{i}$ and $\mathbf{j}$ are **unit vectors** (vectors of length 1 unit) in the x and y directions respectively.

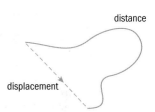

In three dimensions a displacement can be represented as $\begin{pmatrix} x \\ y \\ z \end{pmatrix}$.

The diagram shows a three-dimensional set of axes, x, y, and z.

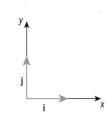

The column vector $\begin{pmatrix} x \\ y \\ z \end{pmatrix}$ can be represented as $x\mathbf{i} + y\mathbf{j} + z\mathbf{k}$,

where $\mathbf{i}$, $\mathbf{j}$, and $\mathbf{k}$ are unit vectors in the x, y, and z directions respectively.

For example, $\begin{pmatrix} 3 \\ 1 \\ -2 \end{pmatrix}$ can be represented as $3\mathbf{i} + \mathbf{j} - 2\mathbf{k}$.

This vector means

3 units in the direction of the x axis,

1 unit in the direction of the y axis and

-2 units in the direction of the z axis

(or 2 units in the direction of the $-z$ axis).

Example 1

a) Write down the displacement from $A(2, 6)$ to $B(5, 7)$ as a column vector.

b) Write down the displacement from $P(2, 1, 3)$ to $Q(7, 2, -1)$ in terms of the unit vectors $\mathbf{i}$, $\mathbf{j}$, $\mathbf{k}$.

a) $\overrightarrow{AB} = \begin{pmatrix} 5-2 \\ 7-6 \end{pmatrix} = \begin{pmatrix} 3 \\ 1 \end{pmatrix}$

Show the increase in both coordinates as two positive components.

b) $\overrightarrow{PQ} = (7-2)\mathbf{i} + (2-1)\mathbf{j} + (-1-3)\mathbf{k} = 5\mathbf{i} + \mathbf{j} - 4\mathbf{k}$

Show the decrease in the z-coordinate by using a negative component.

Example 2

In the diagram, $OABCDEFG$ is a cuboid with $OA = 10\,\text{cm}$, $AB = 6\,\text{cm}$, $OD = 5\,\text{cm}$.

The unit vectors $\mathbf{i}$, $\mathbf{j}$, $\mathbf{k}$ are in the directions OA, OC, OD respectively.

The point P is the mid-point of AB and Q lies on GF such that $GQ = 4QF$.

Express each of the vectors $\overrightarrow{OP}$ and $\overrightarrow{OQ}$ in terms of $\mathbf{i}$, $\mathbf{j}$, $\mathbf{k}$.

$\overrightarrow{OP} = \overrightarrow{OA} + \overrightarrow{AP} = 10\mathbf{i} + 3\mathbf{j} + 0\mathbf{k} = 10\mathbf{i} + 3\mathbf{j}$

10 units in x-direction, 3 units in y-direction and 0 units in z-direction

$\overrightarrow{OQ} = 8\mathbf{i} + 6\mathbf{j} + 5\mathbf{k}$

$GQ = \dfrac{4}{5}\,GF = \dfrac{4}{5} \times 10 = 8$

Note: The x, y, and z axes are not always shown in exactly the same way with the same orientations, but they are always perpendicular to one another.

A **displacement vector** represents a change in position.
For example, in the diagram the displacement from A to B is

$$\vec{AB} = \begin{pmatrix} 5 \\ -2 \\ 1 \end{pmatrix} \text{ or } 5\mathbf{i} - 2\mathbf{j} + \mathbf{k}$$

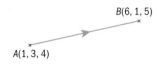

A **position vector** describes the displacement from the origin to a point.
For example, in the diagram the position vector of P is

$$\vec{OP} = \begin{pmatrix} 2 \\ 3 \\ 1 \end{pmatrix} \text{ or } 2\mathbf{i} + 3\mathbf{j} + \mathbf{k}$$

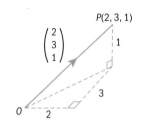

A point with coordinates (x, y, z) has position vector $\begin{pmatrix} x \\ y \\ z \end{pmatrix}$ or $x\mathbf{i} + y\mathbf{j} + z\mathbf{k}$.

If two points, A and B, have coordinates (x_1, y_1, z_1) and (x_2, y_2, z_2) respectively, then

$$\vec{AB} = \begin{pmatrix} x_2 - x_1 \\ y_2 - y_1 \\ z_2 - z_1 \end{pmatrix} \text{ or } (x_2 - x_1)\mathbf{i} + (y_2 - y_1)\mathbf{j} + (z_2 - z_1)\mathbf{k}$$

Example 3

If A has coordinates $A(3, -2, 4)$ and B has coordinates $B(5, -3, 2)$, find as column vectors

a) $\vec{AB}$ b) $\vec{BA}$.

a) $\vec{AB} = \begin{pmatrix} 5-3 \\ -3--2 \\ 2-4 \end{pmatrix} = \begin{pmatrix} 2 \\ -1 \\ -2 \end{pmatrix}$

More commonly we think of $\vec{AB}$ as being the coordinates of A subtracted from the coordinates of B.

b) $\vec{BA} = \begin{pmatrix} 3-5 \\ -2--3 \\ 4-2 \end{pmatrix} = \begin{pmatrix} -2 \\ 1 \\ 2 \end{pmatrix}$

Subtract the coordinates of B from the coordinates of A.

Note: $\vec{AB} = -\vec{BA}$. This is true in every case.

Example 4

The position vector of point A is $3\mathbf{i} - \mathbf{j} + 2\mathbf{k}$.

$\overrightarrow{AB} = 2\mathbf{i} + 5\mathbf{j} - 3\mathbf{k}$

Find the position vector of B.

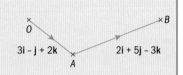

Position vector of $B = \overrightarrow{OB}$

$\overrightarrow{OB} = (3\mathbf{i} + 2\mathbf{i}) + (-\mathbf{j} + 5\mathbf{j}) + (2\mathbf{k} - 3\mathbf{k})$

$\quad = 5\mathbf{i} + 4\mathbf{j} - \mathbf{k}$

$\overrightarrow{OB} = \overrightarrow{OA} + \overrightarrow{AB}$

Exercise 9.1

1. Write down, using column vector notation, the displacement from

    a) $(1, 5)$ to $(3, 4)$
    b) $(-3, 4)$ to $(2, -1)$
    c) $(1, 0, 3)$ to $(5, 0, 8)$
    d) $(5, 2, 0)$ to $(0, 6, -3)$.

2. Write down, using unit vector notation, the displacement from

    a) $(3, 1)$ to $(6, 0)$
    b) $(8, 6)$ to $(5, 7)$
    c) $(1, 0, -2)$ to $(2, 3, -1)$
    d) $(3, 3, 5)$ to $(0, 0, 0)$.

3. The diagram shows a cube $OABCDEFG$ of edge 6 cm.

    Express each of the vectors $\overrightarrow{OE}$, $\overrightarrow{OF}$, $\overrightarrow{EG}$, and $\overrightarrow{CE}$ in terms of $\mathbf{i}, \mathbf{j}, \mathbf{k}$.

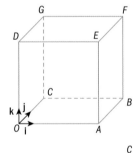

4. The diagram shows a rectangle $ABCD$.

    A is the point $(2, 5, 3)$, B is the point $(4, 4, 4)$, and D is the point $(5, 9, 1)$.

    a) Find the coordinates of C.
    b) Find the coordinates of the mid-point of CD.
    c) Find the column vector $\overrightarrow{DB}$.

5. Find, as column vectors, $\overrightarrow{AB}$ and $\overrightarrow{BA}$ in each of the following cases:

    a) $A(3, 1, 8)$ $B(4, 3, 2)$
    b) $A(-1, -1, 0)$ $B(2, 0, -3)$
    c) $A(2, 0.5, 1.5)$ $B(3, 2.5, 1)$
    d) $A\left(1, \dfrac{3}{2}, -2\right)$ $B\left(\dfrac{5}{2}, 1, 0\right)$.

6. Find, using unit vector notation, $\overrightarrow{AB}$ and $\overrightarrow{BA}$ in each of the following cases:

    a) $A(2, 4, -7)$ $B(3, 5, -2)$
    b) $A(4, -2, 8)$ $B(12, -6, 1)$
    c) $A(-3, -3, -3)$ $B(1, 1, 1)$
    d) $A(t + 2, 2t, -t)$ $B(t, t, t)$.

7. The position vector of a point A is $6\mathbf{i} + 2\mathbf{j} - \mathbf{k}$ and the position vector of a point B is $-2\mathbf{i} + \mathbf{j} + 2\mathbf{k}$. Find

   a) $\overrightarrow{BA}$ b) $\overrightarrow{AB}$.

8. The position vector of a point A is $\begin{pmatrix} x \\ y \\ z \end{pmatrix}$. The position vector of a point B is $\begin{pmatrix} 5 \\ 1 \\ -3 \end{pmatrix}$.

   Given that $\overrightarrow{AB} = \begin{pmatrix} 3 \\ 3 \\ -5 \end{pmatrix}$, find the values of x, y, and z.

9.2 The magnitude of a vector

The magnitude of a vector is the length of the line representing the vector. It is denoted by $|\overrightarrow{AB}|$ and is sometimes called the modulus of the vector.

To calculate the magnitude of a vector we use Pythagoras' theorem.

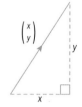

> In two dimensions, the magnitude of vector $\begin{pmatrix} x \\ y \end{pmatrix}$ or $x\mathbf{i} + y\mathbf{j}$ is given by $\sqrt{x^2 + y^2}$.

We write $|\overrightarrow{AB}| = \sqrt{x^2 + y^2}$.

> In three dimensions, the magnitude of vector $\begin{pmatrix} x \\ y \\ z \end{pmatrix}$ or $x\mathbf{i} + y\mathbf{j} + z\mathbf{k}$ is given by $\sqrt{x^2 + y^2 + z^2}$.

We write $|\overrightarrow{AB}| = \sqrt{x^2 + y^2 + z^2}$.

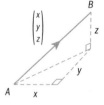

> Vectors are equal if they have the same direction and magnitude.
>
> A negative vector has the same magnitude as the positive vector, but the opposite direction.
>
> A unit vector is a vector of magnitude 1. To find a unit vector we divide the vector by its magnitude, so the unit vector in the direction of a vector $\mathbf{a}$ is given by $\dfrac{\mathbf{a}}{|\mathbf{a}|}$.

Remember: $\mathbf{i}, \mathbf{j}, \mathbf{k}$ are already unit vectors.

To multiply a vector by a scalar we multiply each component by the scalar. Any vector obtained by multiplying another vector by a scalar will be parallel to the original vector.

For example: $2\begin{pmatrix} 3 \\ 4 \\ 5 \end{pmatrix} = \begin{pmatrix} 6 \\ 8 \\ 10 \end{pmatrix}$ or $2(3\mathbf{i} + 4\mathbf{j} + 5\mathbf{k}) = 6\mathbf{i} + 8\mathbf{j} + 10\mathbf{k}$.

$\begin{pmatrix} 6 \\ 8 \\ 10 \end{pmatrix}$ has the same direction as $\begin{pmatrix} 3 \\ 4 \\ 5 \end{pmatrix}$ but twice the magnitude (length).

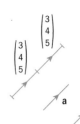

Parallel vectors are vectors which have the same or opposite direction.

Parallel vectors must be scalar multiples of each other.

Example 5

a) Find the magnitude of the vector $\begin{pmatrix} 3 \\ 4 \\ 6 \end{pmatrix}$.

b) Hence write down a unit vector in the direction of $\begin{pmatrix} 3 \\ 4 \\ 6 \end{pmatrix}$.

a) Magnitude of $\begin{pmatrix} 3 \\ 4 \\ 6 \end{pmatrix} = \sqrt{3^2 + 4^2 + 6^2} = \sqrt{61}$ ⟵ Find the length of the vector.

b) Unit vector $= \dfrac{1}{\sqrt{61}} \begin{pmatrix} 3 \\ 4 \\ 6 \end{pmatrix}$ ⟵ This is equivalent to dividing each component of the vector by $\sqrt{61}$.

Example 6

a) Find the value of k for which the vectors $\begin{pmatrix} 2 \\ -1 \\ 6 \end{pmatrix}$ and $\begin{pmatrix} 8 \\ -4 \\ k \end{pmatrix}$ are parallel.

b) Find the value of t for which the vectors $\begin{pmatrix} 5+t \\ 4 \\ 3-3t \end{pmatrix}$ and $\begin{pmatrix} 3 \\ 4 \\ 9 \end{pmatrix}$ are equal.

a) For the vectors to be parallel, they must be scalar multiples of each other.

$\begin{pmatrix} 8 \\ -4 \\ k \end{pmatrix} = 4 \begin{pmatrix} 2 \\ -1 \\ 6 \end{pmatrix}$ ⟵ Find the scalar multiple.

$k = 4 \times 6 = 24$ ⟵ Use the third component to find k.

b) For the vectors to be equal, all three components must be equal.

$5 + t = 3$ and $3 - 3t = 9$

$t = -2$ ⟵ Equate each component.

Exercise 9.2

1. Calculate the magnitude of
    a) $5\mathbf{i} + 3\mathbf{j}$
    b) $2\mathbf{i} - 4\mathbf{j} + 5\mathbf{k}$
    c) $\begin{pmatrix} 1 \\ -7 \\ 0 \end{pmatrix}$
    d) $\begin{pmatrix} -3 \\ -9 \end{pmatrix}$
    e) $\begin{pmatrix} 10 \\ -5 \\ 10 \end{pmatrix}$.

2. Find a unit vector in the direction of
    a) $2\mathbf{i} - 3\mathbf{j} + \mathbf{k}$
    b) $\begin{pmatrix} 1 \\ -1 \\ 1 \end{pmatrix}$
    c) $-\mathbf{i} + 2\mathbf{j} - 2\mathbf{k}$
    d) $\begin{pmatrix} -1 \\ 2 \\ -3 \end{pmatrix}$.

3. State which of the following vectors are parallel to $\begin{pmatrix} 1 \\ -4 \\ -7 \end{pmatrix}$.

    a) $\begin{pmatrix} 2 \\ -8 \\ -16 \end{pmatrix}$
    b) $\frac{1}{3}\mathbf{i} - \frac{4}{3}\mathbf{j} - \frac{7}{3}\mathbf{k}$
    c) $\begin{pmatrix} 3 \\ -12 \\ -21 \end{pmatrix}$
    d) $p(-5\mathbf{i} + 20\mathbf{j} - 30\mathbf{k})$

4. Find the values of c for which the following vectors are parallel.
    a) $\mathbf{i} - 2\mathbf{j} + 4\mathbf{k}$ and $0.5\mathbf{i} - c\mathbf{j} + 2\mathbf{k}$
    b) $\begin{pmatrix} -15 \\ 36 \\ 21 \end{pmatrix}$ and $\begin{pmatrix} -20 \\ 48 \\ c \end{pmatrix}$
    c) $\begin{pmatrix} 2t \\ -4t \\ 5t \end{pmatrix}$ and $\begin{pmatrix} 3t \\ -6t \\ ct \end{pmatrix}$

5. If the following pairs of vectors are equal, find the values of a, b, and c.
    a) $a\mathbf{i} + b\mathbf{j} + c\mathbf{k}$ and $3\mathbf{i} - 4\mathbf{j} + \mathbf{k}$
    b) $(a - 1)\mathbf{i} + (b + 2)\mathbf{j} + c\mathbf{k}$ and $-b\mathbf{i} + (a - 1)\mathbf{j} + 2c\mathbf{k}$

6. The points A, B, C, and D have position vectors $\mathbf{i} + \mathbf{j} + \mathbf{k}$, $2\mathbf{i} + 3\mathbf{j}$, $3\mathbf{i} + 5\mathbf{j} - 2\mathbf{k}$, and $-\mathbf{j} + \mathbf{k}$ respectively.
    a) Determine which of the following pairs of lines are parallel.
        i) AB and CD
        ii) BC and CD
        iii) BC and AD
    b) Determine which of the lines AB, BC, CD, and DA is longest.

7. The points $A(-1, 0, 3)$, $B(1, -2, 2)$, $C(x, y, z)$, and $D(-2, 4, 5)$ form a parallelogram.
    a) Find the coordinates of C.
    b) Find a unit vector in the direction of AD.
    c) Find a vector of magnitude 5 units which has the same direction as AB.

9.3 Addition and subtraction of vectors: a geometric approach

Consider three points A, B, and C.

$$\overrightarrow{AB} + \overrightarrow{BC} = \overrightarrow{AC}$$

This is the triangle law of vector addition.

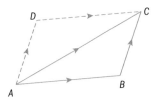

Alternatively, in the parallelogram $ABCD$

$$\overrightarrow{AD} = \overrightarrow{BC}$$

so $\overrightarrow{AB} + \overrightarrow{AD} = \overrightarrow{AC}$ where $\overrightarrow{AC}$ is the diagonal of the parallelogram. This is the parallelogram law of vector addition.

More generally, vector addition can be interpreted as moving from a start-point to an end-point by any route.

For example,

$$\overrightarrow{AB} + \overrightarrow{BC} + \overrightarrow{CD} + \overrightarrow{DE} + \overrightarrow{EF} = \overrightarrow{AF}$$

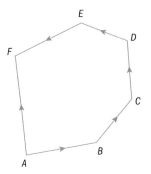

Vector subtraction is also important.

For the vector triangle

$$\overrightarrow{AB} = \overrightarrow{AO} + \overrightarrow{OB}$$
$$= -\mathbf{a} + \mathbf{b}$$

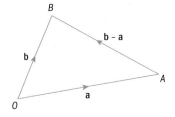

$\boxed{\overrightarrow{AB} = \mathbf{b} - \mathbf{a} \text{ where } \mathbf{a}, \mathbf{b} \text{ are the position vectors of } A, B \text{ respectively.}}$

Note: This explains why to find $\overrightarrow{AB}$ we use the position vector of B minus the position vector of A. In practice this means that to find $\overrightarrow{AB}$ we subtract the coordinates of A from the coordinates of B.

Example 7

a) A is a point with coordinates $(1, 0, 3)$ and B is a point with coordinates $(3, -2, 6)$.
Find, using column vectors,

i) $\overrightarrow{AB}$

ii) the position vector of the point C, which is the mid-point of AB.

b) $OADB$ is a parallelogram.
Find the position vector of the point D.

c) What do your answers to **(a)** and **(b)** tell you?

a) i) $\overrightarrow{AB} = \begin{pmatrix} 3 \\ -2 \\ 6 \end{pmatrix} - \begin{pmatrix} 1 \\ 0 \\ 3 \end{pmatrix} = \begin{pmatrix} 2 \\ -2 \\ 3 \end{pmatrix}$

▶ Continued on the next page

ii) $\overrightarrow{OC} = \overrightarrow{OA} + \overrightarrow{AC}$

but $\overrightarrow{AC} = \frac{1}{2}\overrightarrow{AB} = \begin{pmatrix} 1 \\ -1 \\ \frac{3}{2} \end{pmatrix}$

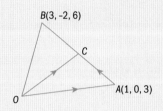

so $\overrightarrow{OC} = \begin{pmatrix} 1 \\ 0 \\ 3 \end{pmatrix} + \begin{pmatrix} 1 \\ -1 \\ \frac{3}{2} \end{pmatrix} = \begin{pmatrix} 2 \\ -1 \\ \frac{9}{2} \end{pmatrix}$

b) $\overrightarrow{OD} = \overrightarrow{OA} + \overrightarrow{AD}$

$= \begin{pmatrix} 1 \\ 0 \\ 3 \end{pmatrix} + \begin{pmatrix} 3 \\ -2 \\ 6 \end{pmatrix}$

$\overrightarrow{AD} = \overrightarrow{OB}$

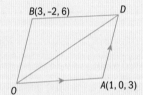

$\overrightarrow{OD} = \begin{pmatrix} 4 \\ -2 \\ 9 \end{pmatrix}$

c) $\overrightarrow{OD} = 2\overrightarrow{OC}$ so C is the mid-point of OD.

Therefore the diagonals of the parallelogram bisect each other.

Example 8

The points A, B, and C are such that

$\overrightarrow{OA} = 2\mathbf{i} + 6\mathbf{j} + 3\mathbf{k}$, $\overrightarrow{OB} = \mathbf{i} + 2\mathbf{j} + 7\mathbf{k}$, and $\overrightarrow{OC} = 4\mathbf{i} + 14\mathbf{j} - 5\mathbf{k}$.

Show that the vectors $\overrightarrow{BA}$ and $\overrightarrow{AC}$ are in the same direction and hence that A, B, C lie on the same straight line.

$\overrightarrow{BA} = \overrightarrow{OA} - \overrightarrow{OB}$

$= (2\mathbf{i} + 6\mathbf{j} + 3\mathbf{k}) - (\mathbf{i} + 2\mathbf{j} + 7\mathbf{k})$ ← $\overrightarrow{BA}$ = position vector of A minus position vector of B

$= \mathbf{i} + 4\mathbf{j} - 4\mathbf{k}$

$\overrightarrow{AC} = \overrightarrow{OC} - \overrightarrow{OA}$

$= (4\mathbf{i} + 14\mathbf{j} - 5\mathbf{k}) - (2\mathbf{i} + 6\mathbf{j} + 3\mathbf{k})$ ← $\overrightarrow{AC}$ = position vector of C minus position vector of A

$= 2\mathbf{i} + 8\mathbf{j} - 8\mathbf{k}$

$\overrightarrow{AC} = 2\overrightarrow{BA}$ so $\overrightarrow{BA}$ and $\overrightarrow{AC}$ are parallel.

But A lies on both $\overrightarrow{BA}$ and $\overrightarrow{AC}$, so A, B, C lie on a straight line.

Exercise 9.3

1. Here are three vectors:

$$\mathbf{r} = \begin{pmatrix} 3 \\ 1 \\ 2 \end{pmatrix} \qquad \mathbf{s} = \begin{pmatrix} 2 \\ 0 \\ 3 \end{pmatrix} \qquad \mathbf{t} = \begin{pmatrix} -1 \\ -2 \\ 0 \end{pmatrix}.$$

Find, as a single column vector,

    a) $\mathbf{r} + \mathbf{s} + \mathbf{t}$ **b)** $2\mathbf{r} + 3\mathbf{s}$ **c)** $\mathbf{s} - \mathbf{t}$

    d) $\mathbf{r} + 2\mathbf{s} - \mathbf{t}$ **e)** $3(\mathbf{t} - 2\mathbf{r})$ **f)** $\frac{1}{2}(3\mathbf{r} - \mathbf{s} + \mathbf{t})$.

2. Two points A and B have position vectors $\mathbf{a} = (4\mathbf{i} - 2\mathbf{j} - 3\mathbf{k})$ and $\mathbf{b} = (3\mathbf{i} - 4\mathbf{j} - \mathbf{k})$. Find the vector which has magnitude 6 and is in the same direction as $\overrightarrow{AB}$.

3. $OABCD$, shown in the diagram, is a square-based pyramid.

    a) Write each of the following vectors in terms of $\mathbf{i}$, $\mathbf{j}$, and $\mathbf{k}$.

        i) $\overrightarrow{OA}$ **ii)** $\overrightarrow{OD}$ **iii)** $\overrightarrow{DO}$

    b) Hence find the vector $\overrightarrow{DA}$ by using vector addition.

    c) Find each of these vectors in the form $x\mathbf{i} + y\mathbf{j} + z\mathbf{k}$.

        i) $\overrightarrow{OB}$ **ii)** $\overrightarrow{DB}$ **iii)** $\overrightarrow{CD}$

    d) Find $|\overrightarrow{CD}|$.

4. Show that the points P, Q, and R, with position vectors $\begin{pmatrix} 2 \\ -1 \\ -8 \end{pmatrix}$, $\begin{pmatrix} 0 \\ -5 \\ -2 \end{pmatrix}$ and $\begin{pmatrix} -3 \\ -11 \\ 7 \end{pmatrix}$, lie on a straight line.

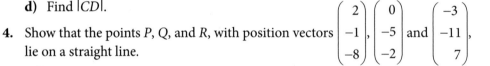

5. $\overrightarrow{OA} = 3\mathbf{i} - 2\mathbf{j} + \mathbf{k}$

   $\overrightarrow{OB} = 4\mathbf{i} + \mathbf{j} - 2\mathbf{k}$

   Find the vectors $\overrightarrow{OR}$ and $\overrightarrow{OQ}$ such that

    a) R is the mid-point of AB

    b) Q is the mid-point of RB.

6. $OABCDEFG$, shown in the diagram, is a prism.

   The cross section $OABC$ is a trapezium and is such that OC is parallel to AB.

   $OC = 6\,\text{cm}$, $OA = 4\,\text{cm}$, $AB = 3\,\text{cm}$, $OE = 2\,\text{cm}$

    a) Write each of these as a column vector.

        i) $\overrightarrow{OC}$ **ii)** $\overrightarrow{OE}$ **iii)** $\overrightarrow{EF}$

    b) Use $\overrightarrow{FC} = \overrightarrow{FE} + \overrightarrow{EO} + \overrightarrow{OC}$ to write $\overrightarrow{FC}$ as a column vector.

    c) Write each of these as a column vector.

        i) $\overrightarrow{CG}$ **ii)** $\overrightarrow{FB}$ **iii)** $\overrightarrow{OG}$ **iv)** $\overrightarrow{BE}$

    d) Find a unit vector in the direction of $\overrightarrow{AD}$.

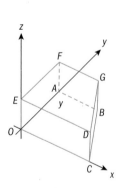

Addition and subtraction of vectors: a geometric approach

7. The points P and Q have position vectors $\overrightarrow{OP} = 2\mathbf{i} - \mathbf{j} - 8\mathbf{k}$ and
$\overrightarrow{OQ} = 5\mathbf{i} - \mathbf{j} + 4\mathbf{k}$ respectively.
The point X lies on PQ such that $PX = \frac{2}{3}PQ$.
Find the position vector of X, $\overrightarrow{OX}$.

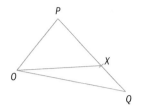

9.4 The vector equation of a straight line

We are familiar with writing the equation of a two-dimensional
straight line in one of the forms $y = mx + c$ or $ax + by = c$.
These equations are said to be written in **Cartesian form**.

We can also write the equation of a line in vector form. The
vector equation of a line is a way of expressing the position
vector of any point on the line.

To find the vector equation of a line we need to know:

   i)   the position vector, say **a**, of any point A which lies
       on the line,

   ii) any vector, say **b**, which runs in the direction of the line.

The position vector (**r**) of any point P on the line can then be
expressed in the form $\mathbf{r} = \mathbf{a} + t\mathbf{b}$, where t is a scalar.

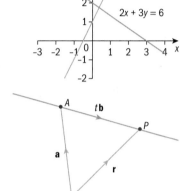

The vector equation of a straight line passing through a point
with position vector **a** and with direction vector **b** is
$$\mathbf{r} = \mathbf{a} + t\mathbf{b}$$
where t is a scalar.

Note: The direction vector b can
have any magnitude but it must
be parallel to the line.
The value of t will be positive
for points on one side of A and
negative for points on the other
side of A.

Example 9

a) Write down a vector equation for the straight line through
the point $(3, 1)$ with direction vector $\begin{pmatrix} 4 \\ -2 \end{pmatrix}$.

b) Find a vector equation for the line through the points $L(1, 2)$
and $M(5, 3)$.

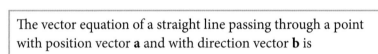

a) $\mathbf{a} = \begin{pmatrix} 3 \\ 1 \end{pmatrix}$ This is the position vector of a point on the line.

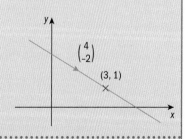

▶ Continued on the next page

$$\mathbf{b} = \begin{pmatrix} 4 \\ -2 \end{pmatrix}$$

A vector equation of the line is therefore $\mathbf{r} = \begin{pmatrix} 3 \\ 1 \end{pmatrix} + t\begin{pmatrix} 4 \\ -2 \end{pmatrix}$

This gives the direction of the line.

b) $\mathbf{a} = \overrightarrow{OL} = \begin{pmatrix} 1 \\ 2 \end{pmatrix}$

Alternatively we could use $\overrightarrow{OM} = \begin{pmatrix} 5 \\ 3 \end{pmatrix}$.

$$\mathbf{b} = \overrightarrow{LM} = \begin{pmatrix} 5 \\ 3 \end{pmatrix} - \begin{pmatrix} 1 \\ 2 \end{pmatrix} = \begin{pmatrix} 4 \\ 1 \end{pmatrix}$$

Alternatively we could use $\overrightarrow{ML} = \begin{pmatrix} -4 \\ -1 \end{pmatrix}$ or any multiple of $\overrightarrow{LM}$ or $\overrightarrow{ML}$.

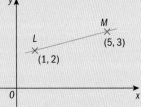

A vector equation of the line is therefore $\mathbf{r} = \begin{pmatrix} 1 \\ 2 \end{pmatrix} + t\begin{pmatrix} 4 \\ 1 \end{pmatrix}$

Remember: Include 'r =' as part of the equation.

We can also use vectors to write the equation of a line in three dimensions, as Example 10 shows.

Example 10

Points A and B have position vectors $\begin{pmatrix} 3 \\ 1 \\ -4 \end{pmatrix}$ and $\begin{pmatrix} -2 \\ 5 \\ 0 \end{pmatrix}$ respectively.

a) Find, in vector form, an equation of the straight line that passes through A and B.

b) Show that the point $C(-12, 13, 8)$ lies on this line.

a) $\overrightarrow{AB} = \begin{pmatrix} -2 \\ 5 \\ 0 \end{pmatrix} - \begin{pmatrix} 3 \\ 1 \\ -4 \end{pmatrix} = \begin{pmatrix} -5 \\ 4 \\ 4 \end{pmatrix}$

$\overrightarrow{AB}$ = position vector of B minus position vector of A

A vector equation of the line is $\mathbf{r} = \begin{pmatrix} 3 \\ 1 \\ -4 \end{pmatrix} + t\begin{pmatrix} -5 \\ 4 \\ 4 \end{pmatrix}$

b) $\begin{pmatrix} 3 \\ 1 \\ -4 \end{pmatrix} + t\begin{pmatrix} -5 \\ 4 \\ 4 \end{pmatrix} = \begin{pmatrix} -12 \\ 13 \\ 8 \end{pmatrix}$

For C to lie on the line there must be a value of t which, when substituted into the equation, gives the position vector of C.

▶ Continued on the next page

$3 - 5t = -12$

$1 + 4t = 13$ ← Write down an equation from each component.

$-4 + 4t = 8$

Solving each equation gives $t = 3$. ← This shows that $t = 3$ gives the position vector of C.

So $C(-12, 13, 8)$ lies on the line that passes through A and B.

Note: We must show that the three equations are consistent and the same value of t satisfies each equation. If this is not true, then the point does not lie on the line.

Example 11

A line has a vector equation given by $\mathbf{r} = (3\mathbf{i} + \mathbf{j} + \mathbf{k}) + \mu(2\mathbf{i} - \mathbf{j} - \mathbf{k})$.

a) Show that this line intersects the x-axis.

b) The point P lies on this line and has coordinates $(-5, a, b)$. Find the values of a and b.

a) If the line intersects the x-axis then there will be a value of μ which gives a position vector of the form $c\mathbf{i} + 0\mathbf{j} + 0\mathbf{k}$. ← Any point on the x-axis has position vector with y and z components equal to zero.

$(3\mathbf{i} + \mathbf{j} + \mathbf{k}) + \mu(2\mathbf{i} - \mathbf{j} - \mathbf{k}) = c\mathbf{i} + 0\mathbf{j} + 0\mathbf{k}$

$(3 + 2\mu)\mathbf{i} + (1 - \mu)\mathbf{j} + (1 - \mu)\mathbf{k} = c\mathbf{i} - 0\mathbf{j} - 0\mathbf{k}$ ← Rewrite the equation so we can compare the i, j, and k components.

$3 + 2\mu = c$ ← Compare the i components.

$1 - \mu = 0$ so $\mu = 1$ ← Compare the j and k components.

$\Rightarrow 3 + 2 \times 1 = c$ ← Substitute $\mu = 1$ into $3 + 2\mu = c$ to find the value of c.

$\Rightarrow c = 5$

When $\mu = 1$ and $c = 5$, $\mathbf{r} = 5\mathbf{i} - 0\mathbf{j} - 0\mathbf{k}$ ← Substitute $\mu = 1$ and $c = 5$ into the equation of the line.

so the point $(5, 0, 0)$ lies on the line and on the x-axis.

Therefore the line intersects the x-axis.

b) $(3\mathbf{i} + \mathbf{j} + \mathbf{k}) + \mu(2\mathbf{i} - \mathbf{j} - \mathbf{k}) = -5\mathbf{i} + a\mathbf{j} + b\mathbf{k}$ ← Equate the RHS of the equation of the line with the position vector of P.

$(3 + 2\mu)\mathbf{i} + (1 - \mu)\mathbf{j} + (1 - \mu)\mathbf{k} = -5\mathbf{i} + a\mathbf{j} + b\mathbf{k}$

$3 + 2\mu = -5$ ← Collect terms in i, j, and k and compare components to find the value of μ.

$\mu = -4$

When $\mu = -4$, ← Substitute $\mu = -4$ into the equation of the line.

$\mathbf{r} = (3\mathbf{i} + \mathbf{j} + \mathbf{k}) + (-4)(2\mathbf{i} - \mathbf{j} - \mathbf{k})$

$= -5\mathbf{i} + 5\mathbf{j} + 5\mathbf{k}$

$a = 5, b = 5$ ← State the values of a and b.

Exercise 9.4

1. **a)** The vector equation of the line l_1 is $\mathbf{r} = \begin{pmatrix} 2 \\ 3 \\ -1 \end{pmatrix} + t \begin{pmatrix} 4 \\ -5 \\ 10 \end{pmatrix}$.

    Write down the coordinates of the points A, B, and C on this line which correspond to the following values of t.

    i) $t = 0$ ii) $t = 1$ iii) $t = -3$

    b) The vector equation of the line l_2 is $\mathbf{r} = (2\mathbf{i} + \mathbf{k}) + \mu(-3\mathbf{i} + 3\mathbf{j} - 2\mathbf{k})$.

    Write down the coordinates of the points P, Q, and R on this line which correspond to the following values of μ.

    i) $\mu = 2$ ii) $\mu = \dfrac{1}{2}$ iii) $\mu = -4$

2. In two dimensions, find, using column vectors, a vector equation for

    a) the line through the point $(4, -1)$ in the direction of the vector $\begin{pmatrix} -3 \\ 5 \end{pmatrix}$

    b) the line joining the points $(2, 5)$ and $(6, -1)$

    c) the x-axis

    d) the line joining the points with position vectors $\begin{pmatrix} 2 \\ -5 \end{pmatrix}$ and $\begin{pmatrix} 4 \\ 3 \end{pmatrix}$

    e) the line $y = 3x$

    f) the line $x + y = 1$.

3. In three dimensions, find, using unit vectors, a vector equation for

    a) the line through $A(2, 1, 5)$ which is parallel to the vector $-\mathbf{i} - \mathbf{j} + \mathbf{k}$

    b) the line through $B(3, 2, 4)$ which is parallel to CD where $\overrightarrow{OC} = 6\mathbf{i} - 2\mathbf{j} + 3\mathbf{k}$
    and $\overrightarrow{OD} = 5\mathbf{i} + 2\mathbf{j} - \mathbf{k}$

    c) the y-axis

    d) the line joining the points $(-3, 7\ 0)$ and $(-5, 10, 2)$

    e) the line through the origin in the direction $\mathbf{i} + \mathbf{j} + \mathbf{k}$

    f) the line through the point with position vector $-3\mathbf{i} + 2\mathbf{j} + 6\mathbf{k}$, in the direction of the z-axis.

4. Points A, B, and C have position vectors

    $$\mathbf{a} = \begin{pmatrix} 1 \\ 2 \\ 3 \end{pmatrix}, \qquad \mathbf{b} = \begin{pmatrix} -1 \\ 2 \\ -1 \end{pmatrix}, \qquad \mathbf{c} = \begin{pmatrix} 4 \\ 2 \\ 5 \end{pmatrix}.$$

    Find

    a) a vector equation for the line AB

    b) a vector equation for the line through A parallel to BC

    c) a vector equation for the line through the mid-point of AB and the mid-point of AC.

5. Determine which of the points $P(6, 7, -13)$, $Q(-3, -11, 14)$, and $R(0, -6, 5)$ lie on the straight line with equation $\mathbf{r} = \begin{pmatrix} 1 \\ -3 \\ 2 \end{pmatrix} + t \begin{pmatrix} 2 \\ 4 \\ -6 \end{pmatrix}$.

6. Points A and B have position vectors $-2\mathbf{i} + \mathbf{j}$ and $4\mathbf{i} + 4\mathbf{k}$.
   a) Find an equation for the line that passes through A and B.
   b) Show that this line does not intersect the x-axis.

7. Points A and B have position vectors $\mathbf{a} = \begin{pmatrix} 2 \\ -1 \\ 3 \end{pmatrix}$ and $\mathbf{b} = \begin{pmatrix} 1 \\ 1 \\ -1 \end{pmatrix}$.

   a) Find an equation for the line AB.
   b) Find the values of α and β if the point $C(3, \alpha, \beta)$ lies on the line AB.

8. a) Find the vector equation of the line joining the points $(2, 3, 4)$ and $(1, 4, 3)$.
   b) Find the x- and y-coordinates on this line where the z-coordinate is 0.

9.5 Intersecting lines

In two dimensions, there are three possible relationships between two straight lines.

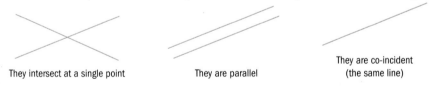

| They intersect at a single point | They are parallel | They are co-incident (the same line) |

> Two lines intersect if they have one point in common.
> Two lines are parallel if they have the same direction but no point in common.
> Two lines are coincident if they have the same direction and an infinite number of points in common.

In three dimensions there is a fourth possible relationship between two lines.

> Lines that are not parallel, do not intersect, and are not coincident are called skew lines. They have different directions and no point in common.

For example, in the cuboid shown, the straight lines AC and GD are skew.

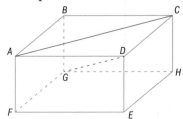

Example 12

Two lines l_1 and l_2 are given by the vector equations

$$\mathbf{r} = \begin{pmatrix} 5 \\ 1 \\ -1 \end{pmatrix} + t \begin{pmatrix} 2 \\ 1 \\ 5 \end{pmatrix} \text{ and } \mathbf{r} = \begin{pmatrix} 13 \\ -6 \\ 2 \end{pmatrix} + \mu \begin{pmatrix} -3 \\ 4 \\ 1 \end{pmatrix} \text{ respectively.}$$

Show that the lines intersect and find the position vector of the point of intersection.

We can see that either the lines must intersect, or else they are skew. Their directions, given by $\begin{pmatrix} 2 \\ 1 \\ 5 \end{pmatrix}$ and $\begin{pmatrix} -3 \\ 4 \\ 1 \end{pmatrix}$,

are not the same and are not multiples of each other, so the lines are not parallel or coincident.

$$\begin{pmatrix} 5 \\ 1 \\ -1 \end{pmatrix} + t \begin{pmatrix} 2 \\ 1 \\ 5 \end{pmatrix} = \begin{pmatrix} 13 \\ -6 \\ 2 \end{pmatrix} + \mu \begin{pmatrix} -3 \\ 4 \\ 1 \end{pmatrix}$$

For the lines to intersect there must be a value of t and a value of μ which give the same position vector from each equation.

$$5 + 2t = 13 - 3\mu \quad (1)$$
$$1 + t = -6 + 4\mu \quad (2)$$
$$-1 + 5t = 2 + \mu \quad (3)$$

Equation (1) − 2 × equation (2) gives

Solve two of the equations to find a pair of values which satisfies those two equations.

$$3 = 25 - 11\mu$$
$$\mu = 2$$

Substituting $\mu = 2$ into equation (2):

$$1 + t = -6 + 4 \times 2$$
$$t = 1$$

Checking equation (3):

Check that these values satisfy the third equation.

When $t = 1$ and $\mu = 2$, $-1 + 5t = 4$
$$2 + \mu = 4$$

So this pair of values also satisfies the third equation and we have shown that the lines intersect.

If we solve two of the equations to get a value for each of the scalars but then, on checking, these values do not satisfy the third equation, then the lines do not intersect.

$$\mathbf{r} = \begin{pmatrix} 5 \\ 1 \\ -1 \end{pmatrix} + 1 \begin{pmatrix} 2 \\ 1 \\ 5 \end{pmatrix} = \begin{pmatrix} 7 \\ 2 \\ 4 \end{pmatrix}$$

Find the position vector of the point of intersection by substituting the value of t into the equation of l_1 (or alternatively the value of μ into the equation of l_2).

If we were asked to find the **coordinates** of the point of intersection, we would give the answer in the form (7, 2, 4).

Example 13

Determine whether the following pairs of lines intersect. If they do not intersect, determine whether they are parallel or skew.

a) $\mathbf{r} = \begin{pmatrix} 4 \\ 5 \\ 2 \end{pmatrix} + t \begin{pmatrix} 2 \\ -1 \\ 3 \end{pmatrix}, \quad \mathbf{r} = \begin{pmatrix} 5 \\ 0 \\ 4 \end{pmatrix} + \mu \begin{pmatrix} -4 \\ 2 \\ -6 \end{pmatrix}$

b) $\mathbf{r} = \begin{pmatrix} 4 \\ 3 \\ 1 \end{pmatrix} + t \begin{pmatrix} 3 \\ 1 \\ 2 \end{pmatrix}, \quad \mathbf{r} = \begin{pmatrix} 2 \\ 1 \\ 6 \end{pmatrix} + \mu \begin{pmatrix} 5 \\ -2 \\ 4 \end{pmatrix}$

a) By inspection we see that

$\begin{pmatrix} -4 \\ 2 \\ -6 \end{pmatrix} = -2 \begin{pmatrix} 2 \\ -1 \\ 3 \end{pmatrix}$

> Firstly check to see if the lines are parallel.

> It is easy to check whether the lines are parallel by looking at the direction vectors, so it can save work to do this first.

The lines are therefore parallel and do not intersect.

b) The lines are not parallel.

> Check the direction vectors in order to conclude this.

For intersection

$4 + 3t = 2 + 5\mu$ (1)
$3 + 1t = 1 - 2\mu$ (2)
$1 + 2t = 6 + 4\mu$ (3)

Equation (1) − 3 × equation (2) gives

$-5 = -1 + 11\mu$

$\mu = -\dfrac{4}{11}$

> Solve two of the equations for t and μ values.

Substituting $\mu = -\dfrac{4}{11}$ into equation (1):

$4 + 3t = 2 + 5 \times -\dfrac{4}{11}$

$t = -\dfrac{14}{11}$

Checking equation (3):

$1 + 2t = 1 + 2 \times -\dfrac{14}{11} = -\dfrac{17}{11}$

$6 + 4\mu = 6 + 4 \times -\dfrac{4}{11} = \dfrac{50}{11}$

> Check to see if the values found also satisfy the third equation.

$1 + 2t \neq 6 + 4\mu$

There are no values which satisfy all three equations, so there is no point in common for these two lines.

The lines are not parallel so the lines must be skew.

Exercise 9.5

1. State whether the following pairs of lines are parallel or not. You may assume that the lines are not coincident.

a) $\mathbf{r} = \begin{pmatrix} 2 \\ -5 \\ 3 \end{pmatrix} + t \begin{pmatrix} -1 \\ 3 \\ 2 \end{pmatrix}$, $\mathbf{r} = \begin{pmatrix} 3 \\ -6 \\ 5 \end{pmatrix} + \mu \begin{pmatrix} 4 \\ -12 \\ -8 \end{pmatrix}$

b) $\mathbf{r} = \begin{pmatrix} -1 \\ 3 \\ 1 \end{pmatrix} + t \begin{pmatrix} 3 \\ 8 \\ 6 \end{pmatrix}$, $\mathbf{r} = \begin{pmatrix} 4 \\ -7 \\ -4 \end{pmatrix} + \mu \begin{pmatrix} 1.5 \\ 4 \\ 3 \end{pmatrix}$

c) $\mathbf{r} = \begin{pmatrix} 0 \\ 0 \\ 1 \end{pmatrix} + s \begin{pmatrix} 1 \\ -1 \\ -3 \end{pmatrix}$, $\mathbf{r} = \begin{pmatrix} 2 \\ 1 \\ 0 \end{pmatrix} + t \begin{pmatrix} 0 \\ 3 \\ 5 \end{pmatrix}$

d) $\mathbf{r} = \begin{pmatrix} -3 \\ -2 \\ 1 \end{pmatrix} + s \begin{pmatrix} 2 \\ 1 \\ 1 \end{pmatrix}$, $\mathbf{r} = \begin{pmatrix} 1 \\ 2 \\ -5 \end{pmatrix} + \mu \begin{pmatrix} 1 \\ -1 \\ 3 \end{pmatrix}$

e) $\mathbf{r} = (2\mathbf{i} + \mathbf{k}) + t(-3\mathbf{i} + 3\mathbf{j} - 2\mathbf{k})$, $\mathbf{r} = (-5\mathbf{i} - \mathbf{j} + 2\mathbf{k}) + \mu(6\mathbf{i} + 5\mathbf{j} + 2\mathbf{k})$

f) $\mathbf{r} = (\mathbf{i} - \mathbf{j} + 2\mathbf{k}) + t(2\mathbf{i} + 5\mathbf{j} - \mathbf{k})$, $\mathbf{r} = (2\mathbf{i} - 2\mathbf{j} + 4\mathbf{k}) + \mu(2\mathbf{i} - 5\mathbf{j} + \mathbf{k})$

g) $\mathbf{r} = (5 + 2s)\mathbf{i} + (1 + s)\mathbf{j} + (-1 + 5s)\mathbf{k}$, $\mathbf{r} = (13 - 4t)\mathbf{i} + (-6 - 2t)\mathbf{j} + (2 - 10t)\mathbf{k}$

h) $\mathbf{r} = (s - 1)\mathbf{i} + (2s - 1)\mathbf{j} + (3s + 1)\mathbf{k}$, $\mathbf{r} = \left(1 - \frac{1}{2}s\right)\mathbf{i} + (1 - s)\mathbf{j} + \left(3 - \frac{3}{2}s\right)\mathbf{k}$

2. In each case determine whether the two lines intersect and find the coordinates of any points of intersection.

a) $\mathbf{r} = \begin{pmatrix} 2 \\ 1 \\ 0 \end{pmatrix} + t \begin{pmatrix} 1 \\ 1 \\ -2 \end{pmatrix}$, $\mathbf{r} = \begin{pmatrix} 5 \\ 2 \\ -1 \end{pmatrix} + \mu \begin{pmatrix} 2 \\ 0 \\ 1 \end{pmatrix}$

b) $\mathbf{r} = \begin{pmatrix} 1 \\ 2 \\ 0 \end{pmatrix} + t \begin{pmatrix} -1 \\ 5 \\ 2 \end{pmatrix}$, $\mathbf{r} = \begin{pmatrix} 2 \\ -1 \\ 5 \end{pmatrix} + \mu \begin{pmatrix} 3 \\ -2 \\ 6 \end{pmatrix}$

c) $\mathbf{r} = (2\mathbf{i} + 3\mathbf{j} + 5\mathbf{k}) + s(4\mathbf{i} - \mathbf{j} + 3\mathbf{k})$, $\mathbf{r} = (4\mathbf{i} + 7\mathbf{j} + 2\mathbf{k}) + t(2\mathbf{i} - 2\mathbf{j} + 3\mathbf{k})$

d) $\mathbf{r} = 6s\mathbf{i} + (2 - s)\mathbf{j} + (-3 - s)\mathbf{k}$, $\mathbf{r} = (-4 + 2t)\mathbf{i} + (6 + t)\mathbf{j} - (4 + t)\mathbf{k}$

3. Find the position vector of the common point for each of the following pairs of lines.

a) $\mathbf{r} = \begin{pmatrix} 1 \\ 1 \\ 0 \end{pmatrix} + t \begin{pmatrix} 2 \\ -1 \\ 0 \end{pmatrix}$, $\qquad \mathbf{r} = \begin{pmatrix} 2 \\ 1 \\ 0 \end{pmatrix} + \mu \begin{pmatrix} -1 \\ 1 \\ 0 \end{pmatrix}$

b) $\mathbf{r} = (2\mathbf{i} - \mathbf{j} + \mathbf{k}) + s(3\mathbf{i} + \mathbf{j})$, $\quad \mathbf{r} = (-\mathbf{i} + \mathbf{j} + \mathbf{k}) + \mu(-3\mathbf{i} + 2\mathbf{j})$

4. Points A, B, C, and D have position vectors $2\mathbf{i} + 3\mathbf{j}$, $3\mathbf{i} + 2\mathbf{j}$, $4\mathbf{i} + 6\mathbf{j}$, and $9\mathbf{i} + 6\mathbf{j}$ respectively. Find the position vector of the point of intersection of lines AB and CD.

5. Points P, Q, R, and S have coordinates $(2, 0, 1)$, $(-1, 3, -1)$, $(-5, -1, 2)$, and $(1, 4, 4)$ respectively. Show that the lines PQ and RS are skew.

6. Two lines have equations $\mathbf{r} = (3\mathbf{i} + 2\mathbf{j} - \mathbf{k}) + t(2\mathbf{i} - 3\mathbf{j} + 2\mathbf{k})$ and $\mathbf{r} = (-\mathbf{i} + \mathbf{j} + \mathbf{k}) + \mu(\mathbf{i} + 2\mathbf{j} + c\mathbf{k})$. Find the value of c for which the lines intersect and find the coordinates of the point of intersection.

7. Show that the two lines $\mathbf{r} = \begin{pmatrix} 5 \\ 7 \\ 1 \end{pmatrix} + t \begin{pmatrix} -1 \\ 3 \\ 2 \end{pmatrix}$ and $\mathbf{r} = \begin{pmatrix} 7 \\ 1 \\ -3 \end{pmatrix} + \mu \begin{pmatrix} 1 \\ -3 \\ -2 \end{pmatrix}$ are coincident.

9.6 Scalar products

The scalar product (sometimes called the dot product) of two vectors **a** and **b** is written as **a·b** and is equal to $|\mathbf{a}| \times |\mathbf{b}| \times \cos\theta$ where θ is the angle between **a** and **b**.

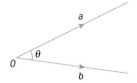

$$\boxed{\mathbf{a \cdot b} = |\mathbf{a}|\,|\mathbf{b}|\,\cos\theta}$$

We often do not know the angle between two vectors, so it is not convenient to work out the scalar product by using the result above.

There is a simpler way to calculate the scalar product.

In the vector triangle shown, $\mathbf{c} = \mathbf{b} - \mathbf{a}$.

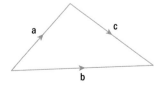

Suppose the two vectors **a** and **b** may be written in column vector form.

Let $\mathbf{a} = \begin{pmatrix} a_1 \\ a_2 \\ a_3 \end{pmatrix}$, $\mathbf{b} = \begin{pmatrix} b_1 \\ b_2 \\ b_3 \end{pmatrix}$.

$$\mathbf{c} = \mathbf{b} - \mathbf{a} = \begin{pmatrix} b_1 - a_1 \\ b_2 - a_2 \\ b_3 - a_3 \end{pmatrix}$$

$$\begin{aligned}
|\mathbf{c}|^2 &= (b_1 - a_1)^2 + (b_2 - a_2)^2 + (b_3 - a_3)^2 \\
&= b_1^2 + a_1^2 - 2a_1 b_1 + b_2^2 + a_2^2 - 2a_2 b_2 + b_3^2 + a_3^2 - 2a_3 b_3 \\
&= a_1^2 + a_2^2 + a_3^2 + b_1^2 + b_2^2 + b_3^2 - 2(a_1 b_1 + a_2 b_2 + a_3 b_3) \\
&= |\mathbf{a}|^2 + |\mathbf{b}|^2 - 2(a_1 b_1 + a_2 b_2 + a_3 b_3)
\end{aligned}$$

From the cosine rule

$$|\mathbf{c}|^2 = |\mathbf{a}|^2 + |\mathbf{b}|^2 - 2|\mathbf{a}||\mathbf{b}|\cos\theta$$

Comparing the results we have $\mathbf{a}\cdot\mathbf{b} = |\mathbf{a}||\mathbf{b}|\cos\theta = a_1 b_1 + a_2 b_2 + a_3 b_3$.

$$\mathbf{a}\cdot\mathbf{b} = a_1 b_1 + a_2 b_2 + a_3 b_3, \text{ where } \mathbf{a} = \begin{pmatrix} a_1 \\ a_2 \\ a_3 \end{pmatrix}, \mathbf{b} = \begin{pmatrix} b_1 \\ b_2 \\ b_3 \end{pmatrix}$$

We now have two ways of calculating the scalar product.

Example 14

Vectors $\overrightarrow{OA}$ and $\overrightarrow{OB}$ are given by $\overrightarrow{OA} = \begin{pmatrix} 3 \\ -1 \\ 2 \end{pmatrix}$ and $\overrightarrow{OB} = \begin{pmatrix} 6 \\ -4 \\ -5 \end{pmatrix}$.

Calculate the scalar product $\overrightarrow{OA} \cdot \overrightarrow{OB}$.

$\overrightarrow{OA} \cdot \overrightarrow{OB} = (3 \times 6) + (-1 \times -4) + (2 \times -5)$ ⟵ Using $a_1 b_1 + a_2 b_2 + a_3 b_3$

$\qquad = 12$

Example 15

The position vectors of points A and B relative to an origin O are given by $\mathbf{a} = 2\mathbf{i} + p\mathbf{j} + 3\mathbf{k}$ and $\mathbf{b} = 3\mathbf{i} - \mathbf{j} + p\mathbf{k}$, where p is a constant.

Calculate $\mathbf{a}\cdot\mathbf{b}$.

$\mathbf{a}\cdot\mathbf{b} = (2 \times 3) + (p \times -1) + (3 \times p)$ ⟵ Using $a_1 b_1 + a_2 b_2 + a_3 b_3$

$\qquad = 2p + 6$

We can use the two ways of calculating a scalar product to provide a method for calculating the angle between two vectors.

$$\mathbf{a}\cdot\mathbf{b} = |\mathbf{a}||\mathbf{b}|\cos\theta = a_1 b_1 + a_2 b_2 + a_3 b_3$$

$$\cos\theta = \frac{a_1 b_1 + a_2 b_2 + a_3 b_3}{|\mathbf{a}||\mathbf{b}|}$$

A special case exists when $a_1b_1 + a_2b_2 + a_3b_3 = 0$
$$\cos\theta = 0$$
$$\theta = 90°$$

This is an important result.

| If $a_1b_1 + a_2b_2 + a_3b_3 = 0$ then $\theta = 90°$, and if $\theta = 90°$ then $a_1b_1 + a_2b_2 + a_3b_3 = 0$ |

Note: Two vectors are perpendicular if their scalar product is zero.

Or

| If $\mathbf{a\cdot b} = 0$ then $\mathbf{a}$ is perpendicular to $\mathbf{b}$, and if $\mathbf{a}$ is perpendicular to $\mathbf{b}$ then $\mathbf{a\cdot b} = 0$ |

Example 16

Find the angle between the two vectors $\begin{pmatrix} 1 \\ -1 \\ 1 \end{pmatrix}$ and $\begin{pmatrix} 2 \\ 1 \\ 4 \end{pmatrix}$.

$\cos\theta = \dfrac{a_1b_1 + a_2b_2 + a_3b_3}{|\mathbf{a}||\mathbf{b}|}$ ← Use the standard result.

$\qquad = \dfrac{(1\times2) + (-1\times1) + (1\times4)}{\sqrt{1^2 + (-1)^2 + 1^2}\;\sqrt{2^2 + 1^2 + 4^2}}$ ← Substitute the values.

$\qquad = \dfrac{5}{\sqrt{3}\sqrt{21}}$

$\theta = 51.0°$ (1 d.p.)   ← Give angle answers in degrees to 1 decimal place.

Example 17

a) Find the value of p if the vectors $p\mathbf{i} - 3\mathbf{k}$ and $2\mathbf{i} + \mathbf{j} + 5\mathbf{k}$ are perpendicular.

b) Show that the vectors $q\mathbf{i} - \mathbf{j} + \mathbf{k}$ and $\mathbf{i} + (q + 1)\mathbf{j} + \mathbf{k}$ are perpendicular.

a) The vectors are perpendicular if the scalar product = 0   ← $a_1b_1 + a_2b_2 + a_3b_3 = 0$

$\qquad (p \times 2) + (0 \times 1) + (-3 \times 5) = 0$

$\qquad\qquad\qquad\qquad 2p - 15 = 0$ ← Simplify and solve the equation.

$\qquad\qquad\qquad\qquad\qquad p = \dfrac{15}{2}$

b) Scalar product $= (q \times 1) + (-1 \times (q + 1)) + (1 \times 1)$ ← Use $\mathbf{a\cdot b} = a_1b_1 + a_2b_2 + a_3b_3$.

$\qquad\qquad\qquad = q - q - 1 + 1 = 0$

Therefore the vectors are perpendicular.   ← Write your conclusion.

Note: This means that the two vectors are perpendicular whatever the value of q is.

Example 18

Prove that the lines $r_1 = (2t+1)i + (1-t)j + (1-t)k$ and $r_2 = (3+\mu)i - (2\mu+1)j + (4\mu+2)k$ are perpendicular.

- -

The equations of the lines can be written as

$\qquad r_1 = i + j + k + t(2i - j - k)$ ⟵   Write each equation in the form $r = a + tb$.
$\qquad r_2 = 3i - j + 2k + \mu(i - 2j + 4k)$

But $(2i - j - k)\cdot(i - 2j + 4k)$

$\qquad = (2 \times 1) + (-1 \times -2) + (-1 \times 4)$ ⟵   Find the scalar product of the direction vectors.
$\qquad = 2 + 2 - 4$
$\qquad = 0$

The scalar product is 0 so the lines are perpendicular.

Note: These two lines are perpendicular but happen to be skew.

Exercise 9.6

1. Find the angle between the following pairs of vectors.
   a) $i + j + k$, $i + 3j + 2k$ b) $2i + j - 3k$, $-i + 2j - 3k$
   c) $2i + 5j - 2k$, $i + 4j + 11k$ d) $i + 2k$, $j + k$
   e) $2i + j - k$, $i - 2j + k$ f) $i - j + 2k$, $-2i + 2j - 4k$

2. Find the angle between the following pairs of vectors.

   a) $\begin{pmatrix} 3 \\ -1 \\ 4 \end{pmatrix}$, $\begin{pmatrix} 0 \\ 0 \\ 4 \end{pmatrix}$ b) $\begin{pmatrix} 2 \\ 5 \\ 0 \end{pmatrix}$, $\begin{pmatrix} 3 \\ -1 \\ 0 \end{pmatrix}$ c) $\begin{pmatrix} 1 \\ 2 \\ 3 \end{pmatrix}$, $\begin{pmatrix} 6 \\ -1 \\ -2 \end{pmatrix}$ d) $\begin{pmatrix} p \\ 2p \\ 5p \end{pmatrix}$, $\begin{pmatrix} 2p \\ 4p \\ 10p \end{pmatrix}$

3. Find which of the following vectors are perpendicular to each other.

   $a = 3i + 6j$ $b = i - 3j - 4k$ $c = 7i + 4j + 5k$ $d = 4i - 2j + 5k$

4. Find the value of p for which the vectors $i - 2j + k$ and $3i + 4j + pk$ are perpendicular to each other.

5. The position vectors of the points A and B, relative to origin O, are given by

   $\overrightarrow{OA} = \begin{pmatrix} 3 \\ 2 \\ 0 \end{pmatrix}$ and $\overrightarrow{OB} = \begin{pmatrix} 2 \\ k \\ 0 \end{pmatrix}$.

   Find the value of k so that
   a) $\overrightarrow{OA}$ and $\overrightarrow{OB}$ are perpendicular   b) $\overrightarrow{OA}$ and $\overrightarrow{OB}$ are parallel.

6. Find the angles in the triangle ABC where the coordinates of A, B, and C are $(1, 0, 1)$, $(2, 2, -2)$, and $(-3, 0, -5)$.

7. Show that the lines $r = \begin{pmatrix} 2 \\ 1 \\ 4 \end{pmatrix} + t \begin{pmatrix} 3 \\ 6 \\ -2 \end{pmatrix}$ and $r = \begin{pmatrix} 7 \\ 21 \\ 3 \end{pmatrix} + \mu \begin{pmatrix} 2 \\ -1 \\ 0 \end{pmatrix}$ are perpendicular and skew.

9.7 The angle between two straight lines

In section 9.6 we learned how to find the angle between two vectors $\mathbf{a} = \begin{pmatrix} a_1 \\ a_2 \\ a_3 \end{pmatrix}$ and $\mathbf{b} = \begin{pmatrix} b_1 \\ b_2 \\ b_3 \end{pmatrix}$ using the result $\cos\theta = \dfrac{a_1b_1 + a_2b_2 + a_3b_3}{|\mathbf{a}||\mathbf{b}|}$.

We now use this result to find the angle between two lines.

> The angle between two straight lines is defined as the angle between their direction vectors.

Example 19

Two lines have equations $\mathbf{r} = \begin{pmatrix} 2 \\ 0 \\ 1 \end{pmatrix} + t\begin{pmatrix} 5 \\ 3 \\ -1 \end{pmatrix}$ and $\mathbf{r} = \begin{pmatrix} 6 \\ 7 \\ -1 \end{pmatrix} + \mu\begin{pmatrix} 1 \\ -5 \\ -3 \end{pmatrix}$.

Find the acute angle between the two lines.

....

$\begin{pmatrix} 5 \\ 3 \\ -1 \end{pmatrix}$ and $\begin{pmatrix} 1 \\ -5 \\ -3 \end{pmatrix}$ give the directions of the lines.

$\cos\theta = \dfrac{(5 \times 1) + (3 \times -5) + (-1 \times -3)}{\sqrt{5^2 + 3^2 + (-1)^2}\ \sqrt{1^2 + (-5)^2 + (-3)^2}}$ ◀—— Use the standard result to find $\cos\theta$.

$= -0.2$ ◀—— A negative cosine results in an obtuse angle.

$\theta = 101.5°$ (to 1 d.p.)

Acute angle between these lines $= 180 - 101.5°$

$= 78.5°$ (1 d.p.)   ◀—— This is the acute angle required.

Note that it can be shown that these two lines do not intersect. The angle between two skew lines is, nevertheless, still defined as the angle between their directions.

Exercise 9.7

1. Find the acute angle between the two straight lines.

   a) $\mathbf{r} = \begin{pmatrix} 5 \\ 1 \\ -1 \end{pmatrix} + t\begin{pmatrix} 2 \\ 1 \\ 5 \end{pmatrix}$, $\mathbf{r} = \begin{pmatrix} 13 \\ -6 \\ 2 \end{pmatrix} + \mu\begin{pmatrix} -3 \\ 4 \\ 1 \end{pmatrix}$

b) $\mathbf{r} = \begin{pmatrix} 1 \\ 2 \\ 0 \end{pmatrix} + s\begin{pmatrix} -2 \\ -1 \\ 3 \end{pmatrix}$, $\mathbf{r} = \begin{pmatrix} 1 \\ 0 \\ -1 \end{pmatrix} + t\begin{pmatrix} 1.5 \\ 4 \\ 3 \end{pmatrix}$

c) $\mathbf{r} = \begin{pmatrix} 0 \\ 5 \\ 2 \end{pmatrix} + t\begin{pmatrix} 1 \\ 2 \\ 3 \end{pmatrix}$, $\mathbf{r} = \begin{pmatrix} -2 \\ 3 \\ 1 \end{pmatrix} + \mu\begin{pmatrix} 2 \\ -1 \\ 4 \end{pmatrix}$

d) $\mathbf{r} = s(\mathbf{i} - \mathbf{j} + \mathbf{k})$, $\mathbf{r} = (4\mathbf{i} + \mathbf{j}) + t(\mathbf{i} + 2\mathbf{j} + 3\mathbf{k})$

e) $\mathbf{r} = (\mathbf{i} - \mathbf{j} + \mathbf{k}) + t(\mathbf{i} - \mathbf{j} + \mathbf{k})$, $\mathbf{r} = \mu(2\mathbf{i} - 3\mathbf{j} + \mathbf{k})$

2. Show that the following two lines are perpendicular.

$\mathbf{r} = \begin{pmatrix} 2 \\ -1 \\ 6 \end{pmatrix} + \lambda\begin{pmatrix} -2 \\ -4 \\ 5 \end{pmatrix}$ and $\mathbf{r} = \begin{pmatrix} 6 \\ 1 \\ -1 \end{pmatrix} + \mu\begin{pmatrix} -11 \\ 3 \\ -2 \end{pmatrix}$

Hint: Show that
$a_1b_1 + a_2b_2 + a_3b_3 = 0$

3. A square-based pyramid $OABCD$ is shown in the diagram.
   Find the angles between the following lines.

   a) OD and AD **b)** CD and AD **c)** OB and BD

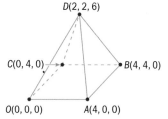

9.8 The distance from a point to a line

The distance from a point to a line is defined as the shortest distance, that is, the distance along a direction perpendicular to the line.

Let Q be the foot of the perpendicular from the point P to the line l, which has equation $\mathbf{r} = \mathbf{a} + t\mathbf{b}$.

To find the distance from point P to line l, the coordinates of Q are found by using the fact that PQ is perpendicular to l. The direction vector from P to any point on the line can be found in terms of t, then the scalar product between this vector and the direction of l can be equated to 0 in order to find the value of t for point Q, the foot of the perpendicular from P to the line.

This value of t can then be substituted into $\mathbf{r} = \mathbf{a} + t\mathbf{b}$ to find the position vector of Q and, hence, the distance PQ.

This process is demonstrated in Example 20.

Example 20

Find the distance of the point $P(3, 3, 1)$ from the line l whose equation is $\mathbf{r} = \begin{pmatrix} -4 \\ 3 \\ 0 \end{pmatrix} + t \begin{pmatrix} 1 \\ 2 \\ -4 \end{pmatrix}$.

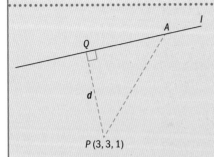

Any point on the line, say Q, has position vector

$$\mathbf{q} = \begin{pmatrix} -4+t \\ 3+2t \\ -4t \end{pmatrix}.$$

> Find the general position vector of a point on the line.

The direction of l is $\begin{pmatrix} 1 \\ 2 \\ -4 \end{pmatrix}$

For Q to be the foot of the perpendicular from P to the line,

$$\overrightarrow{PQ} \cdot \begin{pmatrix} 1 \\ 2 \\ -4 \end{pmatrix} = 0$$

But $\overrightarrow{PQ} = \mathbf{q} - \mathbf{p} = \begin{pmatrix} -4+t \\ 3+2t \\ -4t \end{pmatrix} - \begin{pmatrix} 3 \\ 3 \\ 1 \end{pmatrix} = \begin{pmatrix} -7+t \\ 2t \\ -4t-1 \end{pmatrix}$

> Find $\overrightarrow{PQ}$ in terms of t.

$$\begin{pmatrix} -7+t \\ 2t \\ -4t-1 \end{pmatrix} \cdot \begin{pmatrix} 1 \\ 2 \\ -4 \end{pmatrix} = 0$$

> Substitute for $\overrightarrow{PQ}$ in $\overrightarrow{PQ} \cdot \begin{pmatrix} 1 \\ 2 \\ -4 \end{pmatrix} = 0$.

$$(-7+t) \times 1 + (2t) \times 2 + (-4t-1) \times (-4) = 0$$
$$-7 + t + 4t + 16t + 4 = 0$$
$$t = \frac{1}{7}$$

> Calculate the value of t.

▶ Continued on the next page

$$\text{So } \overrightarrow{PQ} = \begin{pmatrix} -7+\dfrac{1}{7} \\ 2 \times \dfrac{1}{7} \\ -4 \times \dfrac{1}{7} - 1 \end{pmatrix} = \begin{pmatrix} -\dfrac{48}{7} \\ \dfrac{2}{7} \\ -\dfrac{11}{7} \end{pmatrix} = \dfrac{1}{7}\begin{pmatrix} -48 \\ 2 \\ -11 \end{pmatrix}$$ ⟵ Substitute for t in $\overrightarrow{PQ}$.

$$PQ = \dfrac{1}{7}\sqrt{(-48)^2 + 2^2 + (-11)^2} = \dfrac{\sqrt{2429}}{7}$$ ⟵ Calculate the length of PQ.

Note: If we want to find the position vector or coordinates of Q, we can substitute

$$t = \dfrac{1}{7} \text{ into } \mathbf{q} = \begin{pmatrix} -4+t \\ 3+2t \\ -4t \end{pmatrix}.$$

Exercise 9.8

1. The line l passes through the points $A(3, -1, 2)$ and $B(2, 0, 2)$.
   a) Find a vector equation for l.
   b) C is the point $(9, 1, -6)$. Find the coordinates of the point D on l such that CD is perpendicular to l.
   c) Calculate the length of CD.

2. Find the length of the perpendicular from the origin to the following lines.
   a) $\mathbf{r} = \mathbf{i} + \mathbf{j} + \mathbf{k} + t(2\mathbf{i} + 2\mathbf{j} - \mathbf{k})$

   b) $\mathbf{r} = \begin{pmatrix} 2 \\ 1 \\ 2 \end{pmatrix} + t\begin{pmatrix} 3 \\ 4 \\ -5 \end{pmatrix}$

3. Find the distance from the point $(3, -1, -2)$ to the line $\mathbf{r} = (-4\mathbf{i} + 3\mathbf{j}) + t(\mathbf{i} + 2\mathbf{j} - 4\mathbf{k})$.

4. Find the length of the perpendicular from the point with position vector $3\mathbf{i} + 2\mathbf{j} - 6\mathbf{k}$ to the line $\mathbf{r} = t(2\mathbf{i} - \mathbf{j} + 2\mathbf{k})$.

5. In the diagram, *OABCDEFG* is a cuboid.
The unit vectors **i**, **j**, and **k** are in the directions
OC, *OE*, and *OA* respectively.
Point *G* has coordinates (4, 5, 2).

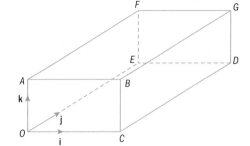

a) Find a vector equation for the line *CF*.

b) Find the shortest distance from
point *D* to the line *CF*.

6. *P* is the point (2, −1, 6). *Q* is the point (3, −2, 7).

a) Find a vector equation for the line *PQ*.

R is the point (−1, 2, 3).

b) Show that the line *OR* is perpendicular to the line *PQ*.

c) Find the distance from the point *P* to the line *OR*.

Summary exercise 9

1. Find the magnitude of the vector $\begin{pmatrix} 2 \\ -1 \\ -3 \end{pmatrix}$.

2. Find the value of k if the vectors $k\mathbf{i} + 4\mathbf{j}$ and $-2\mathbf{i} + 6\mathbf{j}$ are parallel.

3. The sum of the vectors $3\mathbf{i} + 2\mathbf{j} - 4\mathbf{k}$, $p\mathbf{i} - 5\mathbf{j} + 3\mathbf{k}$, and $6\mathbf{i} - 2\mathbf{j} + 7\mathbf{k}$ is $10\mathbf{i} - q\mathbf{j} + 6\mathbf{k}$. Find the values of p and q.

4. The two vectors $(a - 1)\mathbf{i} + (b + 2)\mathbf{j}$ and $-b\mathbf{i} + (a - 1)\mathbf{j}$ are equal. Find the values of a and b.

5. Find the unit vector which is in the direction of the vector $2\mathbf{i} - \mathbf{j} + 2\mathbf{k}$.

6. Express, in column vector form, the vector which has magnitude 10 and which has the same direction as $\begin{pmatrix} 3 \\ 0 \\ -4 \end{pmatrix}$.

7. Find a column vector which is perpendicular to both of the vectors $\begin{pmatrix} 1 \\ 3 \\ -1 \end{pmatrix}$ and $\begin{pmatrix} 3 \\ -1 \\ -1 \end{pmatrix}$.

EXAM-STYLE QUESTIONS

8. *OABCDEFG* is a cube with edge 5 cm.

a) Find in terms of **i**, **j**, and **k**

i) $\overrightarrow{OF}$ ii) $\overrightarrow{AG}$.

b) Find the angle between the diagonals *OF* and *AG*.

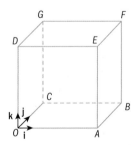

9. The position vectors of two points A and B relative to an origin O are $\mathbf{i} + \mathbf{j}$ and $\mathbf{i} + \mathbf{j} + \sqrt{2}\mathbf{k}$. Find angle AOB.

10. Column vectors v_1, v_2, v_3 are such that $2v_1 - v_2 + v_3 = 0$.

$$v_1 = \begin{pmatrix} 4 \\ -1 \\ 1 \end{pmatrix} \quad v_2 = \begin{pmatrix} 1 \\ 0 \\ -2 \end{pmatrix}$$

Find the unit vector in the direction of v_3.

EXAM-STYLE QUESTIONS

11. The diagram shows triangle ABC in which the position vectors of A, B, and C with respect to O are given by $\overrightarrow{OA} = 2\mathbf{i} - \mathbf{j} + 2\mathbf{k}$, $\overrightarrow{OB} = 3\mathbf{i} + 3\mathbf{j} - \mathbf{k}$, and $\overrightarrow{OC} = -2\mathbf{i} + \mathbf{j} - 4\mathbf{k}$. Find the angles of the triangle.

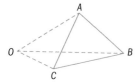

12. The quadrilateral $OABC$ shown in the diagram is such that
$\overrightarrow{OA} = \mathbf{i} + \mathbf{j} + 2\mathbf{k}$,
$\overrightarrow{OB} = 3\mathbf{i} + 5\mathbf{k}$,
$\overrightarrow{OC} = 2\mathbf{i} - \mathbf{j} + 3\mathbf{k}$.

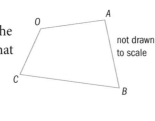

not drawn to scale

a) Find the lengths of the sides of the quadrilateral.

b) What type of quadrilateral is $OABC$?

c) Find the angles of the quadrilateral.

d) Find the distance from point A to the line CB.

13. ABC is a triangle in which the position vectors of A, B, and C are $\begin{pmatrix} 5 \\ 2 \\ -3 \end{pmatrix}$, $\begin{pmatrix} 3 \\ 1 \\ -1 \end{pmatrix}$, and $\begin{pmatrix} 6 \\ -5 \\ 1 \end{pmatrix}$ respectively.
Find the perimeter of the triangle.

EXAM-STYLE QUESTIONS

14. Relative to an origin O, the position vectors of the points A and B are given by

$$\overrightarrow{OA} = \begin{pmatrix} 1 \\ p \\ 1 \end{pmatrix} \text{ and } \overrightarrow{OB} = \begin{pmatrix} p+1 \\ 4 \\ -2 \end{pmatrix}.$$

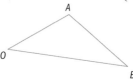

a) Find the value of p for which $\overrightarrow{OA}$ is perpendicular to $\overrightarrow{OB}$.

b) Find the size of angle OAB when $p = -1$.

c) Find the distance from point A to the line which passes through O and B in the case where $p = -1$.

15. With O as origin, the position vectors of the points A and B are $2\mathbf{i} + 4\mathbf{j} + 7\mathbf{k}$ and $-4\mathbf{i} + \mathbf{j} + \mathbf{k}$ respectively.
Find the position vector of the point X on AB which is such that $2\overrightarrow{AX} = \overrightarrow{XB}$.

16. Find the angle which the vector $\mathbf{i} - 2\mathbf{j} + \mathbf{k}$ makes with
a) the x-axis b) the y-axis c) the z-axis.

17. A quadrilateral $ABCE$ has coordinates $A(1, 0, 2)$, $B(-3, -2, 0)$, $C(-3, 2, c)$, and $D(-1, 3, 3)$.
AB is parallel to CD.
a) Find the value of c.
b) Find the unit column vector which is in the direction of BD.

18. The position vectors of points P and Q relative to an origin O are $\begin{pmatrix} 1 \\ 2 \\ 4 \end{pmatrix}$ and $\begin{pmatrix} 2 \\ 4 \\ k \end{pmatrix}$ respectively.

a) Find the value of k for which $\overrightarrow{PQ}$ is

perpendicular to the vector $\begin{pmatrix} -1 \\ 2 \\ -3 \end{pmatrix}$.

b) For the value of k found in part **(a)** find the size of angle POQ.

19. The diagram shows a triangular prism $ABCDEF$.

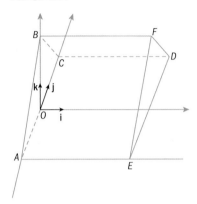

The position vectors of A, B, C, and D with respect to an origin O are given by $\overrightarrow{OA} = -3\mathbf{j}$, $\overrightarrow{OB} = 4\mathbf{k}$, $\overrightarrow{OC} = 3\mathbf{j}$, and $\overrightarrow{OD} = 8\mathbf{i} + 3\mathbf{j}$ respectively.

a) Find the lengths of the edges of the prism.

b) Find the size of angle FAD.

20. The position vectors of points P and Q relative to an origin O are $2\mathbf{i} + \mathbf{j} - 2\mathbf{k}$ and $-4\mathbf{i} + 2\mathbf{j} + 4\mathbf{k}$ respectively. See the diagram.

a) Find unit vectors in the directions of OP and OQ.

b) Find the unit vector in the direction of the line which bisects angle POQ.

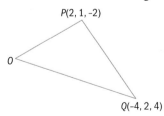

P(2, 1, -2)

O

Q(-4, 2, 4)

21. a) Find the vector equation of the line joining $P(1, 2, -1)$ and $Q(-1, 0, -1)$.

b) Find the coordinates of the foot of the perpendicular from the origin to line PQ.

22. Find the cosine of the acute angle between

the line $\mathbf{r} = \begin{pmatrix} 1 \\ 0 \\ 0 \end{pmatrix} + t \begin{pmatrix} -4 \\ 3 \\ -5 \end{pmatrix}$ and the vector $\begin{pmatrix} 1 \\ 1 \\ 0 \end{pmatrix}$.

EXAM-STYLE QUESTIONS

23. The three lines l_1, l_2, and l_3 have equations

$$\mathbf{r} = \begin{pmatrix} 7 \\ 3 \\ -3 \end{pmatrix} + \lambda \begin{pmatrix} 3 \\ -2 \\ 1 \end{pmatrix}, \mathbf{r} = \begin{pmatrix} 7 \\ -2 \\ 4 \end{pmatrix} + \mu \begin{pmatrix} -2 \\ 1 \\ -1 \end{pmatrix}, \quad \text{and}$$

$$\mathbf{r} = \begin{pmatrix} 1 \\ 0 \\ 0 \end{pmatrix} + v \begin{pmatrix} 0 \\ 1 \\ -1 \end{pmatrix}$$

a) Show that l_1 and l_2 intersect and find the point of intersection.

b) Show that l_1 and l_3 do not intersect and find the acute angle between them.

24. a) Find the point of intersection of the line through the points $(2, 0, 1)$ and $(-1, 3, 4)$ with the line through the points $(-1, 3, 0)$ and $(4, -2, 5)$.

b) Calculate the acute angle between the lines.

25. The point A lies on the line which is parallel to the vector $2\mathbf{i} + \mathbf{j} - \mathbf{k}$ and which passes through the point $(1, 1, 2)$. The point B lies on a different line which is parallel to the vector $\mathbf{i} + \mathbf{j} - 2\mathbf{k}$ and which passes through the point $(1, 1, 4)$. The line AB is perpendicular to both of these lines. Find the equation of the line AB.

26. With respect to the origin O, the lines l and m have equations

$\mathbf{r} = 3\mathbf{i} + 6\mathbf{j} + \mathbf{k} + s(2\mathbf{i} + 3\mathbf{j} - \mathbf{k})$ and
$\mathbf{r} = 3\mathbf{i} - \mathbf{j} + 4\mathbf{k} + t(\mathbf{i} - 2\mathbf{j} + \mathbf{k})$

respectively.

   i) Show that l and m intersect and find the coordinates of the point of intersection.

   ii) Find the angle between the two straight lines l and m.

   iii) Find the distance between the origin and line l.

27. The points A, B, and C have position vectors, relative to the origin O, given by

$$\overrightarrow{OA} = \begin{pmatrix} 2 \\ -2 \\ 1 \end{pmatrix}, \overrightarrow{OB} = \begin{pmatrix} 4 \\ 2 \\ 1 \end{pmatrix}, \overrightarrow{OC} = \begin{pmatrix} 1 \\ 1 \\ 3 \end{pmatrix}.$$

   i) Find angle ACB.

   ii) Find a vector equation for the line CM, where M is the midpoint of the line AB.

   iii) Show that AB and CM are perpendicular.

   iv) Find the position vector of the point N, the foot of the perpendicular from O to CM.

28.

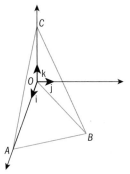

The diagram shows a tetrahedron $OABC$. The edges $\overrightarrow{OA}$ and $\overrightarrow{OC}$ have length 8 units and 6 units respectively. The position vector of B is $5\mathbf{i} + 4\mathbf{j}$.

   i) Express each of the vectors $\overrightarrow{BC}$ and $\overrightarrow{BA}$ in terms of $\mathbf{i}$, $\mathbf{j}$, and $\mathbf{k}$.

   ii) Use a vector product to calculate angle ABC.

   iii) Find the length of the perpendicular from C to AB.

29. With respect to the origin O, the points A and B have position vectors given by

$$\overrightarrow{OA} = \begin{pmatrix} 5p \\ -p \\ -p \end{pmatrix} \text{ and } \overrightarrow{OB} = \begin{pmatrix} p \\ -5p \\ 7p \end{pmatrix}$$

Where p is a constant.

   i) Find a vector equation for the line AB.

   ii) Show that the point C $(4p, -2p, p)$ lie on AB.

   iii) Show that OC is perpendicular to AB.

   iv) Find the position vector of the point D such that $OD = OA$.

30. The straight line **l** has equation $\mathbf{r} = \mathbf{i} + 4\mathbf{j} + \lambda(\mathbf{i} + \mathbf{j} + \mathbf{k})$, relative to an origin O. The point P has coordinates $(2, 3, -1)$.

   i) Find the coordinates of the foot of the perpendicular from P to l.

   ii) Find the distance from P to l.

Chapter summary

Vectors and scalars

- A vector has both magnitude and direction.
- A scalar has magnitude only.

Notation

- Vectors written as $\begin{pmatrix} x \\ y \\ z \end{pmatrix}$ are said to be in column vector form.
- Vectors written as $x\mathbf{i} + y\mathbf{j} + z\mathbf{k}$ are said to be in unit vector form.

Types of vector

- A displacement vector indicates a movement from one point to another.
- A position vector indicates a movement from the origin to a point.
- If A and B are points with position vectors $\mathbf{a}$ and $\mathbf{b}$ then $\overrightarrow{AB} = \mathbf{b} - \mathbf{a}$.

Properties of vectors

- Two vectors are equal if they have the same magnitude and direction.
- Two vectors are parallel if they are scalar multiples of each other.
- A negative vector has the same magnitude as the positive vector but the opposite direction.
- In three dimensions, the magnitude of the vector $\begin{pmatrix} x \\ y \\ z \end{pmatrix}$ or $x\mathbf{i} + y\mathbf{j} + z\mathbf{k}$ is given by $\sqrt{x^2 + y^2 + z^2}$.
- A unit vector is a vector of magnitude 1. The unit vector in the direction of $\mathbf{a}$ is $\dfrac{\mathbf{a}}{|\mathbf{a}|}$.

The scalar product of two vectors

- The scalar product is defined by $\mathbf{a} \cdot \mathbf{b} = |\mathbf{a}||\mathbf{b}|\cos\theta$ where θ is the angle between $\mathbf{a}$ and $\mathbf{b}$.
- The scalar product is usually found by using $\mathbf{a} \cdot \mathbf{b} = a_1 b_1 + a_2 b_2 + a_3 b_3$, where $\mathbf{a} = \begin{pmatrix} a_1 \\ a_2 \\ a_3 \end{pmatrix}$, $\mathbf{b} = \begin{pmatrix} b_1 \\ b_2 \\ b_3 \end{pmatrix}$.
- The angle between two vectors can be found by using the result $\cos\theta = \dfrac{a_1 b_1 + a_2 b_2 + a_3 b_3}{|\mathbf{a}||\mathbf{b}|}$.
- The scalar product of two vectors is zero if the vectors are perpendicular, i.e. if $\mathbf{a} \cdot \mathbf{b} = 0$ then $\mathbf{a}$ is perpendicular to $\mathbf{b}$.

Straight lines

- The vector equation of a straight line passing through a point with position vector $\mathbf{a}$ and direction vector $\mathbf{b}$ is $\mathbf{r} = \mathbf{a} + t\mathbf{b}$, where t is a scalar.
- Two lines intersect if they have one point in common.
- Two lines are parallel if they have the same direction but no point in common.
- Two lines are coincident if they have the same direction and an infinite number of points in common.
- In three dimensions, skew lines are lines that are not parallel but do not intersect.
- The angle between two straight lines is defined as the angle between their direction vectors.

Distance from a point to a line

- To find the shortest distance from a point to a line, first find the coordinates of the foot of the perpendicular from the point to the line.

10 Differential equations

A differential equation is an equation which relates a variable, y, to its derivatives, $\dfrac{dy}{dx}$ or sometimes $\dfrac{d^2 y}{dx^2}$. The behaviour of many real-life systems in both nature and technology can be modelled by differential equations. Finding ways to produce a function which satisfies the relationship in the differential equation has resulted in many scientific advances in recent times. For example, the Navier–Stokes non-linear partial differential equations are used to describe the motion of liquids and gases. As such, these equations are incredibly useful when predicting how a liquid or gas will behave in a wide range of fields – from improving the aerodynamic design of a car, to improving the effective delivery of drugs through the bloodstream.

Objectives

- Formulate a simple statement involving a rate of change as a differential equation, including the introduction if necessary of a constant of proportionality.
- Find by integration a general form of solution for a first-order differential equation in which the variables are separable.
- Use an initial condition to find a particular solution.
- Interpret the solution of a differential equation in the context of a problem being modelled by the equation.

Before you start

You should know how to:

1. Work out the value of c, given a curve and point on the curve,

   e.g. a) $y = 2x + c$ $(-1, 4)$
   $4 = 2(-1) + c$ $c = 6$

   b) $y = e^x + \tan x + c$ $(0, -1)$
   $-1 = 1 + 0 + c$ $c = -2$

Skills check:

1. Work out the value of c, given the equation of a curve and the coordinates of a point on the curve.

   a) $y = -2x + c$ $(3, -2)$
   b) $y = 6x^2 - x + c$ $(-4, -3)$
   c) $y = e^x + c$ $(0, -2)$
   d) $y = \ln x + c$ $(e, 3)$

2. Use trigonometric identities,

   e.g. Express $\sin^3\theta\cos^2\theta$ in terms of $\sin\theta$ only.

   $$\sin^3\theta\cos^2\theta = \sin^3\theta(1-\sin^2\theta)$$
   $$= \sin^3\theta - \sin^5\theta$$

3. Decompose rational functions into partial fractions,

   e.g. $\dfrac{x+3}{x^2+2x}$

   $$\frac{x+3}{x^2+2x} \equiv \frac{A}{x} + \frac{B}{x+2} \Rightarrow A+B=1 \text{ and } 2A=3$$

   $$\Rightarrow A = \frac{3}{2}, \ B = -\frac{1}{2}$$

2. Express $\cos 2\theta$ in terms of $\sin\theta$ only.

3. Express $\dfrac{2}{x^2-1}$ in partial fractions.

10.1 Forming simple differential equations (DEs)

A differential equation (DE) is an equation in which at least one differential expression appears.

In Chapter 5, and previously in P1, you met the simplest differential equation, $\dfrac{dy}{dx} = f(x)$. This has a solution $y = \displaystyle\int f(x)\,dx$, and progressively you learned how to integrate more functions, all of which were in the form $\dfrac{dy}{dx} = f(x)$.

$\dfrac{dy}{dx} = f(x)$ is a first-order differential equation because it only contains the first derivative $\dfrac{dy}{dx}$.

With these skills, you will be able to solve first-order DEs in which the variables are separable, i.e. DEs where the equation can be written in the form $g(y)\dfrac{dy}{dx} = f(x)$.

Note that sometimes one of the functions $f(x)$ or $g(y)$ may be 1.

Since you know that $\dfrac{dy}{dx}$ gives the gradient of the function which satisfies the DE, you can get a feel for the behaviour of the functions which satisfy the DE by sketching a 'needle diagram': we sketch a number of short line segments at certain points on the domain, each with (approximately) the gradient given by the value of $\dfrac{dy}{dx}$ at each point.

Such a needle diagram is shown here. Drawing this by hand is somewhat tedious, but thinking about the patterns is helpful.

This needle diagram shows approximately what the solutions to $\frac{dy}{dx} = 2x$ look like. The gradient is only a function of x, so the needles lying on the same vertical line (i.e. those with the same x-value) will have the same gradient. On the y-axis (where $x = 0$) we see that the gradient is 0, so the needles that cross the y-axis are horizontal. When the value of x is positive, so is the value of the gradient, and thus the needles have a positive gradient which gets steeper as x increases. When the value of x is negative, so is the value of the gradient, and thus the needles have a negative gradient which gets steeper as x decreases.

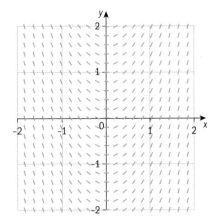

Solving the DE $\frac{dy}{dx} = 2x$ by integrating, as you already know how to do, gives a general solution $y = \int 2x \, dx = x^2 + c$. The solution is a quadratic which is symmetric about the y-axis, with an undetermined constant (which is why there is actually a family of functions which satisfy the differential relationship).

Completing a number of the paths for different values of the constant c lets you see how the needle diagram gives an impression of what the solutions look like. Assigning a specific value to c determines one particular solution. We are often given an initial condition by which we calculate c, the constant of integration.

The graph here shows three particular solutions: $y = x^2 + 1$, $y = x^2$, and $y = x^2 - 2$. However, there are an infinite number of particular solutions within the family that comprises the general solution. For first-order differential equations, each particular solution is distinct (it does not cross with another particular solution at any point).

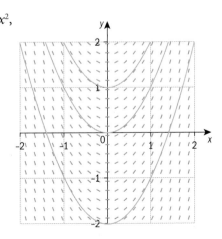

Often the mathematical description of a physical situation can be captured in a DE.

Example 1

A body falling through a fluid experiences resistance, causing it to lose speed at a rate proportional to its speed at that instant. Write down a differential equation satisfied by the speed v at time t, and sketch a needle diagram in the first quadrant to illustrate the solution to this differential equation.

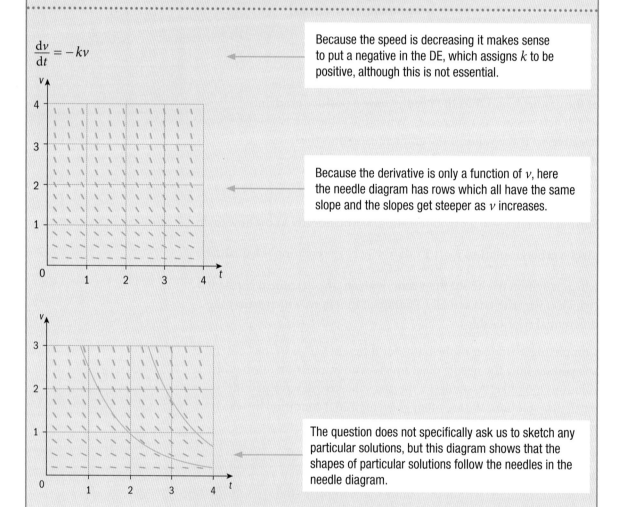

$$\frac{dv}{dt} = -kv$$

Because the speed is decreasing it makes sense to put a negative in the DE, which assigns k to be positive, although this is not essential.

Because the derivative is only a function of v, here the needle diagram has rows which all have the same slope and the slopes get steeper as v increases.

The question does not specifically ask us to sketch any particular solutions, but this diagram shows that the shapes of particular solutions follow the needles in the needle diagram.

Note: You will not be required to draw needle diagrams in examinations, but once you have some experience of drawing them you may find that thinking about what they look like – and therefore what any solution should look like – is a helpful check that your solution 'makes sense', in the same way as you estimate solutions to problems.

Example 2

Water is leaking from a tank. The rate at which the depth of water is decreasing is proportional to the square root of the current depth. Write down a differential equation satisfied by the depth h at time t, and sketch a needle diagram in the first quadrant to illustrate the solution to this differential equation.

$$\frac{\mathrm{d}h}{\mathrm{d}t} = -k\sqrt{h}$$

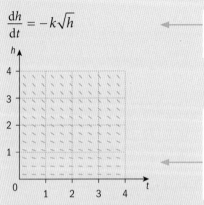

This is similar to the equation from Example 1, but the power of the variable h is different.

Because the derivative is a function of h, again the needle diagram has rows which all have the same slope. However, now the increase in gradient is much slower as h increases, since $\frac{\mathrm{d}h}{\mathrm{d}t}$ is proportional to $\sqrt{h}$, not to h.

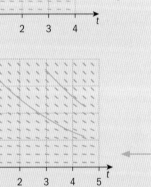

This is reflected in the particular solutions the needle diagram suggests.

Example 3

A differential equation has the form $\frac{\mathrm{d}y}{\mathrm{d}x} = xy$. Sketch a needle diagram in all four quadrants to illustrate the solution of this differential equation.

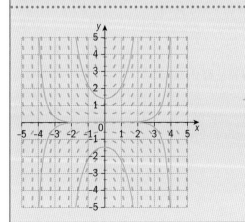

In quadrant 1, we see that the gradient is zero when either x or y is zero, and then increases as either x or y increase.

Also, the function xy has reflective symmetry about both the x- and y-axes, so we can easily draw in all four quadrants.

Example 4

A rectangular tank has a base of $5\,m^2$. Water is flowing into the tank at a constant rate of 10 litres per second ($= 0.01\,m^3 s^{-1}$). At the same time, water is leaking out at a rate proportional to $\sqrt{h}$ where h is the depth of the water at time t. When the depth is 1 metre the water is leaking out at 2 litres per second. Find the differential equation describing the rate of change of h with time t.

....

If water comes in at rate $0.01\,m^3 s^{-1}$ and the base area is $5\,m^2$, the water coming in is increasing the depth at a rate of $0.002\,m\,s^{-1}$.

So the DE is $\dfrac{dh}{dt} = 0.002t - k\sqrt{h}$.

If water is leaking out at 2 litres per second ($0.002\,m^3 s^{-1}$) then this is decreasing the depth at a rate of $\dfrac{0.002}{5} = 0.0004\,m\,s^{-1}$.

When $h = 1$, $k\sqrt{h} = 0.0004 \Rightarrow k = 0.0004$, giving $\dfrac{dh}{dt} = 0.002t - 0.0004\sqrt{h}$.

Exercise 10.1

1. The rate at which body temperature, T, falls is proportional to the difference between the body temperature T and the temperature T_0 of the surroundings. Find a differential equation relating body temperature, T, and time t.

2. A certain substance is formed in a chemical reaction. The mass of the substance formed t seconds after the start of the reaction is x grams. At any time the rate of formation of the substance is proportional to $(30 - x)$. Find a differential equation relating x and t.

3. In studying the spread of a disease, a scientist thinks that the rate of infection is proportional to the product of the number of people infected and the number of people uninfected. If N is the number infected at time t and P is the total number of people in the population, form a differential equation to summarise the scientist's theory.

4. The rate of increase of a population is proportional to its size at the time. Write down a differential equation to describe this situation.

   It also known that when the population was 2 million, the rate of increase was 140 000 per day. Find the constant of proportionality in your DE.

5. Sketch needle diagrams for the DEs in questions 1 to 3.

10.2 Solving first-order differential equations with separable variables

We saw in section 10.1 that when asked to solve $\dfrac{dy}{dx} = 2x$, we can write $y = \displaystyle\int 2x\,dx = x^2 + c$.

To be rigorous in solving this differential equation, we should write

$$\frac{dy}{dx} = 2x \Rightarrow \int 1\,dy = \int 2x\,dx \Rightarrow y = x^2 + c$$

If we need to solve a differential equation such as $\dfrac{dy}{dx} = \dfrac{f(x)}{g(y)}$, we can write

$$g(y)\frac{dy}{dx} = f(x), \text{ giving } \int g(y)\frac{dy}{dx}\,dx = \int f(x)\,dx$$

If $G(y) = \displaystyle\int g(y)\,dy$

then $\dfrac{d}{dx}\{G(y)\} = g(y)\dfrac{dy}{dx}$.

This has solution $\displaystyle\int g(y)\,dy = \int f(x)\,dx.$

> A differential equation which can be rearranged to the form
> $g(y)\dfrac{dy}{dx} = f(x)$ is said to have **separable variables**.

Example 5

Solve $\dfrac{dy}{dx} = 2y$.

This DE is similar to the one in Example 1 except here the constant of proportionality is positive. The solution here is one of exponential growth, whereas the needle diagram in Example 1 showed the solution was in the form of exponential decay.

· ·

$$\frac{dy}{dx} = 2y$$

$$\frac{1}{y}\frac{dy}{dx} = 2$$ ← Separate the variables first.

$$\int \frac{1}{y}\,dy = \int 2\,dx$$ ← Show what you are intending to do.

$$\ln y = 2x + c$$ ← Integrate both sides.

$$y = Ae^{2x}$$ ← Write the solution in its simplest form.

The constant which is added in integration will become a multiplicative constant if there is a log from the integration and you take exponentials (i.e. $A = e^c$) – remember to change the letter when you do this.

Example 6

Solve $\dfrac{dy}{dx} = 2xy$.

$$\dfrac{dy}{dx} = 2xy$$

$$\dfrac{1}{y}\dfrac{dy}{dx} = 2x$$

$$\int \dfrac{1}{y}\, dy = \int 2x\, dx$$

$$\ln y = x^2 + c$$

$$y = Ae^{x^2}$$

This DE is very similar to that in Example 5. Once the variables are separated you essentially always have two integrations to do and it can require any of the integration techniques you learned in P1 or in Chapters 5 and 8 of this book.

Example 7

Solve $\dfrac{dy}{dx} = \cos x$.

$$\dfrac{dy}{dx} = \cos x$$

$$\int 1\, dy = \int \cos x\, dx$$

$$y = \sin x + c$$

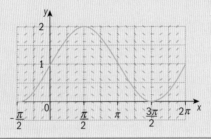

The needle diagram shows the gradient function, and the particular solution with $c = 1$ is shown in blue.

Example 8

Solve $\dfrac{dy}{dx} = \cos^2 y$.

$$\dfrac{dy}{dx} = \cos^2 y$$

$$\sec^2 y\,\dfrac{dy}{dx} = 1$$

$$\int \sec^2 y\, dy = \int 1\, dx$$

$$\tan y = x + c$$

$$y = \tan^{-1}(x + c)$$

You need to get integrable functions on both sides.

Integrate both sides.

If the functional form is very complicated then it is not a requirement to give the function of y explicitly, unless the question specifies this. Here $\tan y = x + c$ is a perfectly acceptable form of the solution to the DE.

Example 9

Solve $\dfrac{dy}{dx} = \dfrac{3x^2 \sin^2 y}{(x^3+2)}$.

. . . .

$\dfrac{dy}{dx} = \dfrac{3x^2 \sin^2 y}{(x^3+2)}$

$\mathrm{cosec}^2 y\, \dfrac{dy}{dx} = \dfrac{3x^2}{(x^3+2)}$ ←   You need to get integrable functions on both sides.

$\displaystyle\int \mathrm{cosec}^2 y\, dy = \int \dfrac{3x^2}{(x^3+2)}\, dx$

$\cot y = -\ln A(x^3+2)$, where $c = \ln A$ ←   Integrate both sides – you need to remember standard forms of integrable functions.

Example 10

Solve $\dfrac{dy}{dx} = \dfrac{2}{x^2-1}$.

. . . .

$\dfrac{dy}{dx} = \dfrac{2}{x^2-1}$

$\dfrac{dy}{dx} = \dfrac{1}{x-1} - \dfrac{1}{x+1}$ ←   You need to use partial fractions to get this into an integrable form.

$\displaystyle\int dy = \int \left(\dfrac{1}{x-1} - \dfrac{1}{x+1}\right) dx$

$y = \ln(x-1) - \ln(x+1) + c$ ←   Here we express c as $\ln A$ in order to combine the terms.

$y = \ln\left\{\dfrac{A(x-1)}{(x+1)}\right\}$

In examinations, you will normally be asked to first express a rational function in partial fractions, and then integrate the partial fractions to solve the DE.

Example 11

Solve $\dfrac{dy}{dx} = y^2 \ln x$.

. . . .

$\dfrac{dy}{dx} = y^2 \ln x$

$\dfrac{1}{y^2}\dfrac{dy}{dx} = \ln x$ ←   Separate the variables.

$\displaystyle\int \dfrac{1}{y^2}\, dy = \int \ln x\, dx$ ←   You will need to integrate this by parts.

$-\dfrac{1}{y} = x\ln x - \displaystyle\int x\dfrac{1}{x}\, dx = x\ln x - x + c$ Remember, you may be asked to use any of the integration techniques you met in Chapter 8 in solving DEs. After separating the variables you always have two integrations to perform.

Exercise 10.2

Find the general solution to the following differential equations.

1. $\dfrac{dy}{dx} = \dfrac{x}{y}$

2. $\dfrac{dy}{dx} = \dfrac{x^3}{y}$

3. $\dfrac{dy}{dx} = \dfrac{x^2 + x + 1}{y^2}$

4. $\dfrac{dy}{dx} = y(x + 3)$

5. $\dfrac{dy}{dx} = \dfrac{2}{\cos y}$

6. $\dfrac{dy}{dx} = \dfrac{1}{xy}$

7. $2y\dfrac{dy}{dx} = 4x^3$

8. $x^2\dfrac{dy}{dx} = 2y^3$

9. $\dfrac{dy}{dx} = y\cos x$

10. $\dfrac{dy}{dx} = 2xe^y$

11. $\dfrac{dy}{dx} = 2x\sec y$

12. $\dfrac{dy}{dx} = \dfrac{y^2}{x(x+2)}$

13. $\dfrac{dy}{dx} = \dfrac{\cos x}{\cos y}$

14. $\dfrac{dy}{dx} = \dfrac{\cos^2 y}{x}$

15. $2\dfrac{dy}{dx} = 4x^3(y^2 - 1)$

16. $\dfrac{dy}{dx} = \dfrac{\cos^2 y}{\cos^2 x}$

17. $\dfrac{dy}{dx} = xe^x \sec y$

18. $2\dfrac{dy}{dx} = \dfrac{y^2 - 1}{x^2 + x}$

19. $\dfrac{dy}{dx} = y^2 \sec^2 x$

20. $\dfrac{dy}{dx} = \dfrac{x(1 + y^4)}{y^3}$

21. $\dfrac{dy}{dx} = \left(\dfrac{x+3}{x^2 + x}\right)y^2$

22. $\left(\dfrac{-2}{y^3}\right)\dfrac{dy}{dx} = \dfrac{e^x}{1 + e^{2x}}$ Use the substitution $u = e^x$.

23. $y\dfrac{dy}{dx} = \sin^3 x\cos^2 x$

24. $y\dfrac{dy}{dx} = \dfrac{27}{(9 + x^2)}$ Use the substitution $x = 3\tan\theta$.

10.3 Finding particular solutions to differential equations

In the last section, the only information you were given was the differential equation, and you could find the general solution: a family of solutions, each member of which is determined by the value of the constant of integration. If you also have a condition which gives a point on the solution curve, then the particular value of the constant of integration can be determined. Such a condition is called an **initial condition**, or sometimes a 'boundary condition'.

> Essentially the DE gives the gradient function at any point in the plane. If you have an initial condition – a starting point – and know what direction to head in, then you can move along the whole function path.

The first two examples use DEs for which we have already produced the general solution, so you can see the extra work required in order to also produce a particular solution.

Example 12

Find the curve which satisfies $\dfrac{dy}{dx} = 2y$ and passes through $(0, 3)$.

$\dfrac{dy}{dx} = 2y$

$\Rightarrow y = Ae^{2x}$ ← From Example 5: this is the general solution.

$x = 0, y = 3 \Rightarrow A = 3$ ← Use the initial condition to find A.

$\Rightarrow y = 3e^{2x}$ ← This is the particular solution.

Example 13

Find the curve which satisfies $\dfrac{dy}{dx} = \dfrac{2}{x^2 - 1}$ and passes through $(2, 0)$.

$\dfrac{dy}{dx} = \dfrac{2}{x^2 - 1}$ ← From Example 10

$y = \ln\left\{\dfrac{A(x - 1)}{(x + 1)}\right\}$

$x = 2, y = 0 \Rightarrow 0 = \ln\left(\dfrac{A}{3}\right) \Rightarrow A = 3$

$y = \ln\left\{\dfrac{3(x - 1)}{(x + 1)}\right\}$ The needle diagram shows the DE with the particular solution through $(2, 0)$.

Example 14

Find the curve which satisfies $\dfrac{dy}{dx} = \dfrac{4xy}{(x^2 + 3)}$ and passes through $(0, 9)$.

$\dfrac{dy}{dx} = \dfrac{4xy}{(x^2 + 3)}$

$\Rightarrow \dfrac{1}{y}\dfrac{dy}{dx} = \dfrac{4x}{(x^2 + 3)}$ ← Separate the variables.

$\Rightarrow \displaystyle\int \dfrac{1}{y}\, dy = \int \dfrac{4x}{(x^2 + 3)}\, dx$ ← Integrate both sides.

$\Rightarrow \ln y = 2\ln(x^2 + 3) + c = \ln A(x^2 + 3)^2$

$\Rightarrow y = A(x^2 + 3)^2$

$\Rightarrow x = 0, y = 9 \Rightarrow A = 1$ ← Use the initial condition to find A.

$\Rightarrow y = (x^2 + 3)^2$

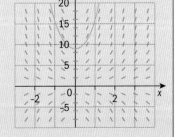

Example 15

Find the curve which satisfies $\dfrac{dy}{dx} = \sin x \sec y$ and passes through $(0, 0)$.

$$\frac{dy}{dx} = \sin x \sec y$$

$$\cos y \, \frac{dy}{dx} = \sin x$$

$$\int \cos y \, dy = \int \sin x \, dx$$

$$\Rightarrow \sin y = -\cos x + c$$

$$x = 0, y = 0 \Rightarrow 0 = -1 + c \Rightarrow c = 1$$

$$\Rightarrow \sin y = -\cos x + 1$$

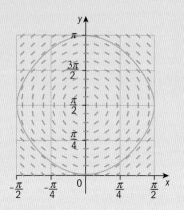

The needle diagram shows the DE with the particular solution through $(0, 0)$. Often the solution curves for DEs are very hard to draw without using a computer package.

Example 16

Find the curve which satisfies $\dfrac{dy}{dx} = xy^2 e^{2x}$ and passes through $(0, 1)$.

$$\frac{dy}{dx} = xy^2 e^{2x}$$

$$\frac{1}{y^2} \frac{dy}{dx} = x e^{2x}$$

Separate the variables.

$$\int \frac{1}{y^2} \, dy = \int x e^{2x} \, dx$$

This will need integration by parts.

$$\Rightarrow -\frac{1}{y} = x \frac{1}{2} e^{2x} - \int \frac{1}{2} e^{2x} \, dx$$

$$\Rightarrow -\frac{1}{y} = \frac{1}{2} x e^{2x} - \frac{1}{4} e^{2x} + c$$

Remember the constant at the end here. This is the general solution.

$$x = 0, y = 1 \Rightarrow -1 = 0 - \frac{1}{4} + c \Rightarrow c = -\frac{3}{4}$$

Use the initial condition to find c.

$$\Rightarrow -\frac{1}{y} = \frac{1}{2} x e^{2x} - \frac{1}{4} e^{2x} - \frac{3}{4} = \frac{-(3 + e^{2x} - 2x e^{2x})}{4}$$

$$\Rightarrow y = \frac{4}{(3 + e^{2x} - 2x e^{2x})}$$

Reorganise to find y explicitly. This is the particular solution.

Example 17

Find the curve which satisfies $y^2 \dfrac{dy}{dx} = x^2 e^{x^3}$ and passes through the origin.

$y^2 \dfrac{dy}{dx} = x^2 e^{x^3}$

⟵ The variables are already separated here.

$\Rightarrow \displaystyle\int y^2 \, dy = \int x^2 e^{x^3} \, dx$

$\Rightarrow \dfrac{1}{3} y^3 = \dfrac{1}{3} e^{x^3} + c$

⟵ Since $\dfrac{d}{dx}(e^{x^3}) = 3x^2 e^{x^3}$

$x = 0, y = 0 \Rightarrow c = -\dfrac{1}{3}$

⟵ Use the initial condition to find the particular solution.

$\Rightarrow y^3 = e^{x^3} - 1$

Exercise 10.3

Find the particular solution to each of the following differential equations, given the prescribed initial condition. The first nine questions in this exercise use the same DEs as those in questions 1–9 of Exercise 10.2, for which you should already have found the general solutions.

1. $\dfrac{dy}{dx} = \dfrac{x}{y}$; $x = 1, y = 2$

2. $\dfrac{dy}{dx} = \dfrac{x^3}{y}$; $x = 1, y = 2$

3. $\dfrac{dy}{dx} = \dfrac{x^2 + x + 1}{y^2}$; $x = 0, y = 3$

4. $\dfrac{dy}{dx} = y(x + 3)$; $x = 0, y = 5$

5. $\dfrac{dy}{dx} = \dfrac{2}{\cos y}$; $x = 0, y = \dfrac{\pi}{2}$

6. $\dfrac{dy}{dx} = \dfrac{1}{xy}$; $x = 1, y = 2$

7. $2y\dfrac{dy}{dx} = 4x^3$; $x = 2, y = 3$

8. $2x^2 \dfrac{dy}{dx} = y^3$; $x = 1, y = 0.5$

9. $\dfrac{dy}{dx} = y\cos x$; $x = \dfrac{\pi}{2}, y = 1$

10. $\dfrac{dy}{dx} = \dfrac{x}{y}$; $x = 2, y = 0$

This is the same DE as Q1, but with a different initial condition.

11. $\dfrac{dy}{dx} = \dfrac{1}{xy}$; $\qquad x = e, y = 1$ $\qquad$ This is the same DE as Q6, but with a different initial condition.

12. A curve passes through the point $(1, 2)$ and its gradient function is $\dfrac{2x + 1}{2y}$.
Find the equation of the curve.

13. A curve is such that $\dfrac{dy}{dx} = \sqrt{\dfrac{y}{x + 1}}$ and the point $(3, 9)$ lies on the curve.
Find the equation of the curve.

14. A curve is such that $x^3 \dfrac{dy}{dx} = \sec y$ and the point $\left(1, \dfrac{\pi}{6}\right)$ lies on the curve.
Find the equation of the curve.

15. A curve is such that $e^y \dfrac{dy}{dx} - 2 \sec^2 x = 10$ and the point $\left(\dfrac{\pi}{4}, 0\right)$ lies on the curve.
Find the equation of the curve.

16. A curve is such that $\sqrt{xy} \dfrac{dy}{dx} = 1$ and the point $(4, 9)$ lies on the curve.
Find the equation of the curve.

17. A curve is such that $\dfrac{dy}{dx} = \dfrac{2e^{-y}x}{x^2 + 1}$ and the point $(0, 0)$ lies on the curve.
Find the equation of the curve.

18. A curve is such that $x(x + 1)\dfrac{dy}{dx} = y$ and the point $(1, 3)$ lies on the curve.
Find the equation of the curve.

19. A curve is such that $e^{-y} \dfrac{dy}{dx} = -e^{3x}$ and the point $(0, 0)$ lies on the curve.
Find the equation of the curve.

20. A curve is such that $x^3 \dfrac{dy}{dx} = \cos^2 y$ and the point $\left(1, \dfrac{\pi}{4}\right)$ lies on the curve.
Find the equation of the curve.

21. A curve is such that $y\dfrac{dy}{dx} = xe^{x^2}$ and the point $(0, 3)$ lies on the curve.
Find the equation of the curve.

10.4 Modelling with differential equations

In this chapter, you have already seen how to set up simple differential equations which describe a real-life context, and also how to find both general and particular solutions of first-order DEs where the variables are separable. This section combines all these building blocks to find solutions to real-life problems (usually in order to determine when certain conditions in the future will be satisfied).

Example 18

A tank is draining in such a way that when the height of water in the tank is h cm, it is decreasing at the rate of $0.5\sqrt{h}$ cm s^{-1}. Initially the water in the tank is at a height of 25 cm.

a) Write down a differential equation which describes this situation.

b) Solve the differential equation to find h as a function of time.

c) What is the height of the water after 10 seconds?

d) How long does it take for the water to reach a height of 5 cm?

e) Sketch a graph of the height against time for $0 \le t \le 20$.

· ·

a) $\dfrac{dh}{dt} = -0.5\sqrt{h}$

⟵ The height is decreasing so $\dfrac{dh}{dt}$ is negative.

b) $\dfrac{1}{\sqrt{h}}\dfrac{dh}{dt} = -0.5$

⟵ Separate the variables.

$\Rightarrow \displaystyle\int \dfrac{1}{\sqrt{h}}\,dh = \int -0.5\,dt$

$\Rightarrow 2\sqrt{h} = -0.5t + c$

⟵ Integrate both sides: this is the general solution.

$t = 0, h = 25 \Rightarrow c = 10$

⟵ Use the initial condition to find the particular solution.

$\Rightarrow 2\sqrt{h} = -0.5t + 10$

$\Rightarrow h = \left(-0.25t + 5\right)^2; \quad t = 20 - 4\sqrt{h}$

⟵ Writing the solution explicitly for h and for t is helpful here as parts (c) and (d) ask for a height and a time.

c) $h = \left(-0.25 \times 10 + 5\right)^2 = 6.25$ cm

d) $t = 20 - 4\sqrt{5} = 11.1$ seconds

e)

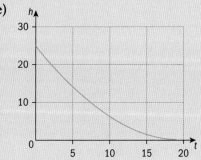

⟵ This is a section of the quadratic curve $h = (-0.25t + 5)^2$.

Example 19

The spread of a disease occurs at a rate proportional to the product of the number of people infected and the number not infected. Initially 50 out of a population of 1050 are infected and the disease is spreading at a rate of 10 new cases per day.

a) If n is the number infected after t days, show that $\dfrac{dn}{dt} = \dfrac{n(1050 - n)}{5000}$.

b) Solve this differential equation to find the number of people infected after t days.

c) How long will it take for 250 people to be infected?

d) Explain why everyone in the population will eventually be infected.

a) $\dfrac{dn}{dt} = kn(1050 - n)$

$t = 0, n = 50, \dfrac{dn}{dt} = 10 \Rightarrow k = \dfrac{1}{5000}$

$\Rightarrow \dfrac{dn}{dt} = \dfrac{n(1050 - n)}{5000}$

n people are infected, and $(1050 - n)$ are not infected.

The initial condition on the rate of change allows you to find the constant of proportionality.

b) $\dfrac{1}{n(1050 - n)} \dfrac{dn}{dt} = \dfrac{1}{5000}$

$\dfrac{1}{n(1050 - n)} \equiv \dfrac{A}{n} + \dfrac{B}{1050 - n} \Rightarrow A = B = \dfrac{1}{1050}$

Use partial fractions to get an integrable form.

$\displaystyle \int \left(\dfrac{1}{n} + \dfrac{1}{1050 - n} \right) dn = \int \dfrac{1050}{5000} \, dt = \int 0.21 \, dt$

$\Rightarrow \ln n - \ln(1050 - n) = 0.21t + c$

It is easier to take all the constant terms into one.

$\Rightarrow \ln \left(\dfrac{n}{1050 - n} \right) = 0.21t + c$

Solve the DE and express in simplest form.

$\Rightarrow \left(\dfrac{n}{1050 - n} \right) = Ae^{0.21t}$

$t = 0, n = 50 \Rightarrow A = 0.05$

$\Rightarrow \left(\dfrac{n}{1050 - n} \right) = 0.05e^{0.21t}$ (*)

Find the particular solution using the initial condition.

c) $n = 250; \quad 0.05e^{0.21t} = \dfrac{250}{800} \Rightarrow e^{0.21t} = 6.25$

$\Rightarrow t = 8.73 \text{ days}$

d) $n = \dfrac{1050 \times 0.05e^{0.21t}}{1 + 0.05e^{0.21t}}$

Rearrange (*) to give *n* explicitly.

$\Rightarrow n = 1050 \times \left(\dfrac{0.05e^{0.21t}}{1 + 0.05e^{0.21t}} \right)$

$n \to 1050$ as $t \to \infty$

As t increases the exponential term dominates in the bracket and this expression tends to 1.

Sometimes the problem will involve a constant of proportionality (given by k in Example 19) and also a constant of integration (given by c in Example 19). Be careful not to mix these up, as they have different values!

Example 20

A stone falls through the air from rest and, t seconds after it was dropped, its speed v satisfies the equation $\dfrac{dv}{dt} = 10 - 0.2v$.

> This is modelling motion under gravity at $10\,\mathrm{m\,s^{-2}}$ with air resistance proportional to speed.

a) Show that $v = 50(1 - e^{-0.2t})$.

b) Calculate the time at which the stone reaches a speed of $20\,\mathrm{m\,s^{-1}}$.

c) Sketch the graph of v against t and hence show that the velocity of the stone will never be more than $50\,\mathrm{m\,s^{-1}}$.

d) Explain what the differential equation tells us would happen if the stone was thrown downwards with a speed of $60\,\mathrm{m\,s^{-1}}$ instead of being dropped.

a) $\dfrac{dv}{dt} = 10 - 0.2v$

$\dfrac{1}{10 - 0.2v}\dfrac{dv}{dt} = 1$ ← Separate the variables.

$\Rightarrow \displaystyle\int \dfrac{5}{50 - v}\,dv = \int 1\,dt$

$\Rightarrow -5\ln(50 - v) = t + c$ ← Find the general solution.

$\Rightarrow \ln(50 - v) = -0.2t + C$ ← Where $C = -0.2c$

$\Rightarrow 50 - v = Ae^{-0.2t}$

$t = 0, v = 0 \Rightarrow A = 50$ ← The initial conditions provide the particular solution.

$\Rightarrow v = 50(1 - e^{-0.2t})$

b) $20 = 50(1 - e^{-0.2t}) \Rightarrow -0.2t = \ln 0.6$

$\Rightarrow t = 2.55$

> The expression in the bracket, $(1 - e^{-0.2t})$, approaches 1 as t increases, but $e^{-0.2t}$ cannot become negative so the bracketed expression will never exceed 1. Hence the speed cannot exceed $50\,\mathrm{m\,s^{-1}}$.

c)

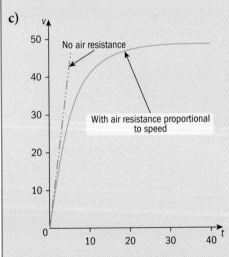

> This sketch also shows the behaviour in the 'free-fall' model with no air resistance. Until the stone starts to travel quickly the free-fall model is quite a good approximation, as air resistance is negligable at a slow speed.

d) If v is 60 initially then there will be a negative rate of change for the speed, so the stone will slow down – again approaching a terminal velocity of $50\,\mathrm{m\,s^{-1}}$, because that speed is where the air resistance exactly matches the gravitational force.

Example 21

The size of a colony of pests, n, which fluctuates during the year, is modelled by an entomologist with the differential equation $\frac{dn}{dt} = 0.2n(0.2 - \cos t)$, where t is the number of weeks from the start of the observations. There are 400 pests in the colony initially.

Did you know?

An entomologist is a scientist who studies insects.

a) Solve the differential equation to find n in terms of t.

b) Find how many pests there are after 3 weeks.

c) Show that the number of pests reaches a minimum after approximately 9.6 days, and find the number of pests at that time.

d) Find how many pests there are after 2π weeks.

e) Explain why the model predicts that the number of pests will grow infinitely large over time.

··

a) $\frac{dn}{dt} = 0.2n(0.2 - \cos t)$

$\frac{5}{n}\frac{dn}{dt} = 0.2 - \cos t$ ⟵ Separate the variables.

$\int \frac{5}{n}\,dn = \int (0.2 - \cos t)\,dt$

$\Rightarrow 5\ln n = 0.2t - \sin t + c$ ⟵ Integrate both sides.

$\Rightarrow n = Ae^{0.2(0.2t - \sin t)}$

$t = 0, n = 400 \Rightarrow A = 400$ ⟵ Use the initial condition to find the particular solution.

$\Rightarrow n = 400e^{0.2(0.2t - \sin t)}$

b) $t = 3 \Rightarrow n = 400e^{0.2(0.6 - \sin 3)} = 438.44... = 438$ ⟵ Substitute the given time.

c) $\frac{dn}{dt} = 0 \Rightarrow \cos t = 0.2 \Rightarrow t = 1.369$ weeks ≈ 9.6 days ⟵ Solve for a turning point.

$\frac{dn}{dt} = 0.2n(0.2 - \cos t) \Rightarrow \frac{d^2n}{dt^2} = 0.2\frac{dn}{dt}(0.2 - \cos t) + 0.2n\sin t$ ⟵ Find the second derivative to determine whether the turning point is a maximum or a minimum.

$\frac{d^2n}{dt^2}$ is positive for $t = 1.369$ (the first term is zero because $\frac{dn}{dt} = 0$ defined the turning point), so it is a minimum.

$n = 400e^{0.2(0.2\times 1.369 - \sin 1.369)} = 347$ ⟵ Find n, given $t = 1.369$ weeks.

▶ Continued on the next page

d) $t = 2\pi, n = 400e^{0.2(0.4\pi - \sin 2\pi)} = 514$

e) The exponential function has a term $0.04\,t$ as well as the bounded periodic term, so as t increases n will grow infinitely large.

> A sketch of the solution function shows the behaviour of the model. While it fluctuates over a period of 2π weeks, there is also an upward trend.

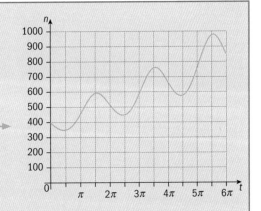

Example 22

A model of the deflation of a sphere of radius r cm assumes that at time t seconds after the start, the rate of decrease of the surface area is proportional to the volume at that time.

When $t = 0$, $r = 20$ cm and $\dfrac{dr}{dt} = -3$.

a) Show that r satisfies $\dfrac{dr}{dt} = -0.0075r^2$.

b) Solve this differential equation, obtaining an expression for r in terms of t.

c) How long will it take for the sphere to deflate to a radius of 10 cm?

d) How much longer will it be before the radius is 5 cm?

....

a) $A = 4\pi r^2 \Rightarrow \dfrac{dA}{dt} = 8\pi r\dfrac{dr}{dt}; \quad V = \dfrac{4}{3}\pi r^3$ ⟵ Use the formulae for a sphere.

$\Rightarrow 8\pi r\dfrac{dr}{dt} = -k\left(\dfrac{4}{3}\pi r^3\right)$ ⟵ Express the stated relationship in terms of r.

$t = 0, r = 20, \dfrac{dr}{dt} = -3$

$\Rightarrow 8\pi \times 20 \times (-3) = -k\left(\dfrac{4}{3}\pi \times 20^3\right)$

$\Rightarrow k = 0.045$ ⟵ Use the initial condition to find the constant of proportionality.

$\Rightarrow \dfrac{dr}{dt} = -0.0075r^2$

b) $\dfrac{-400}{3r^2}\dfrac{dr}{dt} = 1 \Rightarrow \displaystyle\int \dfrac{-400}{3r^2}\,dr = \int 1\,dt$ ⟵ Separate the variables and note that $0.0075 = \dfrac{3}{400}$.

$\Rightarrow \dfrac{400}{3r} = t + c$ ⟵ Integrate both sides.

▶ Continued on the next page

$$t = 0, r = 20 \Rightarrow c = \frac{20}{3} \Rightarrow \frac{400}{3r} = t + \frac{20}{3}$$

Use the initial condition to find the particular solution.

$$\Rightarrow \frac{3r}{400} = \frac{3}{3t + 20}$$

Express r in terms of t.

$$r = \frac{400}{3t + 20}$$

c) $t = \dfrac{400}{3r} - \dfrac{20}{3}$

When $r = 10$,

$$t = \frac{40}{3} - \frac{20}{3} = \frac{20}{3} = 6\frac{2}{3}$$

d) When $r = 5$,

$$t = \frac{80}{3} - \frac{20}{3} = \frac{60}{3} = 20$$

So it takes an extra $13\dfrac{1}{3}$ seconds.

A sketch of the solution function shows the behaviour of the model and why it takes increasingly longer for the radius to decrease by a set amount.

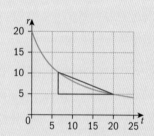

Example 23

In a chemical reaction a substance X reacts with another substance Y. The masses of substances X and Y after time t seconds are x and y grams respectively.

It is given that $\dfrac{dy}{dt} = -0.5xy$ and $x = 20\,e^{-2t}$. Also, we know that when $t = 0$, $y = 50$.

a) Form a differential equation in y and t.

b) Solve this equation to obtain an expression for y in terms of t.

c) Find the mass of Y that remains as t gets very large.

....

a) $\dfrac{dy}{dt} = -0.5xy; \quad x = 20e^{-2t} \Rightarrow \dfrac{dy}{dt} = 10y\,e^{-2t}$

Removing x from the DE.

b) $\dfrac{1}{y}\dfrac{dy}{dt} = -10\,e^{-2t}$

Separate the variables.

$$\int \frac{1}{y}\,dy = \int -10\,e^{-2t}\,dt$$

Integrate both sides.

▶ Continued on the next page

$\Rightarrow \ln y = 5\,e^{-2t} + c$

$t = 0,\, y = 50 \Rightarrow c = \ln 50 - 5$

Use the initial condition to find the particular solution and simplify the expression.

$\Rightarrow \ln y = 5\,e^{-2t} + \ln 50 - 5$

$\Rightarrow \ln\left(\dfrac{y}{50}\right) = 5(e^{-2t} - 1)$

$\Rightarrow y = 50e^{5(e^{-2t} - 1)}$

c) $e^{-2t} \rightarrow -5$ as t gets very large, so $50\,e^{-5}$ grams will remain.

Exercise 10.4

1. A rod has the property that the temperature gradient, $\dfrac{dT}{dx}$, at a distance x cm from the end of the rod being heated is proportional to the distance x.
   The end of the rod is heated to a temperature of $400\,°C$, and when $x = 50$ we know that $\dfrac{dT}{dx} = -8$.

   a) Write down a differential equation which is satisfied by the temperature T at a distance x from the end of the rod.

   b) Solve the differential equation to find an expression for the temperature T in terms of x.

   c) Calculate the temperature at the point 50 cm from the heated end of the rod.

   d) Calculate how far from the heated end of the rod the temperature reaches $20\,°C$.

2. In a chemical reaction a substance X reacts with another substance Y.
   The masses of substances X and Y after time t seconds are x and y grams respectively.
   It is given that $\dfrac{dy}{dt} = -0.1xy$ and $x = 30e^{-3t}$. Also, when $t = 0,\, y = 20$.

   a) Form a differential equation in y and t.

   b) Solve this differential equation to obtain an expression for y in terms of t.

   c) Find the proportion of Y which will remain when t is very large.

3. Carbon-14 occurs in all living creatures at 1 part in a trillion carbon atoms, but when a creature dies, the carbon atoms are no longer exchanged with the atmosphere. Carbon-14 atoms decay at a rate proportional to the amount remaining at any given time.

   a) If m is the mass of carbon-14 in 1 gram of carbon at any time t years after a creature dies, form a differential equation relating m and t and solve it.

   b) Taking $m = 1$ when $t = 0$, show that the differential equation is satisfied by $m = e^{-kt}$.

c) Carbon-14 is used to date fossils. It has a half-life of 5730 years (every 5730 years the mass of carbon-14 per gram of carbon in the fossil reduces by 50%). Find the value of k in the expression for m.

d) A fossil is found to have 0.008 grams of carbon-14 per gram of carbon. Calculate how old the fossil is.

4. A reservoir in the shape of a cuboid has a base area of 60 000 m². Water is seeping from the reservoir at a rate proportional to the depth, and it is known that the reservoir loses water at a rate of 3000 litres per minute when the depth is 10 metres.

The depth of water in the reservoir is 8 metres when a heavy storm starts, causing the reservoir to be filled at a constant rate of 1500 litres per second.

a) Write down a differential equation which is satisfied by the depth of water, d metres, at time t minutes after the storm starts.

b) Solve the differential equation.

c) If the storm lasts for two and a half hours, calculate the depth of water in the reservoir when the storm ends.

5. A colony of bacteria grows at a rate proportional to the size of the colony at any time.

a) Write down a differential equation satisfied by the number of bacteria N (in millions) after t hours.

Initially the colony has 2 million bacteria, and after 6 hours it has 2.5 million bacteria.

b) Solve the differential equation to find an expression for N in terms of t.

c) The space in which the colony is housed has room for 10 million bacteria. Calculate the time at which the space reaches saturation.

6. A body moving through a liquid experiences a resistance to motion which causes it to lose speed at a rate proportional to its speed at any time.

a) Write down a differential equation satisfied by the speed $v\,\mathrm{m\,s^{-1}}$ after time t seconds.

Initially the body has a speed of $25\,\mathrm{m\,s^{-1}}$, and after 10 seconds it has a speed of $20\,\mathrm{m\,s^{-1}}$.

b) Solve the differential equation to find an expression for v in terms of t.

c) Calculate how long it takes for the body to lose half of its initial speed.

7. A disease spreads at a rate proportional to the product of the number of people, n, infected and the number of people not yet infected. The population has size P. Initially 5% of the population is infected.

a) Write down a differential equation which is satisfied by n and the time t.

After 3 days it is found that 10% of the population is infected.

b) Solve this differential equation to find the number of people infected after t days.

c) How long will it take for half the people to be infected?

d) Explain why everyone in the population will eventually be infected.

Summary exercise 10

1. Find the general solution to the following differential equations.

a) $\dfrac{dy}{dx} = \dfrac{x^2}{y^3}$

b) $\dfrac{dy}{dx} = 2x\cos^2 y$

c) $\dfrac{dy}{dx} = e^{3x}\sec y$

d) $x\tan y\,\dfrac{dy}{dx} = 1$

e) $\dfrac{dy}{dx} = e^{3x-y}$

f) $\dfrac{dy}{dx} = \dfrac{xy}{x^2+1}$

g) $(x^2+6)\dfrac{dy}{dx} = 8xy$

h) $\sin 3\theta\,\dfrac{d\theta}{dx} = (x+2)\cos 3\theta$

2. Find the particular solution to the following differential equations, with given initial conditions.

a) $\dfrac{dy}{dx} = \dfrac{3x^2+x}{4y^3}$; $x=2, y=1$

b) $\dfrac{dy}{dx} = 2y^2\sin x$; $x=\pi, y=1$

c) $\dfrac{dy}{dx} = e^{3x-y}$; $x=0, y=0$

d) $\dfrac{dv}{dt} = e^{-t}\sqrt{v}$; $t=0, v=9$

e) $\dfrac{dy}{dx} = xe^{x-y}$; $x=0, y=0$

f) $\ln y\,\dfrac{dy}{dx} = \dfrac{x-1}{x^2+x}$; $x=1, y=1$

EXAM-STYLE QUESTIONS

3. A curve is such that $\dfrac{dy}{dx} = \dfrac{y}{\sqrt{x+1}}$ and the point $(1, 1)$ lies on the curve. Find the equation of the curve.

4. A curve is such that $\dfrac{dy}{dx} = xe^x\sec y$ and the point $\left(0, \dfrac{\pi}{6}\right)$ lies on the curve. Find the equation of the curve.

5. A curve is such that $ye^y\cos^2 x\,\dfrac{dy}{dx} = 1$ and the point $\left(\dfrac{\pi}{4}, 0\right)$ lies on the curve. Find the equation of the curve.

6. A curve is such that $\sqrt{y}\,\dfrac{dy}{dx} = \dfrac{2}{x^2-1}$ and the point $(2, 0)$ lies on the curve. Find the equation of the curve.

7. A curve is such that $y\dfrac{dy}{dx} = \cos^2 x$ and the point $\left(\dfrac{\pi}{4}, 0\right)$ lies on the curve. Find the equation of the curve.

8. A curve is such that $\left(\dfrac{2y+1}{y^2+1}\right)\dfrac{dy}{dx} = \dfrac{1}{x}$ and the point $(1, 0)$ lies on the curve. Find the equation of the curve.

9. The temperature of a hot body decreases at a rate proportional to the difference between its temperature and the air temperature around it. A body is heated to 90 °C and placed in air which is at 20 °C.

a) Write down a differential equation relating the temperature T of the body and the time t.

After 4 minutes the body has cooled to 60 °C.

b) Solve the differential equation to find an expression for the temperature T in terms of t.

c) Calculate the temperature after 10 minutes.

d) Calculate how long it takes for the body to get down to within 5 °C of the air temperature.

10. The resistance a body experiences passing through a liquid is proportional to the square of its speed at any time. The body has a speed of 50 m s^{-1} entering the liquid.

a) Write down a differential equation relating the speed v of the body and the time t.

After 4 seconds the body has slowed to $30\,\mathrm{m\,s^{-1}}$.

b) Solve the differential equation to find an expression for the speed v in terms of t.

c) Calculate the speed after 10 seconds.

d) Calculate how long it takes for the body to slow down to $10\,\mathrm{m\,s^{-1}}$.

11. The spread of a rumour in a large group of people is thought to reach new people at a rate proportional to the product of the number of people who have heard the rumour and the number who have not heard it.

In a population of 2000 people, initially 50 people have heard the rumour. Two hours later, 300 people have heard it.

a) Write down a differential equation relating the number of people, N, who have heard the rumour with the time t.

b) Solve the differential equation to find an expression for N in terms of t.

c) Calculate how long it takes for the majority of people to have heard the rumour.

12. a) Show that $\dfrac{\mathrm{d}v}{\mathrm{d}t} = v\dfrac{\mathrm{d}v}{\mathrm{d}x}$ where $v = \dfrac{\mathrm{d}x}{\mathrm{d}t}$.

The rate at which a body loses speed, v, is proportional to its displacement, x, from a fixed point.

b) Show that the differential equation

$$v\frac{\mathrm{d}v}{\mathrm{d}x} = -kx$$ describes this situation.

The body has speed $10\,\mathrm{m\,s^{-1}}$ as it passes through the fixed point and just reaches a point 5 metres from the fixed point as it comes to rest.

c) Solve the differential equation to find an equation relating v and x, showing that $k = 4$.

d) Find the speed of the body when it is 3 metres from the fixed point.

13. A tank contains a solution of a mineral in water. Initially there is 600 litres of water with $12\,\mathrm{kg}$ of the dissolved mineral. The mixture is drained at a rate of 30 litres per minute and simultaneously pure water is added at a rate of 30 litres per minute. The tank is stirred continuously to keep the mixture uniform at all times.

a) Form a differential equation which is satisfied by the mass, m kg, of the mineral in the solution at time t minutes.

b) Solve the differential equation to find an expression for m in terms of t.

When the solution contains $8\,\mathrm{kg}$ of the mineral another $4\,\mathrm{kg}$ is added, and the process is repeated.

c) Find the first time at which the solution contains $8\,\mathrm{kg}$ of the mineral.

14. The variables x and y are related by the differential equation

$$\frac{\mathrm{d}y}{\mathrm{d}x} = \frac{4ye^{2x}}{5 + e^{2x}}.$$

Given that $y = 36$ when $x = 0$, find an expression for y in terms of x.

15. The variables x and θ satisfy the differential equation

$$\frac{\mathrm{d}x}{\mathrm{d}\theta} = (x + 4)\sin^2 3\theta,$$

and it is given that $x = 0$ when $\theta = 0$. Solve the differential equation and calculate the value of x when $\theta = \frac{1}{4}\pi$, giving your answer correct to 3 significant figures.

16. A large field of area $3\,\text{km}^2$ is becoming infected with a soil disease. At time t years the area infected is $x\,\text{km}^2$ and the rate of growth of the infected area is given by the differential equation $\dfrac{dx}{dt} = kx(3-x)$, where k is a positive constant. It is given that when $t=0$, $x=0.5$ and that when $t=2$, $x=2$.

   i)   Solve the differential equation and show that $k = \dfrac{1}{6}\ln 10$.

   ii) Find the value of t when there is only 0.5 km^2 of the field not infected giving your answer to 3 significant figures.

Note: Question 17 is beyond the scope of the Cambridge International syllabus but may aid in your understanding.

---

17. A man walks along a pier pulling a toy boat by a rope of length L. The man keeps the rope taut and horizontal. Initially the rope is at right angles to the pier, with the boat at A and the man at B. A little while later the man is at E and the boat is at C. D is the point on the pier such that CD is perpendicular to the pier. $CD = x$, $BD = y$.

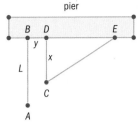

pier

The path followed by the boat is called a tractrix, and has the property that the rope is always tangential to the path of the boat.

a) If $y = \text{f}(x)$ is the path of the boat, show that $\dfrac{dy}{dx} = \dfrac{-\sqrt{L^2 - x^2}}{x}$.

b) Use the substitution $x = L\cos\theta$ to show that the solution to the differential equation is
$$\text{f}(x) = \frac{L\ln x}{(L - \sqrt{(L^2 - x^2)})} - \sqrt{(L^2 - x^2)}$$
[You may use without derivation that $\displaystyle\int \text{cosec}\,\theta\,d\theta = \ln|\text{cosec}\,\theta - \cot\theta|.$]

Chapter summary

Differential equations

- A differential equation is an equation in which at least one differential expression appears.

- A first-order differential equation only contains a first derivative $\left(\dfrac{dy}{dx} \text{ or equivalent}\right)$.

- The differential equation has separable variables if the equation can be written in the form $g(y)\dfrac{dy}{dx} = f(x)$.

- The general solution is obtained by integrating both sides of $\displaystyle\int g(y)\,dy = \int f(x)\,dx$ and including $+c$ on one side.

- A particular solution is obtained by using an initial condition, or boundary condition, to calculate the value of c in the general solution which means the solution satisfies that condition.

- All the techniques used in integration can be required to solve the two integrals obtained when the variables are separated.

11 Complex numbers

We can use computers to generate beautiful images from complex numbers, such as the Mandelbrot set, shown on the left. However, complex numbers are also widely used in real-life applications. For example, they are used in the study of electrical circuits and the flow of fluids around objects, as well as in the technology of digital cameras and mobile phones. They are also used in the design of aeroplane wings.

Objectives

- Understand the idea of a complex number, recall the meaning of the terms real part, imaginary part, modulus, argument and conjugate, and use the fact that two complex numbers are equal if and only if both real and imaginary parts are equal.

- Carry out operations of addition, subtraction, multiplication and division of two complex numbers expressed in Cartesian form $x + iy$.

- Use the result that for a polynomial equation with real coefficients, any non-real roots occur in conjugate pairs.

- Represent complex numbers geometrically by means of an Argand diagram.

- Carry out operations of multiplication and division of two complex numbers expressed in polar form $r\,(\cos\theta + i\sin\theta) \equiv r\,e^{i\theta}$.

- Find the two square roots of a complex number.

- Understand in simple terms the geometrical effects of conjugating a complex number and of adding, subtracting, multiplying and dividing two complex numbers.

- Illustrate simple equations and inequalities involving complex numbers by means of loci in an Argand diagram, e.g. $|z - a| < k$, $|z - a| = |z - b|$, $\arg(z - a) = \alpha$.

Before you start

You should know how to:

1. Use the trigonometric identities to simplify expressions,

e.g. $(1 + \cos\theta)^2 + \sin^2\theta$

$$= 1 + 2\cos\theta + \cos^2\theta + \sin^2\theta$$

$$= 1 + 2\cos\theta + 1$$

$$= 2 + 2\cos\theta = 2(1 + \cos\theta)$$

2. Manipulate and rationalise surds,

e.g. $\left(3\sqrt{2}\right)^2 + \left(2\sqrt{3}\right)^2$

$$= \left(3\sqrt{2}\right)\left(3\sqrt{2}\right) + \left(2\sqrt{3}\right)\left(2\sqrt{3}\right)$$

$$= 9 \times 2 + 4 \times 3$$

$$= 30$$

e.g. $\dfrac{2 - \sqrt{3}}{1 + \sqrt{3}} = \dfrac{2 - \sqrt{3}}{1 + \sqrt{3}} \times \dfrac{1 - \sqrt{3}}{1 - \sqrt{3}}$

$$= \dfrac{2 - 3\sqrt{3} + \left(\sqrt{3}\right)^2}{1 + \sqrt{3} - \sqrt{3} - \left(\sqrt{3}\right)^2}$$

$$= \dfrac{5 - 3\sqrt{3}}{1 - 3}$$

$$= \dfrac{3\sqrt{3} - 5}{2}$$

3. Work out the discriminant for a quadratic equation,

e.g. $2x^2 - 5x + 3 = 0$

$$a = 2, b = -5, c = 3$$

Discriminant: $b^2 - 4ac = (-5)^2 - 4 \times 2 \times 3 = 1$

Skills check:

1. Simplify

a) $(1 + \sin\theta)^2 + \cos^2\theta$

b) $\sin^2\theta - \sin^2\theta\cos^2\theta$

c) $\cos^4\theta + 2\sin^2\theta\cos^2\theta + \sin^4\theta$.

2. Show that

a) $\dfrac{2\sqrt{3} - 3}{\sqrt{3}} = 2 - \sqrt{3}$

b) $\dfrac{4 + \sqrt{2}}{1 + \sqrt{2}} = -2 + 3\sqrt{2}$

c) $\dfrac{1}{2\sqrt{3} - 2} = \dfrac{\sqrt{3} + 1}{4}$.

3. Work out the discriminant for

a) $x^2 - 5x + 2 = 0$

b) $2x^2 - 7x - 4 = 0$

c) $3x - 5x^2 - 4 = 0$.

11.1 Introducing complex numbers

In P1 we saw that if the discriminant, $b^2 - 4ac$, of the quadratic equation $ax^2 + bx + c = 0$ is negative, then the equation has no real roots. This is because there is no real number which, when squared, gives a negative real number.

The ancient Greeks were puzzled by the fact that there was no real number which satisfies the equation $x^2 = -1$. However, it was not until 1545 that the Italian mathematician Gerolamo Cardano thought about the possibility of complex numbers when he solved the equation $x(10 - x) = 40$ to give the solutions $x = 5 \pm \sqrt{-15}$. Later in the 16th century Rafael Bombelli, another Italian mathematician who is generally regarded as the inventor of complex numbers, developed this concept to enable him to solve cubic and quartic equations.

Before we discuss complex numbers in more detail, we introduce the symbol i which we will use to denote the non-real number $\sqrt{-1}$.

It follows that $i^2 = -1$.

$$\sqrt{-1} = i \implies i^2 = -1$$

As such, $\sqrt{-9} = \sqrt{9 \times -1} = \sqrt{9} \times \sqrt{-1} = 3i$.

More generally, $\sqrt{-b^2} = bi$. Numbers of the form bi, where b is real, are called **imaginary numbers**.

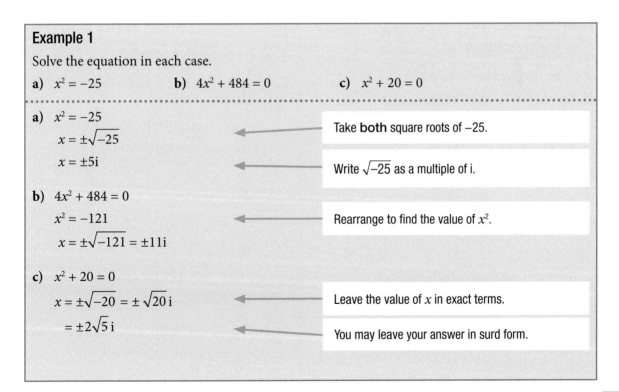

Example 1

Solve the equation in each case.

a) $x^2 = -25$ 　　　　 b) $4x^2 + 484 = 0$ 　　　　 c) $x^2 + 20 = 0$

a) $x^2 = -25$

$x = \pm\sqrt{-25}$ 　　　　　　　　　　 Take **both** square roots of −25.

$x = \pm 5i$ 　　　　　　　　　　 Write $\sqrt{-25}$ as a multiple of i.

b) $4x^2 + 484 = 0$

$x^2 = -121$ 　　　　　　　　　　 Rearrange to find the value of x^2.

$x = \pm\sqrt{-121} = \pm 11i$

c) $x^2 + 20 = 0$

$x = \pm\sqrt{-20} = \pm\sqrt{20}\,i$ 　　　　　　　　 Leave the value of x in exact terms.

$= \pm 2\sqrt{5}\,i$ 　　　　　　　　　　 You may leave your answer in surd form.

Example 2

Simplify:

a) i^3 **b)** i^4 **c)** i^9 **d)** $(-i)^6$ **e)** $(2i)^5$.

a) $i^3 = i^2 \times i = -1 \times i = -i$ Replace i^2 with -1.

b) $i^4 = i^2 \times i^2 = -1 \times -1 = 1$

c) $i^9 = i^4 \times i^4 \times i = 1 \times 1 \times i = i$ Write i^9 as a product of powers of i each of which you can replace with 1, -1, i or $-i$. Use $i^4 = 1$ here.

d) $(-i)^6 = i^6 = i^4 \times i^2 = 1 \times -1 = -1$

e) $(2i)^5 = 32 \times i^5 = 32 \times i^4 \times i = 32i$ Expand the brackets and simplify.

It is useful to remember:
$i^2 = -1$, $i^3 = -i$, $i^4 = 1$.   Use the ordinary laws of algebra to write the expression without brackets, then simplify.

Exercise 11.1A

1. Which of the following equations have no real roots?

   a) $x^2 + 5x + 2 = 0$

   b) $2x^2 - 4x + 1 = 0$

   c) $x^2 + 5x + 7 = 0$

   d) $3x^2 - x + 2 = 0$

   e) $2x^2 + 5x - 1 = 0$

2. Solve the equation in each case.

   a) $x^2 = -4$ **b)** $x^2 = -100$ **c)** $x^2 = -13$ **d)** $x^2 = -75$

3. Simplify the expression in each case.

   a) i^5 **b)** i^7 **c)** i^{11} **d)** i^{25} **e)** i^{64}

4. Simplify the expression in each case.

   a) $(-i)^{14}$ **b)** $(2i)^7$ **c)** $(3i)^3$ **d)** $(-2i)^{10}$

5. Write down the square roots of

   a) 64   **b)** -64 **c)** -19 **d)** -32.

6. Solve the equation in each case.

   a) $x^2 + 49 = 0$ **b)** $x^2 + 23 = 5$ **c)** $2x^2 + 9 = 5$ **d)** $(5x)^2 = -125$

7. Simplify the following expressions.

   a) $\left(\sqrt{2}\,i\right)^2$ **b)** $\left(-\sqrt{7}\,i\right)^4$ **c)** $\left(\frac{1}{3}i\right)^3$ **d)** $\left(-\frac{1}{2}i\right)^8$

8. We wish to find two numbers, x and y, whose sum is 10 and whose product is 40.

   a) Write down two equations in x and y which fully describe this problem.

   b) Show that $x^2 - 10x + 40 = 0$.

   c) Show that this equation has no real roots.

At the beginning of this chapter we discussed how the mathematician Cardano solved the quadratic equation $x(10 - x) = 40$ to give the solutions $x = 5 \pm \sqrt{-15}$. Our experience of solving quadratic equations by using the

formula $x = \dfrac{-b \pm \sqrt{b^2 - 4ac}}{2a}$ shows us that our answer is made up of two

parts, $\dfrac{-b}{2a}$ and $\dfrac{\sqrt{b^2 - 4ac}}{2a}$. The first of these expressions will always be real (provided the coefficients a, b, and c are real). However, we have learned that the second expression might not be real; it might be imaginary and therefore expressed as bi (where b is real). This leads us to our definition of a complex number.

> A number of the form $a + b$i, where a and b are real and $i^2 = -1$, is called a **complex number**.

Note: The 'b' used in the formula for solving a quadratic equation should not be confused with the 'b' used in the complex number $a + b$i.

Note: Complex numbers may be written in the form $a + b$i or in the form $a + i b$.

A number such as 2i is said to be imaginary, whereas a number such as $5 + 2$i is said to be complex. We use the letter z to denote a complex number, $z = x + i y$, where x and y are both real numbers.

We can also see from the quadratic formula, provided that a, b and c are real, that the roots occur in pairs, one root of the form $a + b$i and the other of the form $a - b$i. For example, if one root is $3 + 2$i then the other root will be $3 - 2$i. Pairs of complex numbers like this are referred to as **complex conjugates** of each other.

Note: All imaginary numbers are also complex numbers as they can be written as $0 + b$i. Similarly, all real numbers are complex numbers as they can be written in the form $a + 0$i.

> The complex conjugate of $a + b$i is $a - b$i, where a and b are real numbers.
>
> The complex conjugate of a complex number z is denoted by z^*.

Note: In this book we use $a + b$i when we know the values of the real and imaginary parts (e.g. $3 + 7$i) and $x + i y$ when we do not know the values of the real and imaginary parts. However, this is just a matter of preference and you may like to write it differently.

Example 3

Solve the equation $2z^2 - z + 3 = 0$.

Using the quadratic formula,

$$z = \frac{-b \pm \sqrt{b^2 - 4ac}}{2a}$$

$$z = \frac{-(-1) \pm \sqrt{(-1)^2 - 4 \times 2 \times 3}}{2 \times 2}$$ ← Substitute $a = 2$, $b = -1$ and $c = 3$ into the formula.

$$z = \frac{1 \pm \sqrt{1 - 24}}{4}$$ ← Simplify the expression.

$$z = \frac{1 \pm \sqrt{-23}}{4}$$

$$z = \frac{1}{4} \pm \frac{\sqrt{23}}{4}i$$ ← Here we leave the answer in the form $a \pm bi$, but we could separate the two values and write $z = \frac{1}{4} + \frac{\sqrt{23}}{4}i$ and $\frac{1}{4} - \frac{\sqrt{23}}{4}i$.

Alternatively, completing the square,

$$2z^2 - z + 3 = 0$$

$$2\left(z^2 - \frac{1}{2}z\right) + 3 = 0$$

$$2\left\{\left(z - \frac{1}{4}\right)^2 - \frac{1}{16}\right\} + 3 = 0$$

$$2\left(z - \frac{1}{4}\right)^2 - \frac{1}{8} + 3 = 0$$

$$2\left(z - \frac{1}{4}\right)^2 = -\frac{23}{8}$$

$$\left(z - \frac{1}{4}\right)^2 = -\frac{23}{16}$$

$$\left(z - \frac{1}{4}\right) = \pm\sqrt{-\frac{23}{16}}$$

$$z = \frac{1}{4} \pm \sqrt{-\frac{23}{16}}$$

$$z = \frac{1}{4} \pm \frac{\sqrt{23}}{4}i$$

Exercise 11.1B

1. Express each of the following complex numbers in the form $a + bi$.

   a) $5 + \sqrt{-1}$ b) $4 - \sqrt{-6}$ c) $-2 + \sqrt{-40}$ d) $-1 - \sqrt{-81}$

2. Write down the complex conjugate of the following.

   a) $4 - 2i$ b) $-5 + i$ c) $-9 - 7i$ d) $3i - 6$

3. Write down the conjugate of the complex number in each case.

    a) $z = 10 + 12i$ **b)** $z^* = 3 - 3i$

    c) $z = 5i$ **d)** $z^* = 8$

4. Find whether the following equations have real or complex roots.

    a) $z^2 - 6z + 7 = 0$

    b) $5z^2 + z + 2 = 0$

    c) $z^2 + 8z = 0$

    d) $4 + 3z - 3z^2 = 0$

    e) $4z^2 + 5z + 3 = 1$

    f) $1 - z^2 = 4 + z$

5. Solve the following quadratic equations by using the quadratic formula.

    a) $z^2 - 4z + 5 = 0$

    b) $4z^2 + 2z + 1 = 0$

    c) $2z^2 - 9z - 3 = 0$

    d) $2 - z - 3z^2 = 0$

    e) $5z - 3z^2 = 1 + 4z$

    f) $10z^2 + 5z + 20 = 15$

    g) $z^2 + iz - 1 = 0$

6. Solve the following quadratic equations by completing the square.

    a) $z^2 - 2z + 3 = 0$

    b) $3z^2 + 6z + 10 = 0$

    c) $2z^2 - 8z + 7 = 0$

    d) $9z^2 - 27z + 3 = 0$

    e) $(z + 1)^2 - (2z + 2)^2 = 3$

    f) $z(z + 6) = -10$

    g) $z^2 - 2iz - 5 = 0$

7. Solve $iz^2 + z + 2i = 0$.

    (You may leave your answers in the form $\frac{p}{q}$, where p is a real number and q is an imaginary number.)

11.2 Calculating with complex numbers

We are about to see that when adding, subtracting, and multiplying complex numbers we use the normal rules of algebra. The division of complex numbers is carried out in a similar way to that of dividing surds.

Suppose we have two complex numbers, $z_1 = a + bi$ and $z_2 = c + di$. We add them in the following way:

$$z_1 + z_2 = (a + bi) + (c + di) = (a + c) + (b + d)i$$

> It is common to use $z_1, z_2, z_3\ldots$ if we want to use more than one complex number.

To add two complex numbers we add their real parts and add their imaginary parts.

We subtract them in the following way:

$$z_1 - z_2 = (a + bi) - (c + di) = (a - c) + (b - d)i$$

To subtract two complex numbers we subtract their real parts and subtract their imaginary parts.

We multiply them in the following way:

$$z_1 z_2 = (a + bi)(c + di)$$
$$= ac + bci + adi + bdi^2$$
$$= (ac - bd) + (bc + ad)i$$

To multiply two complex numbers we use the normal rules of algebra and simplify the answer using $i^2 = -1$.

We divide them in the following way:

$$\frac{z_1}{z_2} = \frac{a + bi}{c + di}$$

$$= \frac{a + bi}{c + di} \times \frac{c - di}{c - di}$$

$$= \frac{(a + bi)(c - di)}{(c + di)(c - di)}$$

$$= \frac{ac + bci - adi - bdi^2}{c^2 + cdi - cdi - d^2 i^2}$$

$$= \frac{(ac + bd) + (bc - ad)i}{c^2 + d^2}$$

$$= \frac{(ac + bd)}{c^2 + d^2} + \frac{(bc - ad)}{c^2 + d^2}i$$

To divide two complex numbers we multiply the numerator and the denominator of the fraction by the complex conjugate of the denominator.

Example 4

The complex numbers z_1 and z_2 are given by $z_1 = 5 + 2i$ and $z_2 = 3 + 4i$.

Find $z_1 + z_2$, $z_1 - z_2$, $z_1 z_2$ and $\dfrac{z_1}{z_2}$.

$z_1 + z_2 = 5 + 2i + 3 + 4i$ ⟵ Add the real parts and the imaginary parts separately.

$= 8 + 6i$

$z_1 - z_2 = (5 + 2i) - (3 + 4i)$ ⟵ Use brackets as you would when subtracting one algebraic expression from another.

$= 5 + 2i - 3 - 4i$

$= 2 - 2i$

$z_1 z_2 = (5 + 2i)(3 + 4i)$ ⟵ Expand the brackets and simplify.

$= 15 + 6i + 20i + 8i^2$

$= 7 + 26i$ ⟵ Use $i^2 = -1$.

$\dfrac{z_1}{z_2} = \dfrac{5 + 2i}{3 + 4i}$

$= \dfrac{5 + 2i}{3 + 4i} \times \dfrac{3 - 4i}{3 - 4i}$ ⟵ Multiply the numerator and denominator by $(3 - 4i)$.

$= \dfrac{(5 + 2i)(3 - 4i)}{(3 + 4i)(3 - 4i)}$

$= \dfrac{15 + 6i - 20i - 8i^2}{9 + 12i - 12i - 16i^2}$ ⟵ Expand the brackets and simplify.

$= \dfrac{23 - 14i}{9 + 16}$

$= \dfrac{(23 - 14i)}{25}$ or $\dfrac{23}{25} - \dfrac{14}{25}i$

Exercise 11.2

1. Simplify

   a) $(2 + i) + (3 + 5i)$ b) $(7 + 6i) + (4 - 2i)$

   c) $(1 - 3i) + (-4 + 3i)$ d) $(6 - 5i) + (-3 + 8i)$

   e) $(-2 + 3i) + (3 + 2i)$ f) $(1 + i) + (-2 - 9i)$.

2. Simplify

   a) $(4 + 3i) - (2 + i)$ b) $(5 + 2i) - (3 - 4i)$

   c) $(7 - i) - (-2 + 5i)$ d) $(5 - 5i) - (-6 + 3i)$

   e) $(-3 + 6i) - (3 + 4i)$ f) $(-1 - 2i) - (-3 - 8i)$.

3. Simplify
   a) $2(3 + 4i) + 3(5 + i)$
   b) $2(1 - 6i) - (4 - 2i)$
   c) $3(1 - 3i) + 2(-4 - 3i)$
   d) $5(6 + 5i) + 2(-3 + 8i)$
   e) $6(2 - 3i) - 3(3 + 5i)$
   f) $4(1 + 4i) - 3(2 + 7i)$.

4. Express in the form $x + iy$, where x and y are real.
   a) $(5 + i)(2 + i)$
   b) $(3 + 2i)(5 - 4i)$
   c) $i(7 - i)$
   d) $(3 - 4i)(-1 + 2i)$
   e) $(-1 + 6i)(2 + 2i)$
   f) $3(1 - 3i)(-4 + 3i)$

5. Express in the form $x + iy$, where x and y are real.
   a) $\dfrac{5 + 4i}{2 + i}$
   b) $\dfrac{2 - 7i}{1 - 2i}$
   c) $\dfrac{5 + 6i}{5 - 6i}$
   d) $\dfrac{2i}{8i}$
   e) $\dfrac{9 - 3i}{9 + 3i}$
   f) $\dfrac{2i}{3 + 4i}$
   g) $\dfrac{6 + 2i}{4i}$
   h) $\dfrac{2 - \sqrt{3}i}{2 + \sqrt{3}i}$

6. The complex number z is given by $z = \sqrt{2} - i$.
   a) Express z^2 in the form $x + iy$, where x and y are real.
   b) The complex conjugate of z is denoted by z^*. Express the following in the form $x + iy$, where x and y are real.
      i) $z + z^*$
      ii) $2(z - z^*)$
      iii) $5zz^*$
      iv) $\dfrac{z}{iz^*}$

7. The complex numbers z_1 and z_2 are given by $z_1 = 5 - 12i$ and $z_2 = 3 - 4i$.
   Express the following in the form $x + iy$ where x and y are real.
   a) $z_1 + z_2$
   b) $z_2 - z_1$
   c) $z_1 z_2$
   d) $\dfrac{z_2}{z_1}$

8. Express the following in the form $x + iy$, where x and y are real.
   a) $(2 - i)^2$
   b) $(2 - i)^3$
   c) $(2 - i)^4$
   d) $(2 - i)^7$

9. If $z_1 = 1 + 2i$ and $z_2 = 2 - 3i$, express the following in the form $x + iy$ where x and y are real.
   a) $\dfrac{1}{z_1 + z_2}$
   b) $\dfrac{1}{z_1} + \dfrac{1}{z_2}$

10. Express in the form $x + iy$, where x and y are real.
    a) $\dfrac{1 + i}{1 - i}$
    b) $\dfrac{2}{3 + 5i}$
    c) $\dfrac{2i - 1}{1 + 2i}$
    d) $\dfrac{1}{a + ib}$
    e) $\dfrac{1}{4 + 3i} + \dfrac{12}{4 - 3i}$

11. $u = a + ib$
    Find the real part and the imaginary part of
    a) u^2
    b) $u - u^*$
    c) $u^3 - (u^*)^3$.

12. Express in the form $x + iy$, where x and y are real.
    a) $(\cos\theta + i\sin\theta)^2$
    b) $\dfrac{1}{\cos\theta + i\sin\theta}$
    c) $\dfrac{\cos\theta + i\sin\theta}{\cos\theta - i\sin\theta}$

11.3 Solving equations involving complex numbers

We can order real numbers by placing them on a number line, but because complex numbers have both a real part and an imaginary part, we cannot order them in the same way.

If two real numbers are equal they would occupy the same position on the number line. For two complex numbers to be equal, their real parts must be the same **and** their imaginary parts must be the same. So if for example $x + iy = 5 - 2i$, it follows that $x = 5$ and $y = -2$.

Two complex numbers are equal if both the real parts are equal and the imaginary parts are equal.

Example 5

a) Find the value of the real number p such that $p + (2 - 3i)(1 + 5i)$ is an imaginary number.

b) Find the values of the real numbers x and y such that $4(x + iy) = -2y - 3ix - 5(3 + 2i)$.

a) $p + (2 - 3i)(1 + 5i) = p + (2 - 3i + 10i - 15i^2)$ Expand the brackets and simplify.

$\qquad\qquad\qquad\qquad = p + 7i + 17$

$\qquad\qquad\qquad\qquad = (p + 17) + 7i$ Write in the form $x + iy$.

$\qquad\qquad\qquad p + 17 = 0$

$\qquad\qquad\qquad\qquad p = -17$ Equate the real part to 0 and solve to find p.

b) $4x + 4iy = -2y - 3ix - 15 - 10i$

$4x + 4iy = (-2y - 15) + (-3x - 10)i$ Collect the real parts and the imaginary parts on the RHS.

Comparing real parts, $4x = -2y - 15$

Comparing imaginary parts, $4y = -3x - 10$

$4x + 2y = -15 \qquad (1)$ Solve the simultaneous equations.

$3x + 4y = -10 \qquad (2)$

$2 \times$ equation (1) – equation (2):

$5x = -20$

$x = -4$

Substituting $x = -4$ into equation (1):

$4 \times -4 + 2y = -15$

$y = 0.5$

$x = -4, y = 0.5$

Example 6

The complex number $3 + 2i$ is a root of the quadratic equation $z^2 - (5 + 2i)z + a + bi = 0$, where a and b are real.

a) Find the values of a and b.

b) **Explain** why the two roots are not complex conjugates of each other.

a) As $z = 3 + 2i$ is a root,

$(3 + 2i)^2 - (5 + 2i)(3 + 2i) + a + bi = 0$ Substitute for z in the equation.

$(9 + 12i + 4i^2) - (15 + 16i + 4i^2) + a + bi = 0$ Expand and then simplify.

$-6 - 4i + a + bi = 0$

$-6 + a + (-4 + b)i = 0$ Compare the real part on the LHS with the real part on the RHS, and similarly compare the imaginary parts.

$-6 + a = 0$ and $-4 + b = 0$

$a = 6, b = 4$

b) The coefficients of the quadratic equation $z^2 - (5 + 2i)z + a + bi = 0$ are not all real so the roots will not be complex conjugates.

We have seen that the square roots of a negative number can be expressed as imaginary numbers, that is, the solutions of $z^2 = -a$ are $z = \pm\sqrt{a}\,i$.

> We can use the method of comparing real parts and imaginary parts to find the square roots of a complex number, $a + bi$.

Example 7 demonstrates how we do this.

Example 7

Find the square roots of the complex number $7 - 24i$.

As with any other number we expect to get two square roots of any complex number.

To find the square roots, solve the equation

$z^2 = 7 - 24i$, where $z = x + iy$, where x and y are real.

$(x + iy)^2 = 7 - 24i$ Substitute $x + iy$ for z.

$x^2 + 2xyi + i^2y^2 = 7 - 24i$ Expand the bracket and simplify.

$x^2 + 2xyi - y^2 = 7 - 24i$

$x^2 - y^2 = 7$ (1)   Compare real and imaginary parts to get simultaneous equations.

$2xy = -24$ (2)

▶ Continued on the next page

From (2) $y = \dfrac{-12}{x}$ and substitute for y in equation (1):

$x^2 - \left\{\dfrac{-12}{x}\right\}^2 = 7$

$x^2 - \dfrac{144}{x^2} = 7$

$x^4 - 144 = 7x^2$

$x^4 - 7x^2 - 144 = 0$

$(x^2 - 16)(x^2 + 9) = 0$

$x^2 = 16$ or $x^2 = -9$, but x is real so $x = \pm 4$ only.

Substitute $x = 4$ and $x = -4$ into (2):

$y = \dfrac{-12}{\pm 4} \Rightarrow y = \mp 3$

$x = 4,\ y = -3$ or $x = -4,\ y = 3$

The square roots of $7 - 24\mathrm{i}$ are $4 - 3\mathrm{i}$ and $-4 + 3\mathrm{i}$.

> Eliminate one variable to get a quartic equation in x.

> Solve the equation to get the possible values of x.

> Find the corresponding y values.

> Note that these square roots may be written in the form $\pm(4 - 3\mathrm{i})$.

Exercise 11.3A

1. Find the square roots of the following.
   a) -400
   b) -60
   c) $-3 + 4\mathrm{i}$
   d) $3 + 4\mathrm{i}$
   e) $-5 + 12\mathrm{i}$
   f) $21 + 20\mathrm{i}$
   g) $-21 - 20\mathrm{i}$
   h) $35 - 12\mathrm{i}$
   i) $36\mathrm{i}$

2. a) Find the square roots of $3 - 4\mathrm{i}$.
   b) Hence, solve the equation $z^2 + (2 + \mathrm{i})z + 2\mathrm{i} = 0$.

3. Solve the equation $z + 2z^* = \dfrac{15}{2 + \mathrm{i}}$.

4. Express the square roots of i and $-\mathrm{i}$ in the form $x + \mathrm{i}y$, where x and y are real.

5. Solve the following equations.
   a) $z^2 = 25\mathrm{i}$
   b) $z^2 = -16\mathrm{i}$
   c) $3z^2 + 12 = 0$

6. Find the real numbers a and b which satisfy the equation $(2 + 3\mathrm{i})(5 + b\mathrm{i}) = a + 11\mathrm{i}$.

7. The complex number z is defined by $\dfrac{z - 1}{z + 1 - 2\mathrm{i}} = 1 + \mathrm{i}$.

   Express z in the form $x + \mathrm{i}y$, where x and y are real.

8. Solve the equations, expressing your solutions in the form $x + \mathrm{i}y$, where x and y are real.
   a) $z = \mathrm{i}z + 1$
   b) $z^2 = \mathrm{i}z + 1$

9. Solve the equation $(4 + 5\mathrm{i})z - (1 + \mathrm{i})z^* = 15 + 7\mathrm{i}$, expressing z in the form $x + \mathrm{i}y$, where x and y are real.

10. Given that $(a - 3\mathrm{i})^2 = 16 - b\mathrm{i}$, where a and b are real, find the values of a and b.

We now turn our attention to solving polynomial equations involving solutions which may be complex numbers.

We saw earlier in this chapter that if one root of a quadratic equation with real coefficients is complex then the other root will be its conjugate. We now extend this result to any polynomial with real coefficients.

> If $z = x + iy$ is a root of a polynomial equation with real coefficients, then $z^* = x - iy$ is also a root of the polynomial equation, where z^* is the complex conjugate of z.

This result means that the number of complex roots of a polynomial equation with real coefficients will always be even.

Example 8

Given that $2 + i$ is a root of the equation $z^3 - z^2 - 7z + 15 = 0$, find the other two roots.

Since one complex root is $2 + i$ and the coefficients of the equation are all real, the conjugate $2 - i$ is also a root.

$[z - (2 + i)][z - (2 - i)]$ is a factor. ← Write down the product of the known factors.

$z^2 - (2 + i)z - (2 - i)z + (2 + i)(2 - i)$

$z^2 - 4z + (4 + 2i - 2i - i^2)$ ← Multiply out and simplify.

$z^2 - 4z + 5$ is a factor.

To find the third root (which must be real),

$z^3 - z^2 - 7z + 15 = (\text{linear factor}) \times (z^2 - 4z + 5)$

By considering the term in z^3 and the constant term, we can see that the linear factor must be $(z + 3)$.

The other two roots are -3 and $2 - i$.

In this case, we can find the other factor by inspection. An alternative method is to carry out a long division of $(z^3 - z^2 - 7z + 15)$ by $(z^2 - 4z + 5)$.

Exercise 11.3B

1. A cubic equation with real coefficients has three roots. Two of the roots are 2 and $1 + \sqrt{3}i$.

   a) Write down the third root.

   b) Find the equation which has these three roots. Write it in the form $az^3 + bz^2 + cz + d = 0$.

2. a) Find the roots of the equation $z^2 - 2\sqrt{2}z + 4 = 0$, giving your solutions in the form $x + iy$, where x and y are real.

   b) Showing all your working, verify that each root is also a root of the equation $z^8 - 256 = 0$.

3. Solve the following equations, giving your solutions in the form $x + iy$, where x and y are real.

   a) $(z + 2i)^2 = -9$ b) $z^2 = (z + i)^2$ c) $z^3 = (z + i)^3$

4. Solve the simultaneous equations

   $3z + w = 11$ $2iz + 5w = 8 - 9i$.

5. Solve the simultaneous equations

   $(1 - i)z + iw = 7 + 4i$ $iz + (1 - i)w = -7 - 3i$.

6. a) Write down one solution of the equation $z^3 = 1$.

   b) Hence write down one factor of $z^3 - 1$.   c) Solve the equation $z^3 - 1 = 0$.

7. Solve the equation $z^3 = 27$.

8. The complex number u is defined by $u = (1 + i)^4$.

   a) Express u in the form $x + iy$, where x and y are real.

   b) Solve the equation $z^4 = -4$.

9. The complex number $5 + 2i$ is a root of the equation $2z^3 - 15z^2 + 8z + 145 = 0$. Find the other roots.

10. a) Verify that $-1 + 2i$ is a root of the equation $z^3 + z - 10 = 0$.

    b) Find the other two roots of the equation.

11. a) Verify that $2 + 3i$ is a root of the equation $z^4 - 4z^3 + 12z^2 + 4z - 13 = 0$.

    b) Find the other roots of the equation.

12. The polynomial $x^5 + 2x^3 + 10x^2 + x + 10$ is denoted by p(x).

    a) Find a real root of the equation p(x) = 0.

    b) Showing all your working, verify that i is a root of the equation p(x) = 0.

    c) Find all the roots of the equation p(x) = 0.

11.4 Representing complex numbers geometrically

Any complex number $z = x + iy$ can be represented by using a two-dimensional set of coordinate axes where the horizontal axis is used as the 'real axis' (Re) and the vertical axis is used as the 'imaginary axis' (Im).

So, for example, the complex number $3 + 2i$ is represented by the point (3, 2), as shown in the diagram.

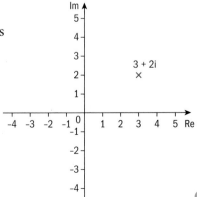

Such a diagram is called an **Argand diagram** and is named after the Swiss mathematician Jean-Robert Argand (1768–1822).

It follows that the point representing $z^* = x - iy$ is the reflection in the real axis of the point representing $z = x + iy$.

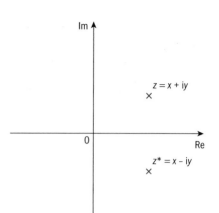

The position of the point representing z^*, the complex conjugate of z, on the Argand diagram is found by reflecting the point representing z in the real axis.

Example 9

$u = 2 - 3i$

On a sketch of an Argand diagram, show the points representing the complex numbers u, u^*, u^2, $u + u^*$, $u - u^*$, uu^* and $\dfrac{u}{u^*}$.

$u^* = 2 + 3i$ ←   Calculate the complex numbers listed.

$u^2 = (2 - 3i)(2 - 3i) = -5 - 12i$

$u + u^* = (2 - 3i) + (2 + 3i) = 4$

$u - u^* = (2 - 3i) - (2 + 3i) = -6i$

$uu^* = (2 - 3i)(2 + 3i) = 13$

$\dfrac{u}{u^*} = \dfrac{2 - 3i}{2 + 3i} \times \dfrac{2 - 3i}{2 - 3i} = \dfrac{-5 - 12i}{13} = -\dfrac{5}{13} - \dfrac{12}{13}i$

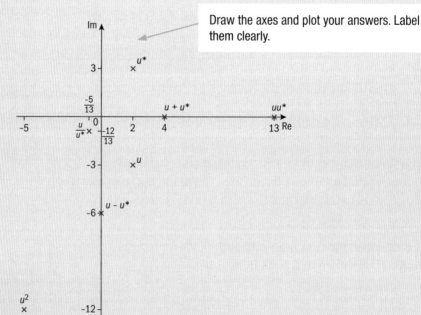

Draw the axes and plot your answers. Label them clearly.

We will now consider how to represent the addition and subtraction of two complex numbers on an Argand diagram. To do this, it is helpful to consider the complex number $z = x + iy$ as a point in the Argand diagram with position vector $\begin{pmatrix} x \\ y \end{pmatrix}$.

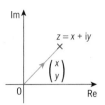

If $z_1 = a + ib$ and $z_2 = c + id$, then
$$z_1 + z_2 = (a + ib) + (c + id)$$
$$= (a + c) + i(b + d)$$

The point representing $z_1 + z_2$ can be constructed by adding the vectors representing z_1 and z_2; that is, the point representing $z_1 + z_2$ can be found by drawing the diagonal of the parallelogram formed by the vectors representing z_1 and z_2. This is illustrated on the diagram.

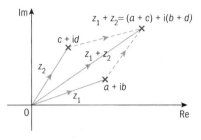

Similarly,
$$z_1 - z_2 = (a - c) + i(b - d)$$

The point representing $z_1 - z_2$ can be constructed by adding the vectors representing z_1 and $-z_2$, then drawing the vector representing $z_1 - z_2$ from the origin.

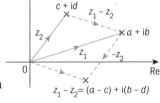

We will look at how we can represent the multiplication and division of complex numbers geometrically later in the chapter. However, at this point, it is worth considering what happens when we multiply a complex number by i.

If $z = x + iy$, then $iz = i(x + iy) = ix - y$, and so the point (x, y) representing z is mapped onto the point $(-y, x)$. The diagram shows that this is equivalent to a rotation of 90° anticlockwise about the origin.

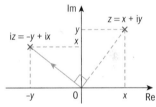

Multiplying a complex number z by the imaginary number i is geometrically equivalent to rotating the point representing z by 90° anticlockwise about the origin.

Exercise 11.4

1. **Sketch** an Argand diagram and show the points A, B, C, and D representing the complex numbers $2 + 3i$, $-1 - 4i$, $3 - i$, and $5 + 0i$ respectively.

2. The complex number u is defined by $u = 5 + 3i$.
   Show, on a sketch of an Argand diagram, the points P, Q, R, and S representing u, u^*, $2u$, and $-u$ respectively.

3. Write down the complex numbers z_1, z_2, z_3, z_4, z_5, z_6, and z_7 represented on the Argand diagram by the points A, B, C, D, E, F, and G respectively.

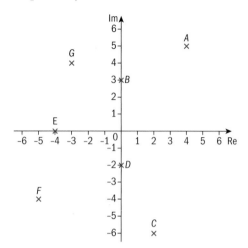

4. The complex number w is defined by $w = (2 + i)^2$.

   a) Express w, iw, $\frac{w}{i}$, and $\frac{25i}{w}$ in the form $x + iy$, where x and y are real.

   b) Sketch an Argand diagram and show the points A, B, C, and D representing w, iw, $\frac{w}{i}$, and $\frac{25i}{w}$ respectively.

5. The complex numbers z_1 and z_2 are defined by $z_1 = 3 + i$ and $z_1 z_2 = -11 + 13i$.

   a) Find z_2, expressing your answer in the form $x + iy$, where x and y are real.

   b) Show, on a sketch of an Argand diagram, the points representing z_1, z_2, $z_1 z_2$, and $iz_1 z_2$.

6. The complex numbers z and w are given below.

   In each case, sketch an Argand diagram to show the positions of z, w, $z + w$, and $z - w$.

   a) $z = 4 + 2i$, $w = 1 + 3i$ b) $z = 2 + 5i$, $w = -4 + i$

   c) $z = -5 - 3i$, $w = 2 + 3i$ d) $z = 3 - 4i$, $w = 1 + i$

7. The complex number z is defined by $z = 2 - 2i$.

   Show, on a sketch of an Argand diagram, the points representing z, $\frac{1}{z}$, and $z + \frac{1}{z}$.

11.5 Polar form and exponential form

As we know, the position of a point can be described by using the Cartesian coordinates (x, y) and this can be plotted on a set of coordinate axes.

Alternatively, we can describe the position of a point by giving its distance from the origin (r) together with the angle that the line joining the origin to the point makes with the positive x-axis (θ radians).

This way of describing the position of a point is called **polar coordinate form**. Polar coordinates are written in the form (r, θ) where $r > 0$ and $-\pi < \theta \leq \pi$.

As we saw in section 11.4, complex numbers expressed in Cartesian form $(x + iy)$ can be plotted on an Argand diagram. We can also represent complex numbers in polar form. This is more commonly called **modulus–argument form** when it is being used with complex numbers.

The modulus of a complex number z, written as $|z|$, is the length of OP, where P is the point representing the complex number on an Argand diagram.

The **argument** of a complex number z, written as $\arg(z)$, is the angle (θ radians) between the line $\theta = 0$ and OP.

Hence, $r = \sqrt{x^2 + y^2}$ and $\tan \theta = \dfrac{y}{x}$.

Also, $x = r \cos \theta$ and $y = r \sin \theta$.

From the diagram we see that $x + iy = r \cos \theta + i r \sin \theta = r(\cos \theta + i \sin \theta)$.

$$x + iy = r(\cos \theta + i \sin \theta)$$

Any complex number can be expressed in Cartesian form as $z = x + iy$ or in modulus–argument form as $z = (r, \theta)$.

Use $r = \sqrt{x^2 + y^2}$ and $\tan \theta = \dfrac{y}{x}$ to convert from Cartesian to modulus–argument form.

Use $x = r \cos \theta$ and $y = r \sin \theta$ to convert from modulus–argument form to Cartesian form.

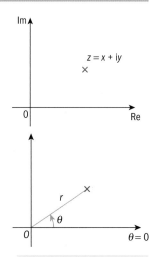

θ may be defined in the interval $-\pi < \theta \leq \pi$ or alternatively in the interval $0 \leq \theta < 2\pi$.

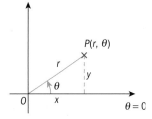

Notes:

$r > 0, -\pi < \theta \leq \pi$

Care needs to be taken to make sure you give the correct argument for a complex number. Always sketch a diagram to check.

Example 10

a) The complex numbers z_1 and z_2 are defined by $z_1 = 1 + i$ and $z_2 = -2 - 4i$.
   Find the modulus and argument of z_1 and z_2.

b) The complex numbers z_3 and z_4 are defined by $z_3 = \left(5, \dfrac{\pi}{3}\right)$ and $z_4 = \left(2, \dfrac{5\pi}{6}\right)$.
   Write z_3 and z_4 in the form $x + iy$, where x and y are real numbers.

....

a) $|z_1| = \sqrt{1^2 + 1^2} = \sqrt{2} = 1.41$ (3 s.f.)

$\arg(z_1) = \tan^{-1}\left(\dfrac{1}{1}\right) = \dfrac{\pi}{4} = 0.785$ (3 s.f.)

$|z_2| = \sqrt{(2)^2 + (4)^2} = \sqrt{20} = 4.47$ (3 s.f.)

$\arg(z_2) = -\left(\pi - \tan^{-1}\left(\dfrac{4}{2}\right)\right) = -2.03$ (3 s.f.)

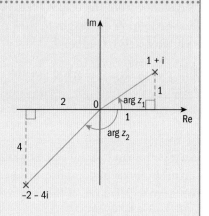

It can be useful to sketch a diagram.

b) $z_3 = 5\left(\cos\dfrac{\pi}{3} + i\sin\dfrac{\pi}{3}\right) = \dfrac{5}{2} + \dfrac{5\sqrt{3}}{2}i = 2.5 + 4.33i$

$z_4 = 2\left(\cos\dfrac{5\pi}{6} + i\sin\dfrac{5\pi}{6}\right) = -\sqrt{3} + 1i = -1.73 + i$

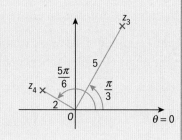

Exponential form is similar to modulus–argument form. A complex number is said to be in exponential form when it is expressed in the form $re^{i\theta}$, where r is the modulus and θ is the argument.

> The derivation of this form is beyond the Cambridge International 9709 syllabus.

For example, the following are all different forms for the same complex number:

(1) $1 + i$ Cartesian form (or real–imaginary form)

(2) $\left(\sqrt{2}, \dfrac{\pi}{4}\right) \equiv \sqrt{2}\left(\cos\dfrac{\pi}{4} + i\sin\dfrac{\pi}{4}\right)$ polar form (or modulus–argument form)

(3) $\sqrt{2}e^{i\frac{\pi}{4}}$ exponential form.

A complex number written in the form $re^{i\theta}$, where r is the modulus and θ is the argument, is said to be in exponential form.

Expressing a complex number in the form $r(\cos\theta + i\sin\theta) \equiv re^{i\theta}$ can help us to multiply or divide two complex numbers more easily.

To show why this is, consider the two complex numbers

$z_1 = (r_1, \theta_1) = r_1(\cos\theta_1 + i\sin\theta_1)$ and $z_2 = (r_2, \theta_2) = r_2(\cos\theta_2 + i\sin\theta_2)$.

$$\begin{aligned}
z_1 \times z_2 &= r_1(\cos\theta_1 + i\sin\theta_1) \times r_2(\cos\theta_2 + i\sin\theta_2) \\
&= r_1 r_2(\cos\theta_1 + i\sin\theta_1)(\cos\theta_2 + i\sin\theta_2) \\
&= r_1 r_2(\cos\theta_1\cos\theta_2 + i\sin\theta_1\cos\theta_2 + i\sin\theta_2\cos\theta_1 + i^2\sin\theta_1\sin\theta_2) \\
&= r_1 r_2((\cos\theta_1\cos\theta_2 - \sin\theta_1\sin\theta_2) + i(\sin\theta_1\cos\theta_2 + \sin\theta_2\cos\theta_1)) \\
&= r_1 r_2(\cos(\theta_1 + \theta_2) + i\sin(\theta_1 + \theta_2))
\end{aligned}$$

So, to multiply two complex numbers in polar form, we multiply their moduli and add their arguments.

If $z_1 = (r_1, \theta_1)$ and $z_2 = (r_2, \theta_2)$, then $z_1 z_2 = (r_1 r_2, \theta_1 + \theta_2)$.

A special case of this can be used to find powers of a complex number.

If $z = (r, \theta)$, then $z^n = (r^n, n\theta)$.

One advantage of writing a complex number in exponential form is that much of our work can be written more concisely. For example, the proof above can be rewritten as

$$z_1 z_2 = r_1 e^{i\theta_1} \times r_2 e^{i\theta_2}$$
$$= r_1 r_2 e^{i(\theta_1 + \theta_2)}$$

Similarly, if we use this notation to divide two complex numbers,

$$\frac{z_1}{z_2} = \frac{r_1 e^{i\theta_1}}{r_2 e^{i\theta_2}} = \frac{r_1}{r_2} e^{i(\theta_1 - \theta_2)}$$

So, to divide two complex numbers in polar form, we divide their moduli and subtract their arguments.

If $z_1 = (r_1, \theta_1)$ and $z_2 = (r_2, \theta_2)$, then $\dfrac{z_1}{z_2} = \left(\dfrac{r_1}{r_2}, \theta_1 - \theta_2\right)$.

This proof using $z_1 = (r_1, \theta_1) = r_1(\cos\theta_1 + i\sin\theta_1)$ and $z_2 = (r_2, \theta_2) = r_2(\cos\theta_2 + i\sin\theta_2)$ is left as an exercise.

Example 11

The complex numbers z and w are defined by $z = 2 + 2i$ and $w = -1 + \sqrt{3}\,i$.

Find the modulus and argument of

a) z^2 b) z^3 c) z^2w^2 d) $\dfrac{w^2}{z^3}$.

a) $z = \left(2\sqrt{2}, \dfrac{\pi}{4}\right)$ ←   Write z in modulus–argument form.

$z^2 = \left\{\left(2\sqrt{2}\right)^2, 2 \times \dfrac{\pi}{4}\right\} = \left(8, \dfrac{\pi}{2}\right)$ ←   Use $(r, \theta)^2 = (r^2, 2\theta)$ and simplify.

$|z^2| = 8,\ \arg(z^2) = \dfrac{\pi}{2}$

b) $z^3 = \left\{\left(2\sqrt{2}\right)^3, 3 \times \dfrac{\pi}{4}\right\} = \left(16\sqrt{2}, \dfrac{3\pi}{4}\right)$ ←   Use $(r, \theta)^3 = (r^3, 3\theta)$ and simplify.

$|z^3| = 16\sqrt{2},\ \arg(z^2) = \dfrac{3\pi}{4}$

Alternatively, exponential form notation may be used.

For example, $z = 2\sqrt{2}\,e^{\frac{i\pi}{4}}$

$z^2 = \left(2\sqrt{2}\,e^{\frac{i\pi}{4}}\right)^2$

$= 8e^{\frac{i\pi}{2}}$ etc.

c) $w = \left(2, \dfrac{2\pi}{3}\right)$

$w^2 = \left(2^2, 2 \times \dfrac{2\pi}{3}\right) = \left(4, \dfrac{4\pi}{3}\right)$ ←   Find w^2 and use your answer for z^2 from part (a).

$z^2w^2 = \left(8, \dfrac{\pi}{2}\right) \times \left(4, \dfrac{4\pi}{3}\right) = \left(8 \times 4, \dfrac{\pi}{2} + \dfrac{4\pi}{3}\right)$ ←   Use $(r_1, \theta_1) \times (r_2, \theta_2) = (r_1\,r_2, \theta_1 + \theta_2)$.

$|z^2w^2| = 32,\ \arg(z^2w^2) = \dfrac{11\pi}{6} = -\dfrac{\pi}{6}$ ←   Write $\arg(z)$ so it lies between $-\pi$ and π.

d) $\dfrac{w^2}{z^3} = \left\{\dfrac{4}{16\sqrt{2}}, \dfrac{4\pi}{3} - \dfrac{3\pi}{4}\right\} = \left\{\dfrac{\sqrt{2}}{8}, \dfrac{7\pi}{12}\right\}$ ←   Use $\dfrac{(r_1, \theta_1)}{(r_2, \theta_2)} = \left(\dfrac{r_1}{r_2}, \theta_1 - \theta_2\right)$ and simplify.

$\left|\dfrac{w^2}{z^3}\right| = \dfrac{\sqrt{2}}{8},\ \arg\left(\dfrac{w^2}{z^3}\right) = \dfrac{7\pi}{12}$

We now turn our attention to understanding the geometrical effect of multiplying and dividing two complex numbers.

If we take a complex number, say $z_1 = (r_1, \theta_1)$, and multiply it by a second complex number $z_2 = (r_2, \theta_2)$, then as $z_2z_1 = (r_2, \theta_2) \times (r_1, \theta_1) = (r_2\,r_1, \theta_2 + \theta_1)$, we see that the effect of multiplying one complex number by another is equivalent to multiplying the modulus of z_1 by the modulus of z_2. This is equivalent to an 'enlargement', centre the origin and scale factor (r_2) equal to the modulus of z_2. Using the same result, we can see that we add the argument of z_2 to the argument of z_1. This is the equivalent of a rotation (anticlockwise) about the origin through an angle (θ_2) equal to the argument of z_2.

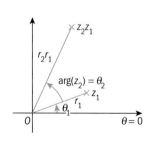

The combined effect of these two aspects results in what is often referred to as a spiral enlargement.

Dividing one complex number by another can be interpreted in a similar way, though the enlargement scale factor will be $\dfrac{1}{r_2}$ and the rotation will be clockwise by an angle θ_2.

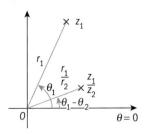

Exercise 11.5

1. Find the modulus and argument of the following complex numbers.

    a) 1   b) −1   c) $1 + i$ d) $1 - i$

    e) $3i$ f) $5 + 12i$ g) $\sqrt{13} - 6i$ h) $\dfrac{\sqrt{2}}{2} + \dfrac{\sqrt{2}}{2}i$

2. $z = \dfrac{1}{2} + \dfrac{1}{2}\sqrt{3}\,i$

    Find the modulus and argument in each case.

    a) z b) z^* c) z^2 d) $(z^*)^2$ e) $(z^2)^*$

    f) z^3 g) z^6 h) $z^3 + z^6$ i) $\dfrac{1}{z}$ j) $\dfrac{2}{z-1}$

3. $z = 4 + 3i$

    a) Find   i) $|z|$ ii) $|z^2|$ iii) $|z^3|$.

    b) Find   i) $\arg z$ ii) $\arg z^2$ iii) $\arg z^3$.

4. Write the following complex numbers, z, in the form $x + iy$.

    a) $|z| = 2$, $\arg(z) = \dfrac{\pi}{2}$ b) $|z| = 5$, $\arg(z) = \pi$ c) $|z| = 4$, $\arg(z) = \dfrac{\pi}{3}$

    d) $|z| = 3$, $\arg(z) = \dfrac{\pi}{6}$ e) $|z| = 10$, $\arg(z) = \dfrac{\pi}{4}$ f) $|z| = \dfrac{1}{2}$, $\arg(z) = -\dfrac{\pi}{2}$

    g) $|z| = 1$, $\arg(z) = -\dfrac{3\pi}{4}$ h) $|z| = 6$, $\arg(z) = \dfrac{2\pi}{3}$ i) $|z| = 7$, $\arg(z) = 0$

5. Show, on a sketch of an Argand diagram, the points representing the complex numbers in question 4.

6. $z = 1 + i$, $w = 1 - \sqrt{3}i$

    Find the modulus and argument of

    a) wz b) $\dfrac{w}{z}$ c) iz d) $\dfrac{1}{w^*}$.

7. Write the following complex numbers in the form $x + iy$.

    a) $e^{\pi i}$ b) $e^{-\frac{\pi i}{2}}$ c) $4e^{\frac{\pi i}{4}}$

8. Express in the form $r e^{i\theta}$.

    a) $1 + i$ b) -3 c) $-1 + 2i$

    d) $7i$ e) $6 + 8i$ f) $6 - 8i$

    g) $-\sqrt{2} - \sqrt{2}i$ h) $\sqrt{2}(5 - i)$ i) $\dfrac{1}{2} + \dfrac{\sqrt{3}}{2}i$

9. If $z_1 = 2\left(\cos\dfrac{\pi}{6} + i\sin\dfrac{\pi}{6}\right)$, $z_2 = 3\left(\cos\dfrac{\pi}{3} + i\sin\dfrac{\pi}{3}\right)$, and $z_3 = 2\left(\cos\dfrac{\pi}{4} + i\sin\dfrac{\pi}{4}\right)$, find, in modulus–argument form,

a) $z_1 z_2$
b) $z_1 z_2 z_3$
c) $\dfrac{z_1}{z_2}$
d) $z_3{}^2$.

10. The complex numbers z and w are defined by

$|z| = 3$, $\arg(z) = \dfrac{\pi}{4}$ and $|w| = 2$, $\arg(w) = \dfrac{\pi}{6}$.

Find the modulus and argument of

a) z^2
b) z^3
c) z^4
d) z^8
e) zw
f) zw^2
g) $(zw)^3$
h) $\dfrac{z}{w}$
i) $\dfrac{1}{z^2}$
j) $w + w^*$
k) $w - w^*$
l) $\dfrac{1}{(wz)^*}$.

11. The complex number z is given by $z = x + iy$.

Prove algebraically that

a) $z + z^* = 2\operatorname{Re}(z)$
b) $z - z^* = 2\operatorname{Im}(z)i$
c) $|\operatorname{Re}(z)| \le |z|$.

12. Illustrate each of the results from question 11 on a separate sketch of an Argand diagram.

11.6 Loci in the Argand diagram

In this section we will see how we can illustrate simple equations and inequalities involving complex numbers as loci in an Argand diagram.

In particular we will consider the locus of z (where z is a variable complex number) in the following three cases, taking into account how the locus changes if the = sign is replaced by an inequality sign.

Case 1 $|z - z_1| = r$, where z_1 is a known complex number and r is real

$|z - z_1|$ is the modulus (or length) of $z - z_1$.
In section 11.4 we saw that $z - z_1$ can be represented as the vector joining the point z_1 to the point representing z. So $|z - z_1| = r$ represents the locus of a point z such that z moves so that its distance from a fixed point z_1 is always r.
z lies on a circle, centre z_1, radius r.

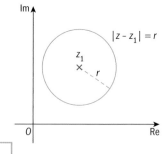

If $|z - z_1| = r$, then z lies on a circle, centre z_1, radius r.

We can also see that if $|z - z_1| < r$, then z lies anywhere inside a circle, centre z_1, radius r;

and if $|z - z_1| > r$, then z lies anywhere outside a circle, centre z_1, radius r.

Example 12

a) Sketch an Argand diagram and show all the points which represent the complex numbers z that satisfy the equation $|z - 1 + i| = 2$.

b) The points in an Argand diagram representing $2 + 5i$ and $-6 + i$ are the ends of a diameter of a circle. Find the equation of the circle, giving your answer in the form $|z - (a + bi)| = k$.

a) $|z - (1 - i)| = 2$ ⟵   Write the equation in the form $|z - z_1| = r$.

z lies on a circle, centre $1 - i$ and radius 2.

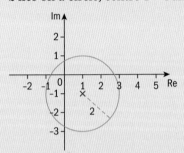

b) The centre of the circle is given by

$$\frac{(2 + -6)}{2} + \frac{(5i + i)}{2} = -2 + 3i$$

The radius of the circle is given by

$$\frac{1}{2}\left\{\sqrt{(-6-2)^2 + (1-5)^2}\right\} = \frac{1}{2}\sqrt{80} = 2\sqrt{5}$$

The equation of the circle is $|z - (-2 + 3i)| = 2\sqrt{5}$.

Use the result that the mid-point of the line joining (x_1, y_1) and (x_2, y_2) is given by

$$\left\{\frac{x_1 + x_2}{2}, \frac{y_1 + y_2}{2}\right\}$$ and that the length of

the line joining (x_1, y_1) and (x_2, y_2) is given by

$$\sqrt{(x_2 - x_1)^2 + (y_2 - y_1)^2}.$$

Example 13

Sketch an Argand diagram and shade the region whose points represent the complex numbers z which satisfy both the inequality $|z - (2 + 3i)| \leq 2$ and the inequality $|z - 4| > 3$.

$|z - (2 + 3i)| \leq 2$

z lies inside or on the circumference of a circle, centre $2 + 3i$ and radius 2.

$|z - 4| > 3$

z lies outside a circle, centre $4 + 0i$ and radius 3.

The set of points which satisfies both conditions lies in the shaded region and on the part of the circle, centre $2 + 3i$, which is shown with a solid line.

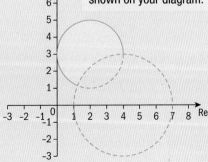

Note that the circle touches the y-axis. This should be clearly shown on your diagram.

Case 2 $|z - z_1| = |z - z_2|$ **where** z_1 **and** z_2 **are known complex numbers**

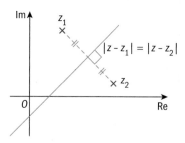

Here the locus of z is such that its distance from the fixed point z_1 is equal to its distance from the fixed point z_2. z will therefore lie on the perpendicular bisector (or mediator) of the line joining z_1 and z_2.

> If $|z - z_1| = |z - z_2|$, then z lies on the perpendicular bisector of the line joining z_1 to z_2.

We can also see that if $|z - z_1| < |z - z_2|$, then z lies anywhere in the region on one side of the perpendicular bisector of the line joining z_1 to z_2 such that the distance from z to z_1 is less than the distance from z to z_2.

Example 14

a) Sketch on an Argand diagram the loci given by $|z - (1 + 2i)| = 5$ and $|z - 5 + i| = |z + 3 - 5i|$.

b) Show that these loci intersect at the point $-2 - 2i$.

. . . .

a) $|z - (1 + 2i)| = 5$

z lies on a circle, centre $1 + 2i$

and radius 5.

$|z - 5 + i| = |z + 3 - 5i|$

$|z - (5 - i)| = |z - (-3 + 5i)|$

z lies on the perpendicular bisector of the line joining $5 - i$ to $-3 + 5i$.

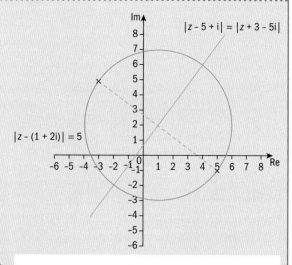

b) The loci will intersect at $-2 - 2i$ if $z = -2 - 2i$ satisfies both equations.

When $z = -2 - 2i$,

$|z - (1 + 2i)| = |-2 - 2i - (1 + 2i)|$

$= |-3 - 4i|$

$= 5$

so $|z - (1 + 2i)| = 5$ is satisfied.

When $z = -2 - 2i$,

$|z - 5 + i| = |-2 - 2i - 5 + i|$

$= |-7 - i|$

$= \sqrt{50}$

Also $|z + 3 - 5i| = |-2 - 2i + 3 - 5i| = |1 - 7i|$

$= \sqrt{50}$

so $|z - 5 + i| = |z + 3 - 5i|$ is satisfied.

Both conditions are satisfied and so these loci intersect at the point $-2 - 2i$.

Substitute $z = -2 - 2i$ into the left-hand side of the equation.

Substitute $z = -2 - 2i$ into each side of the equation in turn.

Make a clear statement in conclusion.

Case 3 $\arg(z - z_1) = \theta$ where z_1 is a known complex number and θ is an angle, measured in radians

As we saw earlier in this chapter, the vector $z - z_1$ can be represented by the vector joining the point z_1 to the point representing z. We can interpret $\arg(z - z_1) = \theta$ as the line joining z_1 to z having argument θ.

The locus of z is therefore a half-line, starting at z_1, at an angle θ with the positive real axis. It is called a half-line as we only want the part of the line which starts at z_1 but which has an infinite length.

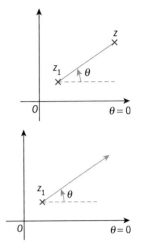

> If $\arg(z - z_1) = \theta$, then the locus of z is a half-line, starting at z_1, at an angle θ with the positive real axis.

Example 15

a) Sketch on an Argand diagram the locus of the complex number z where z satisfies the equation $\arg(z + 2 - i) = \dfrac{2\pi}{3}$.

b) On the same diagram, shade the region whose points represent the complex number z which satisfies $\dfrac{\pi}{2} \leq \arg(z + 2 - i) \leq \dfrac{2\pi}{3}$.

a) $\arg\{z - (-2 + i)\} = \dfrac{2\pi}{3}$ ← Write in the form $\arg(z - z_1) = \theta$.

z lies on a half-line, starting at $-2 + i$, making an angle $\dfrac{2\pi}{3}$ with the positive real axis. ← Identify z_1 and θ.

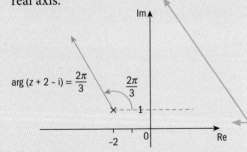

Draw a clear diagram. It is helpful to write a description to accompany your diagram.

b) z may lie anywhere in the shaded region or on its boundaries.

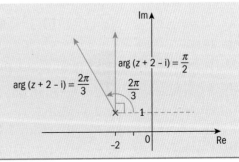

Example 16

a) On a sketch of an Argand diagram, shade the region whose points represent the complex numbers z which satisfy the inequality $|z + 4 - 4\sqrt{3}i| \leq 4$.

b) i) Find the least value of $|z|$ in this region.

ii) Find the greatest value of arg z in this region.

• •

a) $|z - (-4 + 4\sqrt{3}i)| \leq 4$

z lies on or inside a circle,

centre $(-4 + 4\sqrt{3}i)$ and radius 4.

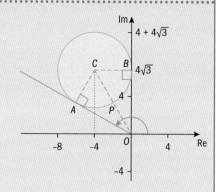

b) i) The least value of $|z|$ is given by the length OP on the diagram.

$$OP = OC - PC$$

But $OC^2 = 4^2 + \left(4\sqrt{3}\right)^2 = 16 + 48 = 64$ ⟵   Use Pythagoras' theorem on triangle OBC.

So $OP = OC - PC = 8 - 4 = 4$

Least value of $|z|$ is 4. ⟵   Use $PC = 4$ as it is a radius of the circle.

ii) The greatest value of arg z is given by the angle OA makes with the positive real axis.

$$\text{Angle } COA = \sin^{-1}\left(\frac{CA}{OC}\right) = \sin^{-1}\left(\frac{4}{8}\right) = \frac{\pi}{6}$$

The greatest value of arg z is $\dfrac{\pi}{6} + \dfrac{\pi}{6} + \dfrac{\pi}{2} = \dfrac{5\pi}{6}$ ⟵   Angle COA = angle $COB = \dfrac{\pi}{6}$

Exercise 11.6

1. On separate sketches of an Argand diagram, show the locus of the complex number z which satisfies

   a) $|z| = 1$ **b)** $|z - 3| = 1$ **c)** $|z + 5| = 2$

   d) $|z - (2 - 2i)| \leq 2$ **e)** $|z + (1 + 3i)| \geq 1$

   f) $|z - a| = a$, where a is a positive real number.

2. **a)** Express, using modulus notation, the equations of the following circles on the Argand diagram.

   i) centre 0, radius 3   **ii)** centre $2 + 2i$, radius 2   **iii)** centre $-4 - 3i$, radius 5

   b) On separate sketches of an Argand diagram, illustrate these circles.

3. On separate sketches of an Argand diagram, show the locus of the complex number z which satisfies the following.

   a) $\arg(z - 1) = \dfrac{\pi}{2}$

   b) $\arg(z + 1) = \dfrac{\pi}{3}$

   c) $\arg(z + i) = -\dfrac{2\pi}{3}$

   d) $\arg\{z - (2 + 3i)\} = \dfrac{3\pi}{4}$

4. On separate sketches of an Argand diagram, show the locus of the complex number z which satisfies the following.

   a) $|z| = |z - 5|$

   b) $|z + i| = |z - 7i|$

   c) $|z + 4| = |z - 2i|$

   d) $|z - (2 + 2i)| = |z - (-2 - 2i)|$

   e) $|z - 5 - 2i| = |z - 1 - 2i|$

5. Show, by shading on separate sketches of an Argand diagram, the region whose points represent the complex number z which satisfies the following.

   a) $\dfrac{\pi}{6} \le \arg z \le \dfrac{\pi}{2}$

   b) $0 \le \arg(z - i) \le \dfrac{\pi}{3}$

   c) $\dfrac{\pi}{4} \le \arg(z - 2) \le \dfrac{2\pi}{3}$

   d) $-\dfrac{\pi}{3} \le \arg\{z - (5 + 3i)\} \le \dfrac{\pi}{2}$

6. Show, by shading on separate sketches of an Argand diagram, the region whose points represent the complex number z which satisfies the following.

   a) $|z - 6| \le 3$

   b) $|z - (1 + 3i)| \ge 1$

   c) $|z - 3| \ge |z - 1|$

   d) $|z - i| \ge |z + 1|$

   e) $|z + (1 + i)| \ge |z - 3 - 3i|$

   f) $|z - 2 + 3i| \le |z - 2i|$

7. Sketch in a diagram the region described by $|z - 2| \le 2$ **and** $-\dfrac{\pi}{6} \le \arg z \le \dfrac{\pi}{6}$.

8. Sketch in a diagram the region described by $|z| \le 3$ **and** $|z| \ge |z - 4|$.

9. a) On a sketch of an Argand diagram, shade the region whose points represent complex numbers satisfying the inequalities $|z| \le 4$ and $\mathrm{Re}(z) \ge \mathrm{Im}(z)$, where $\mathrm{Re}(z)$ and $\mathrm{Im}(z)$ denote the real and imaginary parts of z respectively.

   b) Write down the greatest and least values of $|z|$.

10. a) On a sketch of an Argand diagram, shade the region whose points represent complex numbers satisfying the inequality $|z - \sqrt{5} - 2i| \le 2$.

    b) Calculate the greatest value of $|z|$.

    c) Calculate the greatest value of $\arg z$.

1. Find the values of x and y, where x and y are real, if $\dfrac{1}{x+iy} = \dfrac{4+5i}{6-3i}$.

2. **a)** Show that the two roots of the equation $z^2 + 2z + 10 = 0$ have the same modulus.

    b) Find the argument of each root of the equation $z^2 + 2z + 10 = 0$.

3. **a)** Show that $1, -\dfrac{1}{2} + \dfrac{\sqrt{3}}{2}i,$ and $-\dfrac{1}{2} - \dfrac{\sqrt{3}}{2}i$ are all solutions of the equation $z^3 = 1$.

    b) Hence find $\left\{ -\dfrac{1}{2} + \dfrac{\sqrt{3}}{2}i \right\}^4$ and $\left\{ -\dfrac{1}{2} + \dfrac{\sqrt{3}}{2}i \right\}^5$, giving your answers in the form $x + iy$.

4. The complex number z is given by $z = 1 - \sqrt{5}i$.

    a) Express z^2 in the form $x + iy$ where x and y are real.

    b) Find the value of the real number p such that $z^2 - pz$ is real.

5. Two complex numbers, z and w, satisfy the inequalities $|z - 3 - 3i| \le 2$ and $|w + 1 + i| \le 1$.

    a) On a sketch of an Argand diagram, show the locus of the complex number z and the locus of the complex number w.

    b) Find the least and greatest possible values of $|z - w|$.

6. Show that, for any complex number z, $zz^* + 2(z + z^*)$ is real.

7. **a)** Solve the equation $z^2 + 5z + 4 = 0$.

    b) Hence solve the equation $z^4 + 5z^2 + 4 = 0$.

8. **a)** Sketch in a diagram the locus of the complex number z described by $|z + 3 + 12i| = 3\sqrt{17}$.

    b) Calculate the greatest and least values of $|z|$ for this locus.

9. **a)** Sketch in a diagram the locus of the complex number z described by $|z - 5i| = 3$.

    b) Calculate the greatest and least values of $|z|$ and arg z for this locus.

10. Two complex numbers z and w satisfy the inequalities $|z - 3 - 5i| \le 2$ and $|w - 7 - 10i| \le 2$. Draw an Argand diagram and find the least possible value and the greatest possible value of $|z - w|$.

11. **a)** Verify that $z = 1 + i$ is a root of the equation $z^4 - 6z^3 + 23z^2 - 34z + 26 = 0$.

    b) Write down another root of this equation.

    c) Find all the roots of the equation $z^4 - 6z^3 + 23z^2 - 34z + 26 = 0$.

12. $2 - \sqrt{5}i$ is a root of the equation $z^4 + 4z^3 - 20z^2 + 60z + 27 = 0$. Find the other roots of this equation.

13. **a)** Verify that $z = 1 + 2i$ is a root of the equation $z^3 - 3z^2 + 7z - 5 = 0$.

    b) Solve the equation completely.

14. An equilateral triangle has its vertices on the circle $|z| = 3$. One vertex is at the point such that arg $z = \dfrac{\pi}{6}$. Find the three vertices of the equilateral triangle, expressing your answers in the form $x + iy$, where x and y are real.

15. The complex number w is given by
$$w = -\dfrac{1}{2} + \dfrac{\sqrt{3}}{2}i$$

    a) Find the value of w^2, expressing your answer in the form $x + iy$, where x and y are real.

    b) Verify that $w^4 = w$.

c) i) Write w and w^2 in modulus–argument form.

ii) Explain, using an Argand diagram, why $1 + w + w^2 = 0$.

16. a) Use a sketch of an Argand diagram illustrating the addition of the two complex numbers z_1 and z_2 to show that

$$|z_1 + z_2| \leq |z_1| + |z_2|$$

b) Use a sketch of an Argand diagram illustrating the subtraction of the two complex numbers z_1 and z_2 to show that

$$|z_1 - z_2| \geq |z_1| - |z_2|$$

17. a) Solve the equation $\dfrac{(z-2)}{(z+2-3i)} = 1 - i$,

giving your answer in the form $z = x + iy$ where x and y are real.

b) i) Sketch an Argand diagram showing the set of points representing the complex numbers z which satisfy arg

$(z - 3) = \dfrac{\pi}{3}$

ii) Arg $(z - 3) = \dfrac{\pi}{3}$. Find the least value of $|z|$.

18. i) Verify that the complex number $z_1 = 1 + i$ is a root of the equation $2z^3 - 3z^2 + 2z + 2 = 0$.

ii) Find the other two roots of this equation.

iii) Sketch an Argand diagram showing the set of points representing the complex numbers z which satisfy $|z| = |z - 1 - i|$.

19. $z_1 = -3 + 4i$ is a solution of the equation $z^2 + cz + 25 = 0$.

i) Find the value of c.

ii) Write down the other root of the equation.

iii) Write z_1 in modulus argument form giving arg z_1 in radians correct to 2 decimal places.

iv) Show, on a sketch of an Argand diagram, the points A, B, and C representing the complex numbers z_1, $z_1{}^*$, and $z_1 + z_1{}^*$.

Chapter summary

Introducing complex numbers

- $\sqrt{-1} = i$ and therefore $i^2 = -1$.
- A number of the form $a + bi$, where a and b are real and $i^2 = -1$, is called a complex number.
- The complex conjugate of $z = a + bi$ is $z^* = a - bi$, where a and b are real numbers.

Calculating with complex numbers.

- To add two complex numbers we add their real parts and add their imaginary parts.
- To subtract two complex numbers we subtract their real parts and subtract their imaginary parts.
- To multiply two complex numbers we use the normal rules of algebra and simplify the answer using $i^2 = -1$.
- To divide two complex numbers we multiply the numerator and the denominator of the fraction by the complex conjugate of the denominator.

Solving equations involving complex numbers

- Two complex numbers are equal if both the real parts are equal and the imaginary parts are equal.
- If $z = x + iy$ is a root of a polynomial equation with real coefficients, then $z^* = x - iy$ is also a root of the polynomial equation, where z^* is the complex conjugate of z.
- We can use the method of comparing real parts and imaginary parts to find the square roots of a complex number, $a + bi$.

Representing complex numbers geometrically on an Argand diagram

- The position of z^*, the complex conjugate of z, on the Argand diagram is found by reflecting z in the real axis.

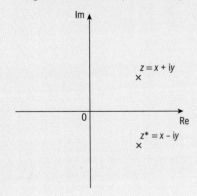

- $z_1 + z_2$ can be represented by drawing the diagonal of the parallelogram formed by adding the vectors representing z_1 and z_2.

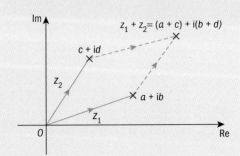

- The points representing $z_1 - z_2$ are constructed by adding the vectors representing z_1 and $-z_2$, then drawing the vector representing $z_1 - z_2$ from the origin.

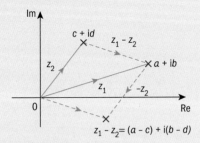

- Multiplying a complex number z, by the imaginary number i is geometrically equivalent to rotating the point representing z by 90° anticlockwise about the origin.

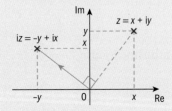

Modulus–argument and exponential forms

- Any complex number can be expressed in Cartesian form as $z = x + iy$ or in modulus–argument (or polar) form as $z = (r, \theta)$.
- For modulus–argument form $r > 0$ and $-\pi \leq \theta \leq \pi$.
- Use $r = \sqrt{x^2 + y^2}$ and $\tan \theta = \dfrac{y}{x}$ to convert from Cartesian to modulus–argument form.

- Use $x = r\cos\theta$ and $y = r\sin\theta$ to convert from modulus–argument form to Cartesian form.
- $x + iy = r(\cos\theta + i\sin\theta)$

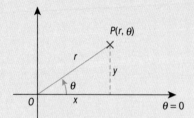

- A complex number written in the form $re^{i\theta}$, where r is the modulus and θ is the argument, is said to be in exponential form.
- If $z_1 = (r_1, \theta_1)$ and $z_2 = (r_2, \theta_2)$, then $z_1 z_2 = (r_1 r_2, \theta_1 + \theta_2)$.
- If $z_1 = (r_1, \theta_1)$ and $z_2 = (r_2, \theta_2)$, then $\dfrac{z_1}{z_2} = \left(\dfrac{r_1}{r_2}, \theta_1 - \theta_2\right)$.

Loci in the Argand diagram

- If $|z - z_1| = r$, then z lies on a circle, centre z_1, radius r.

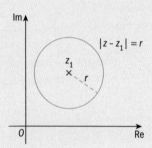

- If $|z - z_1| = |z - z_2|$, then z lies on the perpendicular bisector of the line joining z_1 to z_2.

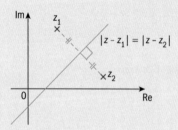

- If $\arg(z - z_1) = \theta$, then the locus of z is a half-line, starting at z_1, at an angle θ with the positive real axis.

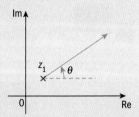

- Inequalities may be represented by shading appropriate regions on an Argand diagram.

Maths in real-life

Electrifying, magnetic and damp: how complex mathematics makes life simpler

If you think of complex numbers in the Argand diagram, you can appreciate that the real and imaginary parts are both separate and interlinked, so it is perhaps not surprising that they should have applications in physical situations where there are separate and linked entities.

In electromagnetism, the electric field and magnetic field can be represented by a single complex-valued field $a + bi$, where a is normally the electric field and b the magnetic field. All the devices shown in these images depend on the use of electromagnetism.

One of the most spectacular applications is in Maglev trains. The Shanghai Maglev train (also known as the Transrapid) is the fastest commercial train in operation as of 2017. It operates by using magnets and electric current to levitate the train above a guide-rail and propel it forwards. With conventional trains there is a lot of friction in the moving parts, but with the Maglev the resistance is almost all related to air resistance. Although a conventional train is also subject to air resistance, the higher speed of the Maglev means that air resistance is much greater. The future development of even faster trains looks possible as designers explore the practicalities of running Maglev trains in a vacuum tube where air resistance would be removed. However, the economics of these projects means that, although technically possible, they may not become practical realities.

You have met simple first-order differential equations in P3; however, physical situations involving forces are described by second-order differential equations, where the solutions sometimes require the use of complex numbers. The simplest model for a standard spring is that it has no resistance (like a projectile) and so continues to oscillate indefinitely once set in motion. But the reality is that the oscillations will die away. In systems like shock absorbers, the designers want the oscillations to die away quickly, so they increase the friction in the system by putting it in a liquid rather than in air.

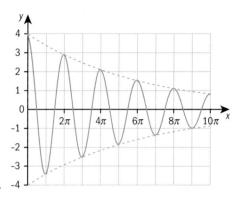

The motion of damped oscillations looks like the graph above: $x = 4e^{-0.05t}\cos t$ (shown in green) is bounded by the curves $x = \pm 4e^{-0.05t}$ shown as dotted blue lines.

Changing the physical characteristics of the system can change the frequency of the oscillations and how quickly they die away, as the following diagrams illustrate.

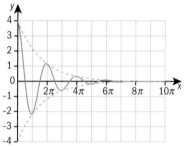
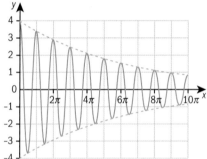
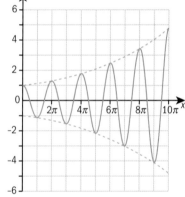

Resonance occurs where the opposite effect is seen: the amplitude of the oscillations grows rather than dies away. Sometimes this is the desired effect but more often it is destructive: feedback in amplification systems, glasses shattering if a singer exactly hits the resonant (or natural) frequency of the glass, people walking in step on a bridge, etc.

The Millennium Bridge in London had to be closed to the public for alterations shortly after it opened when it was found that people walking on the bridge naturally adjusted their walk because the bridge had a natural lateral sway – they adjusted their speed of step unconsciously to match the rhythm of the swaying motion, but this set up resonance and exaggerated the swaying motion.

Exam-style paper 2A – Pure 2

1. **a)** Sketch the graphs of $y = |2x - 4|$ and $y = |4x|$ on the same axes. [2]

   b) Use your sketch to solve the inequality $|2x - 4| < |4x|$. [2]

2. Solve the equation

   $\ln(64 - 3x^4) = 4 \ln x$,

   giving your answer correct to 3 significant figures. [4]

3. The polynomial $2x^4 - 3x^3 - 24x^2 + 13x + 12$ is denoted by p(x).

   a) Find the quotient when p(x) is divided by $x^2 - x - 12$. [3]

   b) Hence solve the equation p(x) = 0. [3]

4. The curve with equation $y = \dfrac{3e^x}{x^2}$ has one stationary point.

   a) Find the exact coordinates of this point. [4]

   b) Determine whether this point is a maximum or a minimum point. [2]

5. **a)** Given that $\sin(\theta + 45°) = 2 \sin \theta$, show that $\tan \theta = \dfrac{1}{2\sqrt{2} - 1}$. [2]

   b) Solve the equation

   $2 \tan^2 \theta - \sec \theta = 4$,

   giving all solutions in the interval $0° < \theta < 360°$. [5]

6. **a)** Show that $(3 \sin x - \cos x)^2$ can be written in the form $5 - 4 \cos 2x - 3 \sin 2x$. [4]

   b) Hence find the exact value of $\displaystyle\int_0^{\frac{1}{2}\pi} (3 \sin x - \cos x)^2 \, dx$. [4]

7. Find the exact value of $\dfrac{dy}{dx}$ when $x = -3$ in each of the following cases:

   a) $y = x^2 \ln(x + 13)$ [4]

   b) $y = \dfrac{3 - 5x}{6 + 5x}$. [3]

8.

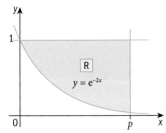

The diagram shows the curve $y = e^{-2x}$. The shaded region R is bounded by the curve and the lines $y = 1$ and $x = p$, where p is a constant.

a) Find the area of R in terms of p. [4]

b) Show that if the area of R is equal to 1, then

$$p = \frac{3}{2} - \frac{1}{2}e^{-2p}$$ [1]

c) Use the iterative formula

$$p_{n+1} = \frac{3}{2} - \frac{1}{2}e^{-2p_n},$$

with initial value $p_1 = 1.5$, to calculate the value of p correct to 2 decimal places. Give the result of each iteration to 4 decimal places. [3]

Exam-style paper 2B – Pure 2

50 marks

1. Solve the inequality $|x| > |x - 1|$. [3]

2. Use the trapezium rule with two intervals to estimate the value of

$$\int_0^1 \frac{2}{4 + e^x}\,dx,$$

giving your answer correct to 2 decimal places. [3]

3.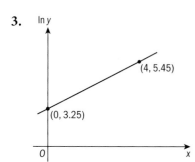

The variables x and y satisfy the equation $y = Ab^x$ where A and b are constants.
The graph of $\ln y$ against x is a straight line passing through the points $(0, 3.25)$
and $(4, 5.45)$ as shown in the diagram.
Find the values of A and b, correct to 1 decimal place. [5]

4. **a)** Find the quotient when the polynomial

$4x^3 + 8x^2 - 45x + 31$

is divided by $2x^2 + 9x - 5$, and show that the remainder is $10x + 6$. [3]

   b) Hence, or otherwise, factorise the polynomial

$4x^3 + 8x^2 - 55x + 25$ [2]

5. **a)** Express $12\cos\theta - 5\sin\theta$ in the form $R\cos(\theta + \alpha)$, where $R > 0$ and $0° \le \alpha \le 90°$,
giving the value of α correct to 2 decimal places. [3]

   b) Hence, solve the equation

$12\cos\theta - 5\sin\theta = 10$,

giving all solutions in the interval $0° \le \theta \le 360°$. [3]

   c) State the least value of $12\cos\theta - 5\sin\theta$ as θ varies. [1]

6. **a)** Show that $\int_0^{\frac{1}{4}\pi} (2 - \cos^2 x)\,dx = \dfrac{3\pi - 2}{8}$. [4]

   b) Show that $\int_0^{\sqrt{15}} \dfrac{x}{x^2 + 5}\,dx = \ln 2$. [5]

7. **a)** By sketching a suitable pair of graphs, show that the equation
$$e^{3x} = 7 - 2x^2$$
has exactly two real roots. [3]

b) Show by calculation that there is a root between 0.6 and 0.7. [2]

c) Show that this root also satisfies the equation
$$x = \frac{1}{3} \ln(7 - 2x^2)$$
[1]

d) Use an iteration process based on the equation in part **(c)**, with a suitable starting value, to find the root correct to 2 decimal places. Give the result of each step of the process to 4 decimal places. [3]

8. The equation of a curve is
$$3x^2 - 2xy + y^2 - 24 = 0$$

a) Show that the tangent to the curve at the point $(2, 6)$ is parallel to the x-axis. [4]

b) Find the equation of the tangent to the curve at the other point on the curve for which $x = 2$, giving your answer in the form $y = mx + c$. [5]

1. Solve the equation

    $7^{x+1} = 7^x + 5,$

    giving your answer correct to 3 significant figures. [4]

2. a) Expand $\dfrac{1}{\sqrt{1-2x}}$ in ascending powers of x, up to and including the term

    in x^3, simplifying the coefficients. [3]

    b) Hence find the coefficient of x^3 in the expansion of $\dfrac{2-3x}{\sqrt{1-2x}}$. [2]

3. The sequence of values given by the iterative formula

    $x_{n+1} = \dfrac{4x_n}{5} + \dfrac{10}{x_n^2},$

    with initial value $x_1 = 3.7$, converges to α.

    a) Use this iterative formula to find α correct to 2 decimal places,
        giving the result of each iteration to 4 decimal places. [3]

    b) State an equation satisfied by α and hence find the exact value of α. [2]

4. $x^2 + y^2 - 4xy + 11 = 0$. Find the two exact values of $\dfrac{dy}{dx}$ when $x = -3$. [6]

5. The polynomial $x^3 - 4x^2 + ax + b$, where a and b are constants, is denoted by p(x).

    It is given that $(x - 2)$ is a factor of p(x) and that when p(x) is divided by
    $(x - 3)$ the remainder is 4.

    a) Find the values of a and b. [5]

    b) When a and b have these values, find the other two linear factors of p(x). [3]

6. a) Express $8 \sin x - 6 \cos x$ in the form $R \sin(x - \alpha)$, where $R > 0$ and

    $0° \leq \alpha \leq 90°$, giving the value of α correct to 2 decimal places. [3]

    b) Hence, solve the equation

    $8 \sin 2\theta - 6 \cos 2\theta = 5,$

    giving all solutions in the interval $0° \leq \theta \leq 180°$. [4]

    c) State the greatest value of $24 \sin 2\theta - 18 \cos 2\theta$ as θ varies. [1]

7. The complex number is defined by $w = 4 + 3i$.

    a) Find the modulus and argument of w. [2]

    b) On a sketch of an Argand diagram, shade the region whose points
        represent the complex numbers which satisfy the inequalities $|z - w| \leq 3$
        and $\arg(z - 1) \leq \dfrac{\pi}{4}$. [5]

    c) Calculate the greatest possible value of $|z|$ in the shaded region. [2]

8. The position vectors of the points A, B, C and D are given by

$$\overrightarrow{OA} = \begin{pmatrix} 2 \\ 3 \\ -1 \end{pmatrix}, \overrightarrow{OB} = \begin{pmatrix} 0 \\ m \\ 2 \end{pmatrix}, \overrightarrow{OC} = \begin{pmatrix} -3 \\ n \\ 1 \end{pmatrix}, \overrightarrow{OD} = \begin{pmatrix} 2 \\ -2 \\ 4 \end{pmatrix},$$

where m and n are constants.

Find

a) the size of angle AOD [4]

b) the value of n such that OC is perpendicular to OD [2]

c) the values of m for which the length of AB is 7. [3]

9. a) Express $\dfrac{25}{y^2(5-y)}$ in partial fractions. [4]

b) Given that $y = 1$ when $x = 0$, solve the differential equation

$$\frac{dy}{dx} = \frac{1}{25}y^2(5-y),$$

obtaining an expression for x in terms of y. [6]

10.

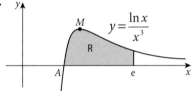

The diagram shows the curve $y = \dfrac{\ln x}{x^3}$ and its maximum point M.
The curve cuts the x-axis at A.

a) Write down the x-coordinate of A [1]

b) Find the exact coordinates of M. [5]

c) Use integration by parts to find the exact area of the shaded region enclosed by the curve, the x-axis and the line $x = e$. [5]

75 marks

1. **a)** Sketch the graphs of $y = |x + 2|$ and $y = |3x - 2|$ on the same axes. [2]

   b) Use your sketch to solve the inequality $|x + 2| \geq |3x - 2|$. [2]

2. **a)** Express $\cos^2 3x$ in terms of $\cos 6x$. [2]

   b) Hence find the exact value of $\displaystyle\int_0^{\frac{1}{18}\pi} \cos^2 3x \, dx$. [4]

3. Use the substitution $u = 1 + e^x$ to show that $\displaystyle\int_0^1 \frac{e^{3x}}{1 + e^x} \, dx = 2.10$ correct to 2 decimal places. [6]

4. Solve the equation $2 \sin 2\theta + \cos 2\theta = 1$ for $0° \leq \theta \leq 360°$. [6]

5. The parametric equations of a curve are
   $$x = \ln(4 - 2t), \quad y = \frac{3t + 5}{t + 1}$$

   a) Express $\dfrac{dy}{dx}$ in terms of t, simplifying your answer. [4]

   b) Find the gradient of the curve at the point where the curve cuts the y-axis. [3]

6. Given that $2 + i$ is a root of the equation $x^3 + px^2 + qx + 15 = 0$, where p, q and r are real constants,

   a) write down the other complex root of this equation [1]

   b) find the value of p and the value of q [4]

   c) find the third root of the equation. [2]

7.

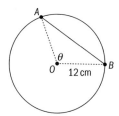

The diagram shows a circle with centre O and radius 12 cm. The angle AOB is θ radians.

The chord AB divides the circle into two regions whose areas are in the ratio $1 : 5$.

It is required to find the length of AB.

   a) Show that $\theta = \dfrac{\pi}{3} + \sin \theta$. [3]

   b) Showing all your working, use an iterative formula based on the equation in part **(a)**, with an initial value of 2, to find θ correct to 2 decimal places.
   Hence find the length of AB correct to 1 decimal place. [5]

8. **a)** Express in partial fractions $\dfrac{4-4x+6x^2}{(1-x)(2+x^2)}$. [5]

    b) Hence obtain the expansion of $\dfrac{4-4x+6x^2}{(1-x)(2+x^2)}$ in ascending powers of x,
    up to and including the term in x^3. [5]

9. A certain substance is formed during a chemical reaction. The mass of
    substance formed t seconds after the start of the reaction is x grams.
    At any time, the rate of formation of the substance is proportional to $(30 - x)$.
    When $t = 0$, $x = 0$ and $\dfrac{\mathrm{d}x}{\mathrm{d}t} = 3$.

    a) Show that x and t satisfy the differential equation $\dfrac{\mathrm{d}x}{\mathrm{d}t} = 0.1\,(30-x)$. [2]

    b) Find, in any form, the solution of this differential equation. [5]

    c) Find x when $t = 10$, giving your answer correct to 1 decimal place. [2]

    d) State what happens to the value of x as t becomes very large. [1]

10.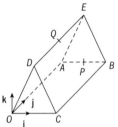

    The diagram shows a prism $OABCDE$ with a horizontal rectangular base $OABC$,
    where $OA = 10$ cm and $OC = 6$ cm. ODC is an isosceles triangle with $OD = CD = 5$ cm.

    Unit vectors $\mathbf{i}$ and $\mathbf{j}$ are parallel to OC and OA. Unit vector $\mathbf{k}$ is vertical.

    The point P is the mid-point of AB and the point Q is the mid-point of DE.

    a) Express each of the vectors $\overrightarrow{OP}$ and $\overrightarrow{PD}$ in terms of $\mathbf{i}$, $\mathbf{j}$, and $\mathbf{k}$. [3]

    b) Find the angle OPD. [3]

    c) Show that the length of the perpendicular from Q to PD is $\dfrac{10}{\sqrt{29}}$ cm. [5]

Answers

The answers given here are concise. However, when answering exam-style questions, you should show as many steps in your working as possible.

1 Algebra

Skills check page 2

1. **a)** 32   **b)** 73   **c)** 47   **d)** 58

2. **a)** 20   **b)** 6   **c)** 22   **d)** 20

Exercise 1.1 page 6

1. **a)** $x = -1$ or $x = 2$ **b)** $x = 1$

   c) $x = \frac{2}{7}$ or $x = \frac{2}{3}$ **d)** $x = \frac{1}{4}$ or $x = \frac{9}{4}$

   e) $x = 3$ or $x = -\frac{1}{3}$ **f)** $x = \frac{4}{3}$ or $x = 4$

   g) $x = -5$ or $x = \frac{3}{5}$ **h)** $x = 1$ or $x = -7$

   i) $x = \frac{3}{2}$ or $x = -\frac{5}{4}$ **j)** $x = 1$ or $x = \frac{1}{2}$

   k) $x = \frac{5}{2}$ or $x = 7$ **l)** $x = -\frac{5}{6}$ or $x = \frac{5}{2}$

2. **a)** $1 < x < 3$ **b)** $x \leq -3$ or $x \geq 5$

   c) $-\frac{5}{3} \leq x \leq 1$ **d)** $x < -3$ or $x > 3$

3. **a)** $x \leq 0$ or $x \geq 6$ **b)** $-\frac{1}{2} < x < 3$

   c) $-1 < x < \frac{1}{3}$ **d)** $x \leq -2$ or $x \geq 3$

   e) $-\frac{1}{3} < x < 5$ **f)** $0 \leq x \leq 1$

   g) $x < -\frac{4}{3}$ or $x > 0$ **h)** $-\frac{1}{2} \leq x \leq 2$

Exercise 1.2 page 7

1. **a)** **i)**

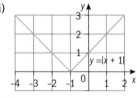

ii)

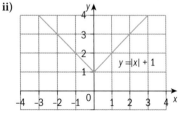

b) **i)**

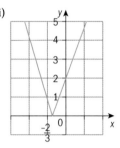

ii)

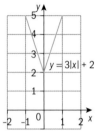

c) **i)**

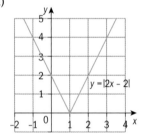

ii)

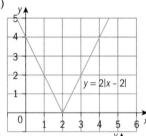

d) **i)**

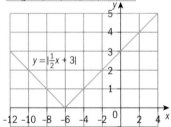

ii)

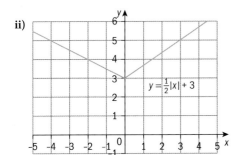

$y = \frac{1}{2}|x| + 3$

e) i)

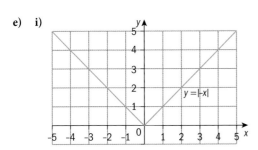

$y = |-x|$

ii)

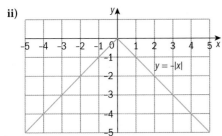

$y = -|x|$

f) i)

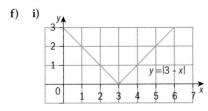

$y = |3 - x|$

ii)

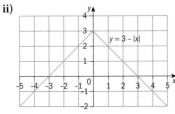

$y = 3 - |x|$

Exercise 1.3 page 9

1. **a)** $x^2 + x + 1$
    b) $x^2 - 3x + 9$
    c) $x^2 - x + 4$

d) $x^2 - x - 1$

e) $2x^2 + 7$

f) $3x^2 - 2x - 2$

g) $2x^2 + 3x + 1,$ remainder 3

2. **a)** -5 **b)** 4   **c)** -8
    d) $-8x - 1$ **e)** 2   **f)** -27

3. Proof

4. **a)** Proof   **b)** $(x - 1)(x - 2)(x - 3)$

5. Proof

6. $x^2 - x + 1$

7. Quotient $= x^2 + x + 1,$ remainder $= -2x + 6$

8. Quotient $= 2x^2 - x - 2,$ remainder $= 9$

9. **a)** Proof   **b)** $(2x - 1)(2x + 3)(3x + 1)$

10. $k = 10$

11. Quotient $= 2x^2 - 2x + 1,$ remainder $= -2$

12. $a = 6$

Exercise 1.4 page 11

1. **a)** 127   **b)** 4   **c)** $-\dfrac{49}{8}$
    d) -1 **e)** 4   **f)** 48

2. $a = 12$ 3. $a = 5$

4. $a = 3$ and $b = -6$

5. **a)** Proof   **b)** $a = -20$

6. Proof

7. $a = -7$ and $b = 20$

8. $a = 2$ 9. $a = \dfrac{1}{6}$

Exercise 1.5 page 14

1. **a)** $(x - 3)(x + 1)(2x - 1)$
    b) $(x - 1)(x - 2)(x - 3)$
    c) $(5x - 1)(x + 1)(x + 2)$
    d) $(x + 4)(x - 2)(2x - 1)$
    e) $(x + 2)(x + 1)(x - 2)$
    f) $(3x + 2)(x + 2)(2x - 1)$

2. **a)** $x = \dfrac{3}{2}$ or $x = -1$ or $x = -4$
    b) $x = \dfrac{1}{2}$ or $x = -2$ or $x = 4$
    c) $x = -1$ or $x = 2$ or $x = 5$
    d) $x = 2$ or $x = -1$ or $x = -4$
    e) $x = 4$ or $x = 5$ or $x = -\dfrac{3}{2}$
    f) $x = -1$ or $x = 2$ or $x = -\dfrac{1}{3}$

3. Proof

4. $(x + 1)(x - 1)(x - 2)(x + 3)$

5. $a = 7$

6. **a)** Proof **b)** $(2x - 5)(x - 3)(2x + 1)$

7. $a = 3$ or $a = -3$

8. $a = 3$ and $b = 8$

9. -63 $(a = 2, b = -3)$

10. $(x - 1)(x + 1)(x + 2)(2x + 1)$

Summary exercise 1 page 15

1. $x = 6$ or $x = -1$

2. $x = -1$ or $x = \dfrac{9}{5}$

3. **a)**

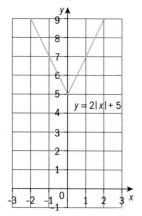

b)

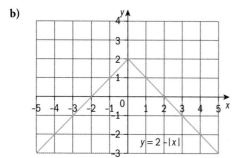

4. $\dfrac{4}{3} < x < 4$

5. $x < 1$ or $x > 3$

6. $1 \le x \le 9$

7. $x \le -\dfrac{5}{2}, \quad x \ge -\dfrac{1}{4}$

8. $x < \dfrac{a}{4}$

9. $2x^3 - 3x^2 + 4x - 3$

10. Quotient $= x^2 - x + 4$, remainder $= 9$

11. **a)** Proof **b)** $(x - 4)(x + 3)(x - 2)$

12. $k = 1$

13. $a = 3, \quad b = 2$

14. **a)** $a = 4, \quad b = -1$

b) $(x + 1)(x - 2)(x + 2)(x + 3)$

15. **a)** $x^2 - 16$

b) $x = -4, x = 4, x = -3, x = 2$

16. **a)** $a = 12, \quad b = 20$

b) $(3x + 2)(x - 2)(2x - 5)$

17. **a)** $A = -14$

b) **i)** $x = 1$ or $x = 4$ or $x = 9$

ii) $x = \pm 1$ or $x = \pm 2$ or $x = \pm 3$

18. **a)** $A = -18, \quad B = 32$

b) **i)** $x = -2$ or $x = 1$ or $x = 16$

ii) $x = \pm 1$ or $x = \pm 2$

19. **i)** Quotient $x^2 + 6$, remainder $8x + 5$

ii) $p = 12, q = -30$

iii) Proof, $x = -1 \pm \sqrt{6}$

2 Logarithms and exponential functions

Skills check page 18

1. **a)** x^5 **b)** x^5 **c)** $x^{\frac{7}{3}}$
2. 2^{ab}

Exercise 2.1 page 21

1. **a)** 614.4 **b)** 972 **c)** 80.1 (3 s.f.)
   d) 2.09×10^{-4}
2. **a)** 252 ml **b)** 7.12 ml
3. \$2.82 million
4. After 12 hours there is $100 \times (0.8)^{12} = 6.9$ mg, so need further injections to be $100 - 6.9 = 93$ mg to bring total to 100 mg.
5. 9 am the next morning is 17 hours later. At 9 am the volume is $3000 \times (0.98)^{17} = 2128$ ml = 2.1 l, so there is enough (>2 litres) for him to drive the car to the garage.

Exercise 2.2 page 24

1. **a)** -4 **b)** $\frac{3}{2}$ **c)** $\frac{7}{2}$ **d)** 8
   e) 3 **f)** $-\frac{5}{3}$ **g)** $\frac{5}{4}$ **h)** -2
   i) $-\frac{1}{2}$ **j)** 5 **k)** -3 **l)** $-\frac{3}{2}$
2. **a)** $256 = 2^8$ **b)** $\frac{1}{27} = 3^{-3}$ **c)** $32 = 4^{2.5}$
   d) $9\sqrt{3} = 3^{2.5}$ **e)** $x = a^q$ **f)** $t = s^u$
3. **a)** $-3 = \log_7\left(\frac{1}{343}\right)$
   b) $9 = \log_{10}(1\,000\,000\,000)$
   c) $-4 = \log_2\left(\frac{1}{16}\right)$ **d)** $2.5 = \log_5\left(25\sqrt{5}\right)$
   e) $-3 = \log_t v$ **f)** $x = \log_p m$
4. **a)** 2 **b)** $\frac{1}{2}$ **c)** $\frac{2}{3}$ **d)** $-\frac{1}{2}$
   e) $5\frac{1}{2}$ **f)** $-\frac{7}{3}$ **g)** 0
5.

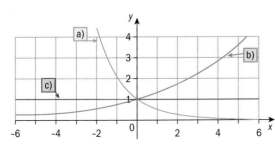

Exercise 2.3 page 28

1. **a)** 1.365 **b)** -1.635 **c)** 0.434
   d) 1.872 **e)** -5.036
2. **a)** 1.152 **b)** 0.878 **c)** 0.461
3. **a)** **i)** 15.1 kg of TNT
   ii) 15 300 000 kg of TNT
   b) 2 890 000 kg of TNT
4. 17.3 grams

Exercise 2.4 page 31

1. **a)** 2.86 **b)** 0.613 **c)** 2.51 **d)** -1
   e) 0.638 **f)** 0.712 **g)** 1.20 **h)** 2.37
   i) 2 **j)** -2 **k)** 6 **l)** -3
2. **a)** 7.64 **b)** 4.96 **c)** 12.1
   d) -1 **e)** 5.45 **f)** 1.31
3. **a)** $x \geq 3.36$ **b)** $x \leq -0.125$ **c)** $x > 1.76$
   d) $x \geq -0.756$ **e)** $x \leq 0.414$ **f)** $x < 4.96$
4. **a)** 0.827 **b)** 1.28 **c)** 2.10
5. **a)** 2.11 **b)** 1.27
6. 14 terms
7. 20 terms
8. 10 terms
9. 12 terms
10. 15 terms
11. 18 terms
12. 9 terms
13. **a)** $\frac{1}{3}$ **b)** $\frac{2}{3}$ **c)** $\frac{1}{2}$
    d) $\frac{4}{99}$ **e)** $\frac{1}{1+e}$
14. **a)** $\frac{2}{5}$ **b)** $\frac{2}{3}, \frac{3}{2}$ **c)** $\frac{6}{5}$ **d)** $\frac{1}{2}, \frac{4}{5}$

6.
a) $\log_5 24$ **b)** $\log_6 70$ **c)** 2
d) $\log_5 36$ **e)** $\log_5 54$ **f)** 0
g) 0 **h)** 2

7.
a) $\log_5 75$ **b)** $\log_6 2160$
c) $\log_{10}\left(\frac{x^2(x+3)}{8}\right)$ **d)** $\log_5(125x^2)$
e) $\log_{10}(1000x)$ **f)** $\log_{10}\left(\frac{800}{x}\right)$

8.
a) $\log x + \log y - 2\log z$
b) $5\log x + 6\log y + 3\log z$
c) $2\log x + \log y - 2\log z$

9.
a) $x + 2y$ **b)** $z + 2y - x$
c) $z - 3x$ **d)** $-3x - 3y$

15. a) 0.248 mg   **b)** 2025

16. 27 years

17. 3 hours and 10 minutes

Exercise 2.5 page 35

1. a) $\log p = \log 3 + t \log b$; $\log p$ and t

   b) $\log y = \log K - 2 \log x$; $\log y$ and $\log x$

   c) $\log y = \log \alpha + \frac{1}{2}\log x$; $\log y$ and $\log x$

2. a) $y = 10x^2$

   b) $V = \frac{4\pi}{3} r^3$

   c) $y = 10^{(0.1+1.3x)}$

3. $y = Ax^n \Rightarrow \log y = \log A + n \log x$
n (the gradient) is approximately 0.6, and the intercept ($\log A$) is approximately 0.75, so A is approximately $10^{0.75} = 5.6$.

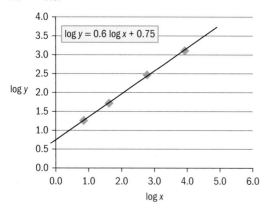

4. $-\ln k$ (the gradient) is approximately -0.675, so k is approximately $e^{0.675} = 2.0$ and the intercept ($\ln A$) is approximately 3.36, so A is approximately $e^{3.36} = 28.8$.

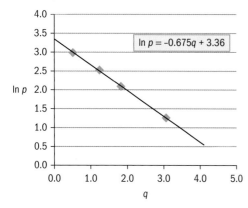

5. n (the gradient) is approximately -0.5, and the intercept ($\log A$) is approximately 3.5, so A is approximately $10^{3.5} = 3200$.

> Note that if you use natural logarithms, the graph will look different but the estimates will be the same – to within the accuracy possible in estimating from a graph.

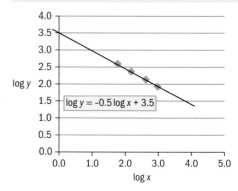

6. n (the gradient) is approximately 0.5, and the intercept ($\log k$) is approximately 2.8, so k is approximately $10^{2.8} = 630$.

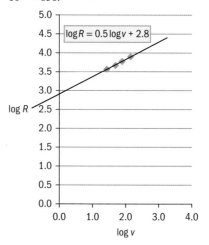

7. n (the gradient) is approximately 1.5, and the intercept ($\log k$) is approximately -0.7, so k is approximately $10^{-0.7} = 0.2$.

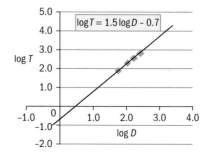

8. ln r (the gradient) is approximately 0.25, and the intercept (ln k) is approximately 6.3, so k is approximately $e^{6.3} = 545$.

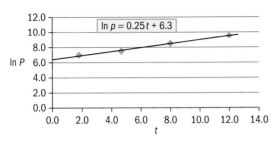

9. gradient $= 3\dfrac{\log 4}{\log 7}$; intercept $= -2\dfrac{\log 4}{\log 7}$

Summary exercise 2 page 37

1. a) -5 b) $\dfrac{3}{2}$ c) $\dfrac{1}{4}$

2. a) $256 = 4^4$ b) $\dfrac{1}{32} = 2^{-5}$ c) $32\sqrt{2} = 2^{5.5}$

3. a) $-3 = \log_6 \dfrac{1}{216}$

b) $7 = \log_{10} 10\,000\,000$

4. a) 4 b) $\dfrac{3}{2}$ c) $\dfrac{7}{4}$

5. a) $\log_4 14$ b) $\log_{10} 204$ c) $\ln 13.5$
    d) 2 e) 2 f) $\log_5 (25x^3)$
    g) $\ln(2x^4 + 3x^3)$ h) $\ln\left(\dfrac{4e^2}{x^2}\right)$

6. $3\log x + 10\log y + 5\log z$

7. $z - 3x - y$

8. a) 2.27 b) -0.748 c) 0.139

9. a) 6 b) 1.03 c) 1.09

10. a) $x \geq 2.34$ b) $x \leq -0.222$ c) $x > 3.38$

11. 10 terms   **12.** 20 terms

13. a) $\dfrac{1}{3}$ b) $\dfrac{1}{33}$ c) 2 d) $x = \dfrac{3}{2 + e^2}$

14. a) 23.6 mg   b) 2108

15. 4 hours 59 minutes

16. $\ln V = \ln A + T \ln r$
ln r (the gradient) is approximately 0.06, so r is approximately $e^{0.06} = 1.06$ (interest rate of 6% pa) and the intercept (ln A) is approximately 8.1, so A is approximately $e^{8.1} = 3300$.

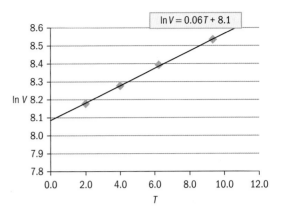

17. $\log P = \log k + t \log r$
$\log r$ (the gradient) is approximately 0.11, so r is approximately $10^{0.11} = 1.29$, and the intercept ($\log k$) is approximately 0.95, so k is approximately $10^{0.95} = 8.9$.

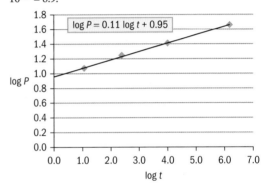

18. gradient $= 2\dfrac{\log 6}{\log 5}$; intercept $= -5\left(\dfrac{\log 6}{\log 5}\right)$

19. $\ln y = \ln A + x \ln k$
ln k (the gradient) is approximately 0.4, so r is approximately $e^{0.4} = 1.49$ and the intercept (ln A) is approximately 0.37, so A is approximately $e^{0.37} = 1.45$.

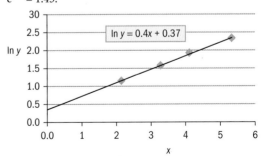

20. $x = 3.54$

21. $x = -12.362$ to 3 d.p.

22. -9.56

3 Trigonometry

Skills check page 40

1. **a)** 0   **b)** $\dfrac{\sqrt{3}}{2}$ **c)** $-\dfrac{1}{\sqrt{3}}\left(=-\dfrac{\sqrt{3}}{3}\right)$

   d) $\dfrac{\sqrt{3}}{2}$ **e)** $\dfrac{\sqrt{3}}{2}$ **f)** -1

2. **a) i)** $\dfrac{\pi}{6}$ **ii)** $-\dfrac{\pi}{4}$

   b) i) $31.8°$ **ii)** $-11.5°$

3. **a)** $26.6°, 206.6°$ **b)** $\dfrac{\pi}{3}, \dfrac{2\pi}{3}, \dfrac{4\pi}{3}, \dfrac{5\pi}{3}$

Exercise 3.1 page 45

1. **a)** $\dfrac{2}{\sqrt{3}}\left(=\dfrac{2\sqrt{3}}{3}\right)$ **b)** 2 **c)** $\dfrac{1}{\sqrt{3}}\left(=\dfrac{\sqrt{3}}{3}\right)$

   d) $\sqrt{2}$ **e)** $-\dfrac{2}{\sqrt{3}}\left(=-\dfrac{2\sqrt{3}}{3}\right)$

   f) $\sqrt{2}$ **g)** $-\dfrac{1}{\sqrt{3}}\left(=-\dfrac{\sqrt{3}}{3}\right)$

   h) $-\dfrac{2}{\sqrt{3}}\left(=-\dfrac{2\sqrt{3}}{3}\right)$

2. **a)** $-\sqrt{2}$ **b)** 2 **c)** $\sqrt{3}$

   d) 1 **e)** $-\dfrac{2}{\sqrt{3}}\left(=-\dfrac{2\sqrt{3}}{3}\right)$ **f)** -1

   g) $-\dfrac{2}{\sqrt{3}}\left(=-\dfrac{2\sqrt{3}}{3}\right)$ **h)** -1

3. **a), b), c), f)** and **h)** are defined.
   d), e) and **g)** are not defined.

4. **a)** 1.60   **b)** 1.06
   c) −1.22   **d)** 1.08
   e) 1.70   **f)** −0.727
   g) −3.24   **h)** −3.24

5. **a)** 45°, 225°   **b)** 45°, 315°
   c) 60°, 120°   **d)** 30°, 150°
   e) 150°, 330°   **f)** 180°

6. **a)** $\dfrac{\pi}{2}$ **b)** $0, 2\pi$ **c)** $\dfrac{3\pi}{4}, \dfrac{7\pi}{4}$

   d) $\dfrac{\pi}{6}, \dfrac{11\pi}{6}$ **e)** $\dfrac{\pi}{2}, \dfrac{3\pi}{2}$ **f)** $\dfrac{4\pi}{3}, \dfrac{5\pi}{3}$

7. **a)** 70.5°, 289.5°
   b) 14.5°, 165.5°

     c) 48.0°, 228.0°
     d) 11.3°, 191.3°
     e) 231.1°, 308.9°
     f) 114.1°, 245.9°

8. **a)** 81.8°, 278.2°   **b)** 81.8°, 98.2°, 261.8°, 278.2°

9.   33.7°, 213.7°

10. 3.87 rad, 5.55 rad

11. **a)** 0°, 180°, 360°
   b) 7.2°, 82.8°, 187.2°, 262.8°
   c) 12.3°, 72.3°, 132.3°, 192.3°, 252.3°, 312.3°

12. **a)** 180°   **b)** 143.1°
   c) 23.9°, 96.1°, 143.9°, 216.1°, 263.9°, 336.1°

13. **a)**

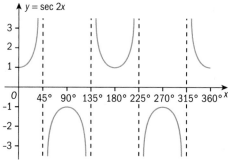

   b)

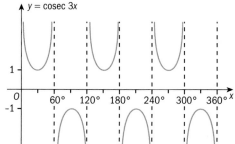

   c)

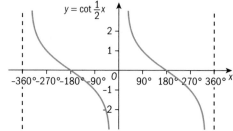

d)

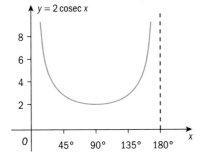

$y = 2\,\text{cosec}\,x$

e)

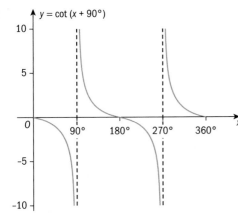

$y = \cot(x + 90°)$

f)

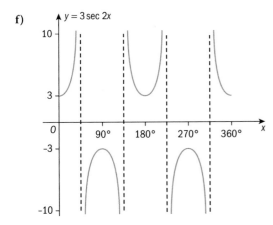

$y = 3\sec 2x$

14. a)

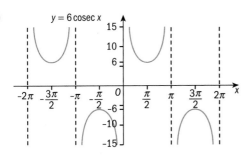

$y = 6\,\text{cosec}\,x$

b)

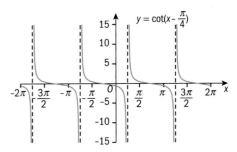

$y = \cot\left(x - \dfrac{\pi}{4}\right)$

c)

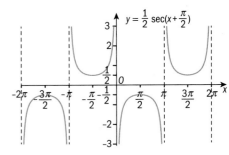

$y = \dfrac{1}{2}\sec\left(x + \dfrac{\pi}{2}\right)$

d)

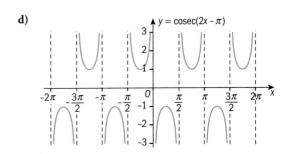

$y = \text{cosec}(2x - \pi)$

Exercise 3.2 page 48

1. **a)** 1 **b)** 1 **c)** $\sin\theta$

  d) 1 **e)** $\sin^3\theta$ **f)** $\tan^2\theta$

2. 26.6°, 116.6°, 206.6°, 296.6°

3. 0.666 rad, 2.48 rad

4. 70.5°, 120°, 240°, 289.5°

5. Proof **6.** Proof **7.** Proof

8. Proof **9.** Proof

10. 18.4°, 135°, 198.4°, 315°

11. 64.3°, 140.1°, 219.9°, 295.7°

12. $\dfrac{3\pi}{4}, \dfrac{7\pi}{4}$ **13.** Proof

14. 45°, 56.3°, 225°, 236.3°

15. 51.8°, 308.2° **16.** Proof

Exercise 3.3 page 53

1. a) $\frac{\sqrt{2}}{4}(\sqrt{3}-1)$ or $\frac{1}{2\sqrt{2}}(\sqrt{3}-1)$ b) $\frac{\sqrt{2}}{4}(\sqrt{3}+1)$

   c) $\frac{1+\sqrt{3}}{1-\sqrt{3}}$ or $-2-\sqrt{3}$ d) $\frac{4+\sqrt{6}-\sqrt{2}}{4-\sqrt{6}+\sqrt{2}}$

2. a) Proof   b) Proof

3. a) $\cos\alpha = \frac{3}{5}$, $\cos\beta = \frac{12}{13}$

   b) $\sin(\alpha+\beta) = \frac{63}{65}$, $\sin(\alpha-\beta) = \frac{33}{65}$, $\cos(\alpha+\beta) = \frac{16}{65}$, $\cos(\alpha-\beta) = \frac{56}{65}$

   c) $\tan(\alpha+\beta) = \frac{63}{16}$, $\tan(\alpha-\beta) = \frac{33}{56}$

4. a) $\frac{\sqrt{2}}{4}(\sqrt{3}+1)$ (or $\frac{1}{2\sqrt{2}}(\sqrt{3}+1)$)

   b) $\frac{\sqrt{2}}{4}(1-\sqrt{3})$ (or $\frac{1}{2\sqrt{2}}(1-\sqrt{3})$)

   c) $\frac{\sqrt{3}-1}{\sqrt{3}+1}$ (or $2-\sqrt{3}$)

   d) $\frac{\sqrt{2}}{4}(1-\sqrt{3})$ (or $\frac{1}{2\sqrt{2}}(1-\sqrt{3})$)

5. a) $\frac{\sqrt{3}}{2}$ b) $\frac{\sqrt{3}}{2}$ c) 1   d) $\sqrt{3}$

6. a) $2\sin\alpha\cos\beta$ b) $2\cos\alpha\cos\beta$ c) $-\sin\beta$

7. Proof   8. Proof

9. $\cot(\theta+\varphi) = \frac{\cot\theta\cot\varphi - 1}{\cot\theta + \cot\varphi}$ 10. Proof

11. a) $\frac{63}{65}$ b) $-\frac{63}{16}$ c) $-\frac{5}{4}$

12. 29.3°, 105.7°, 209.3°, 285.7°

13. 0, π, 2π

14. 45°, 225°

Exercise 3.4 page 57

1. a) 30°, 90°, 150°, 270°
   b) 60°, 90°, 120°, 270°
   c) 60°, 300°
   d) 48.6°, 131.4°, 210°, 330°
   e) 0°, 60°, 120°, 180°, 240°, 300°, 360°
   f) 0°, 45°, 135°, 180°, 225°, 315°, 360°

2. a) 0, 0.42, 2.72, π, 3.56, 5.86, 2π
   b) 0, 1.37, π, 4.91, 2π c) $\frac{\pi}{3}$, $\frac{5\pi}{3}$

3. a) $\frac{7}{25}$ b) $\frac{24}{7}$ c) $-\frac{336}{527}$

4. a) $\frac{240}{289}$ b) $\frac{240}{161}$ c) $\frac{289}{161}$

5. a) $-\frac{3}{\sqrt{10}}$ b) $-\frac{1}{\sqrt{10}}$ c) $-\frac{3}{4}$ d) $-\frac{4}{5}$

6. a) $1 - \sin 2x$ b) $\cos^2 2x$

7. $12\cos^3\theta - 8\cos^5\theta - 4\cos\theta$

8. a) i) $3\sin\theta - 4\sin^3\theta$
    ii) 0°, 45°, 135°, 180°, 225°, 315°, 360°

   b) $\frac{3\tan\theta - \tan^3\theta}{1 - 3\tan^2\theta}$

9. $\frac{\cot^2\theta - 1}{2\cot\theta}$

10. Proof   11. Proof   12. Proof

Exercise 3.5 page 60

1. a) $\sqrt{13}\sin(\theta + 33.69°)$ b) $\sqrt{149}\sin(\theta + 34.99°)$
   c) $\sqrt{5}\sin(\theta + 26.57°)$ d) $41\sin(\theta + 12.68°)$

2. a) i) $\sqrt{89}\sin(\theta - 57.99°)$ ii) $25\sin(\theta - 73.74°)$
    iii) $\sqrt{5}\sin(\theta - 26.57°)$ iv) $\sqrt{19}\sin(\theta - 66.59°)$

   b) i) $\sqrt{89}, -\sqrt{89}$ ii) $25, -25$
    iii) $\sqrt{5}, -\sqrt{5}$ iv) $\sqrt{19}, -\sqrt{19}$

3. a) $\sqrt{2}\cos(\theta + 45°)$ b) $\sqrt{6}\cos(\theta + 35.26°)$
   c) $5\cos(\theta + 36.87°)$ d) $\sqrt{6}\cos(\theta + 24.09°)$

4. a) i) $5\cos(\theta - 53.13°)$ ii) $13\cos(\theta - 67.38°)$
    iii) $\sqrt{3}\cos(\theta - 35.26°)$ iv) $\sqrt{2}\cos(\theta - 45°)$

   b) i) $5, -5$ ii) $13, -13$
    iii) $\sqrt{3}, -\sqrt{3}$ iv) $\sqrt{2}, -\sqrt{2}$

   c) i) 53.13°   ii) 67.38°
    iii) 35.26°   iv) 45°

5. a) $\sqrt{5}\sin(\theta + 63.43°)$ b) 90°, 323.1°

6. a) $2\sin(\theta - 30°)$ b) 60°, 180°   c) 4, 0

7. a) $\sqrt{8}\cos(\theta + 45°)$ b) 24.3°, 245.7°

8. a) $\sqrt{52}\cos(\theta - 56.31°)$ b) 10.2°, 102.4°   c) 57, 5

9. a) 257.6°, 349.8°   b) 163.7°
   c) 53.1°, 323.1°   d) 29.6°, 256.7°

10. a) $\sqrt{5}\sin(2x + 26.57°)$

b) $0, 63.4°, 180°, 243.4°, 360°$

c) $10 + \sqrt{5}, 10 - \sqrt{5}$

11. a) $\sqrt{17}\sin(x + 1.326)$

b) $1.00, 5.77$

c) $16, -1$

12. $79.7°, 153.4°$

Summary exercise 3 page 62

1. $52.5°, 232.5°$

2. $0.841, 2.09, 4.19, 5.44$

3. $23.6°, 156.4°$ **4.** Proof **5.** Proof

6. a) Proof **b)** $\dfrac{1}{1+\sqrt{2}}(=\sqrt{2}-1)$

7. Proof **8.** 1 **9.** Proof

10. a) $13\sin(\theta + 67.38°)$

b) i) 13 **ii)** $22.6°$

11. a) $\sqrt{41}\cos(\theta + 51.34°)$ **b)** $56.9°, 200.5°$ **c)** 20

12. a) $25\sin(\theta + 73.74°)$ **b)** $69.4° < \theta < 323.1°$

13. Proof

14. $0, 56.3°, 180°, 236.3°, 360°$

15. $120°, 240°, 300°$

16. Proof

17. a) $\sin\dfrac{\theta}{2} = \dfrac{t}{\sqrt{1+t^2}}, \cos\dfrac{\theta}{2} = \dfrac{1}{\sqrt{1+t^2}}$

b) Proof **c)** $68.7°, 164.4°$

18. a) Proof **b)** $x = 16.8°, 61.8°, 106.8°, 151.8°$

19. a) Proof **b)** $0°, 30°, 150°, 180°$

20. $60°, 150°, 240°, 330°$

4 Differentiation

Skills check page 68

1. **a)** $12x^3 - \dfrac{1}{x^3}$ **b)** $8 + \dfrac{9}{\sqrt{x}}$

   c) $x^4 - \dfrac{8}{x^5}$ **d)** $-\dfrac{5}{6}x^{\frac{3}{2}}$

2. **a)** $\dfrac{5}{(4-x)^2}$ **b)** $\dfrac{-7}{(2x+3)^{\frac{3}{2}}}$

   c) $\dfrac{3-2x}{(x^2-3x+7)^2}$ **d)** $\dfrac{-30}{\sqrt{(5x+1)^5}}$

Exercise 4.1 page 71

1. **a)** $2e^{2x}$ **b)** $-5e^{-5x}$ **c)** $3e^{3x+9}$

   d) $-e^{8-x}$ **e)** $-7e^x$ **f)** $10x\, e^{x^2}$

   g) $6e^{2x-1}$ **h)** $-8x\, e^{6+x^2}$ **i)** $10x\, e^{5x^2}$

   j) $\dfrac{-3e^{\frac{3}{x}}}{x^2}$ **k)** $\dfrac{3e^{\sqrt{x}}}{\sqrt{x}}$ **l)** $\dfrac{-2}{e^{2x}}$

2. **a)** $-2e^{-2x}$ **b)** e^{5+x} **c)** $27e^{3x}$

   d) $-\dfrac{4e^{\sqrt{x}}}{\sqrt{x}}$ **e)** $3e^x - 8e^{4x}$ **f)** $-4e^{-8x}$

   g) $4e^{4x} - e^{-x}$ **h)** $-6e^{-x} - 12e^{3x}$ **i)** $2e^{2x} + e^x$

   j) $24e^{6x} - 12e^{3x}$ **k)** $-e^{-x}$ **l)** $-4e^{-4x+2} - 14e^{5-2x}$

3. 10 4. $e-3$

5. $(0, 1)$

6. Maximum turning point at $(0, -1)$

Exercise 4.2 page 73

1. **a)** $\dfrac{1}{x}$ **b)** $\dfrac{2x}{x^2-1}$ **c)** $\dfrac{5}{3+5x}$

   d) $\dfrac{3x^2}{x^3+2}$ **e)** $\dfrac{e^x}{e^x-7}$ **f)** $\dfrac{1}{2x}$

   g) $\dfrac{24}{6x-3}$ **h)** $\dfrac{4x+1}{2x^2+x}$ **i)** $-\dfrac{2}{x}$

   j) $\dfrac{3}{x}$ **k)** $-\dfrac{2}{x}$ **l)** $\dfrac{54}{9x-2}$

2. **a)** $\dfrac{8}{x}$ **b)** $\dfrac{3}{x}$ **c)** $\dfrac{2}{2x-1}$

   d) $\dfrac{6}{x}$ **e)** $\dfrac{1}{2x+1}$ **f)** $\dfrac{10}{x}$

   g) $\dfrac{3e^x-2}{3e^x-2x}$ **h)** $\dfrac{18(2x+3)}{x^2+3x}$ **i)** $\dfrac{10}{x(7x-2)}$

   j) $\dfrac{4(x-4)}{x(x-8)}$ **k)** 5 **l)** $-\dfrac{6}{x+3}$

3. 1 4. $\dfrac{3}{5}$

5. $(2, \ln 6)$

6. Minimum turning point at $(1, 1)$

Exercise 4.3 page 75

1. **a)** $e^x(x+1)$ **b)** $2(6x^2 - x - 10)$

   c) $\dfrac{x(3x^2+2)}{\sqrt{x^2+1}}$ **d)** $\dfrac{2+\ln x}{2\sqrt{x}}$

   e) $2x(2x+5)^6(9x+5)$ **f)** $\dfrac{(x-3)(5x-3)}{2\sqrt{x}}$

   g) $y = 2e^{2x}(4x+3)$ **h)** $(x+3)^3(5x+3)$

   i) $x + 2x\ln x$ **j)** $x(x+3)^2(5x+6)$

   k) $e^x(2x-1)^3(2x+7)$ **l)** $\dfrac{1}{(x+1)^2}$

   m) $\dfrac{3e^{\sqrt{x}}(2\sqrt{x}+x+4)}{\sqrt{x}}$ **n)** $4x^2 e^{2x}(2x+3)$

   o) $2e^x\left(\dfrac{1}{x} + \ln 2x\right)$ **p)** $\dfrac{4}{(x+2)^2}$

   q) $\dfrac{17}{(1-5x)^2}$ **r)** $2x\left[1 + \ln(x^2+1)\right]$

2. $\dfrac{1}{e}$

3. $y = 725x - 1200$

4. Proof

5. Minimum turning point at $(-2, -e^{-2})$

Exercise 4.4 page 78

1. **a)** $\dfrac{6}{(x+3)^2}$ **b)** $\dfrac{e^{2x}(2x-1)}{x^2}$

   c) $\dfrac{x(x+8)}{(x+4)^2}$ **d)** $\dfrac{1-x^2}{(1+x^2)^2}$

   e) $\dfrac{6}{(2-x)^2}$ **f)** $y = \dfrac{3x(4e^x - 2xe^x - x)}{(2e^x - x)^2}$

   g) $\dfrac{11}{(2x+1)^2}$ **h)** $\dfrac{-3x-7}{(x+1)^3}$

   i) $\dfrac{6x(4-x)}{(2-x)^2}$ **j)** $\dfrac{4(2x+1)}{(1-x)^4}$

   k) $\dfrac{x^2-4x-1}{(x-2)^2}$ **l)** $\dfrac{5x(2\ln x - 1)}{(\ln x)^2}$

   m) $\dfrac{x^2(3-4x^2)}{(1-2x^2)^{\frac{3}{2}}}$ **n)** $\dfrac{4x^5 + 4x^3 - 6x^2 + 3}{(2x^2+1)^2}$

   o) $\dfrac{2+\sqrt{x}}{2(1+\sqrt{x})^2}$ **p)** $\dfrac{2(x^2-7x-6)}{(2x-7)^2}$

   q) $\dfrac{(3x-2)\sqrt{(x+1)^3}}{2x^2}$ **r)** $\dfrac{-2x(2x^2+3)}{(1+2x^2)^{\frac{3}{2}}}$

2. $x = 0$ and $x = 3$

3. $3x + 4y = 22$

4. $\dfrac{2}{(x-1)^3}$

5. Proof

6. Proof

Exercise 4.5 page 80

1. a) $5\cos 5x$ b) $3\sec^2 \dfrac{x}{2}$

   c) $-2(x+1)\sin(x^2 + 2x)$ d) $8\tan x \sec^2 x$

   e) $3x^2 \cos(x^3 - 7)$ f) $20\sec^2 4x$

   g) $-3\cos^2 x \sin x$

   h) $2\cos 2x \cos x - \sin 2x \sin x$

   i) $x(3x\sec^2 3x + 2\tan 3x)$

   j) $\dfrac{-2x\sin 2x - 2\cos 2x}{x^3}$

   k) $\sin^2 3x\,(9\cos x \cos 3x - \sin x \sin 3x)$

   l) $\dfrac{2\sec^2 \sqrt{x}}{\sqrt{x}}$

2. Proof

3. -3

4. $y = -x + \dfrac{3\pi}{2} - 1$

5. Proof

6. $\left(\dfrac{\pi}{6}, \dfrac{3}{2}\right)$ max, $\left(\dfrac{\pi}{2}, 1\right)$ min, $\left(\dfrac{5\pi}{6}, \dfrac{3}{2}\right)$ max

7. Proof; Minimum value is $\dfrac{\pi}{3} - \sqrt{3}$.

8. Proof

Exercise 4.6 page 84

1. a) $\dfrac{x}{y-1}$ b) $\dfrac{\cos y - 1}{2y + x\sin y}$

   c) $\dfrac{15}{1+9x^2}$ d) $\dfrac{5 - 3x^2 - y^2}{2xy}$

   e) $\dfrac{2x}{1-2y}$ f) $\dfrac{2y+7}{3+2y-2x}$

   g) $\dfrac{y(1-\ln y)}{x}$ h) $e^x\left(\dfrac{1}{(1+x^2)} + \tan^{-1} x\right)$

   i) $\dfrac{x+1}{2-y}$ j) $\dfrac{2x^3 + xy^2}{2y^3 - yx^2}$

   k) $\dfrac{y\cos x - \sin y}{x\cos y - \sin x}$ l) $\dfrac{2}{e^{x+y}} - 1$

   m) $\dfrac{x}{1+x^2} + \tan^{-1} x$ n) $\dfrac{1-2x}{1+3y^2}$

o) $\dfrac{-e^y}{xe^y + 2}$ p) $\dfrac{y\ln y}{y - x}$

q) $\dfrac{1}{2+8x^2}$ r) $\dfrac{2xy^3 + \sin x}{2 - \cos y - 3x^2 y^2}$

2. $-\dfrac{1}{3}$ and $\dfrac{1}{3}$

3. $x + y = 2$

4. Proof

5. Proof

6. $-\dfrac{1}{1 - 2x + 2x^2}$

Exercise 4.7 page 86

1. a) $\dfrac{4}{3t}$ b) $\dfrac{1}{3t^2}$

   c) $-\dfrac{2}{3}\cot t$ d) $\dfrac{1}{t}$

   e) $\dfrac{\sin t}{1-\cos t}$ f) $2t(1-t)^2$

   g) $\dfrac{1}{4t^{\frac{3}{2}}}$ h) $-\tan\theta$

   i) $\dfrac{3}{2t+3}$ j) $-\cosec\theta$

   k) $-\dfrac{b}{a}\cot\theta$ l) $\dfrac{3t^2 - 4}{2t}$

2. Proof   3. Proof

4. Proof   5. Proof

6. 1

7. $x - y - 8 = 0$

8. Proof

9. 0 or $\dfrac{4}{3}$

10. a) $\dfrac{-2\cos 3t}{5\sin 3t}$ b) $(0, 2)$

Summary exercise 4 page 87

1. a) $-5e^{-5x}$ b) $\dfrac{8x}{4x^2 + 5}$

   c) $\dfrac{4}{1+16x^2}$ d) $\dfrac{-1}{(2x-3)^2}$

   e) $\dfrac{\cos x - \sin x}{e^x}$ f) $(x+1)(x-2)^2(5x-1)$

   g) $\dfrac{x(x^2 - 2)}{(x^2 - 1)^{\frac{3}{2}}}$ h) $\dfrac{9}{3x-4}$

   i) $\dfrac{3x^2 - y^2}{2xy - 1}$ j) 14

k) $\dfrac{e^x(x\ln x + \ln x - 1)}{(\ln x)^2}$

l) $\dfrac{3x^3 + 2x}{\sqrt{(1+x^2)}}$

m) $-\dfrac{1}{3t^2}$

n) $y = -\dfrac{e^{\sqrt{x}}}{\sqrt{x}}$

o) $\dfrac{2x^2 + y^2}{y^2 - 2xy}$

p) $\dfrac{4e^{2x} - 9}{e^x}$

q) $2e^{\tan^2 x}\tan x \sec^2 x$

r) $x\,e^{3x-2}(3x+2)$

s) $3\cot\theta$

t) $\dfrac{3}{3x+1}$

u) $\dfrac{12t^2 - 1}{2(t+2)^3}$

v) $1 + \ln x$

w) $\dfrac{1+3x^2}{2\sqrt{x}(1-x^2)^2}$

x) $\dfrac{3x^2 + y^2}{3y^2 - 2xy}$

y) $\dfrac{2}{1+100x^2}$

2. $\left(-1,\ \dfrac{-3}{e^2}\right)$

3. Proof

4. a) $\dfrac{t-1}{t+1}$

b) $(4, 0)$

5. $y = -26x + 60$

6. $-\dfrac{e}{2}$

7. $5x - y + 9 = 0$

8. Proof

9. $\dfrac{-16x}{(1+4x)^2}$

10. $3x - 2y - 1 = 0$

11. $\dfrac{3\pi}{4}$

12. Proof

13. Proof

14. $x - 2y + 4 = 0$

15. $\dfrac{25}{8}$

16. $\left(\dfrac{1}{3e},\ -\dfrac{1}{3e}\right)$

17. Proof

18. Proof

19. Proof

20. Proof

21. a) $\dfrac{5-2x}{3-2y}$ **b)** $-\dfrac{3}{2},\ \dfrac{9}{2}$

22. $(49, -9)$

5 Integration

Skills check page 91

1. a) $2e^{2x+1} - e^{-x}$ b) $4\cos 4x + 6\sin 2x$

 c) $\dfrac{3}{(3x+7)}$

2. Proof

Exercise 5.1 page 94

1. a) $\dfrac{1}{3}e^{3x-1} + c$ b) $\dfrac{1}{4}e^{4x+1} + c$ c) $2e^{0.5x+1} + c$

 d) $\dfrac{1}{\pi}e^{\pi x+3} + c$ e) $\dfrac{1}{4}e^{4(x+1)} + c$ f) $2e^{0.5x+0.5} + c$

 g) $-e^{2-x} + c$ h) $-\dfrac{1}{2}e^{3-2x} + c$ i) $-2e^{0.5-0.5x} + c$

2. a) $\dfrac{1}{3}e^{3} - \dfrac{1}{3}$ b) $\dfrac{1}{4}e - \dfrac{1}{4}$ c) $e - 1$

 d) $e^4 - 1$ e) $4e^3 - 4$ f) $\dfrac{1}{2}e^8 - \dfrac{1}{2}$

 g) $1 - \dfrac{1}{e}$ h) $1 - \dfrac{1}{e^4}$ i) $2e^{0.5} - 2e^{-0.5}$

3. a) 1 b) 8 c) 1 d) 80

 e) 15 f) $\dfrac{4}{5}$ g) $\dfrac{2}{\ln 3}$ h) $\dfrac{7}{4\ln 2}$

 i) $\dfrac{242}{81\ln 3}$

4. $e^3 - 1$ **5.** $\dfrac{1}{2}(1 - e^{-4})$

6. 5.52 **7.** $1 - e^{-k};\ 1$

8. a) $\dfrac{1}{3}$ b) $\dfrac{1}{4}e^{-4}$ c) 1

 d) 1 e) $2e^3$ f) $2e$

9. $\dfrac{\pi}{2}(1 - e^{-6}) = 1.57$

Exercise 5.2 page 98

1. a) $\dfrac{1}{2}\ln|2x - 1| + c;\ x \neq \dfrac{1}{2}$

 b) $\ln|2x - 1| + c;\ x \neq \dfrac{1}{2}$

 c) $\ln|7x + 3| + c;\ x \neq -\dfrac{3}{7}$

 d) $-5\ln|5 - 2x| + c;\ x \neq \dfrac{5}{2}$

 e) $2\ln|1 - 3x| + c;\ x \neq \dfrac{1}{3}$

 f) $-\dfrac{1}{3}\ln|6x + 5| + c;\ x \neq -\dfrac{5}{6}$

 g) $-3\ln|x + 1| + c;\ x \neq -1$

 h) $-\ln|7 - x| + c;\ x \neq 7$

 i) $2\ln|x + e| + c;\ x \neq -e$

 j) $-\dfrac{1}{4}\ln|2x + 1| + c;\ x \neq -\dfrac{1}{2}$

 k) $-\dfrac{4}{3}\ln|9x + 16| + c;\ x \neq -\dfrac{16}{9}$

 l) $-\dfrac{1}{5}\ln|10x - 3| + c;\ x \neq \dfrac{3}{10}$

2. a) $\dfrac{1}{2}\ln\left(\dfrac{5}{3}\right)$ b) $\ln\left(\dfrac{13}{9}\right)$ c) $\ln\left(\dfrac{10}{3}\right)$

 d) $2\ln\left(\dfrac{5}{3}\right)$ e) $\dfrac{9}{2}\ln 2$ f) $7\ln\left(\dfrac{3}{5}\right)$

 g) $2\ln\left(\dfrac{3}{5}\right)$ h) $2\ln\left(\dfrac{15}{19}\right)$ i) $\ln 16 = 2.77$

3. $\ln 2$

4. $2\ln\left(\dfrac{3}{5}\right) = -1.02$

5. $\dfrac{1}{2}\ln 21 = 1.52$

Exercise 5.3 page 101

1. a) $\dfrac{1}{2}\sin 2x + c$ b) $-\dfrac{1}{2}\cos\left(2x + \dfrac{\pi}{4}\right) + c$

 c) $\dfrac{1}{3}\tan\left(3x + \dfrac{\pi}{12}\right) + c$ d) $-\sin(\pi - x) + c$

 e) $-4\sin\left(\dfrac{\pi}{2} - \dfrac{x}{4}\right) + c$ f) $\dfrac{1}{2}\cos\left(2x - \dfrac{\pi}{4}\right) + c$

 g) $-\dfrac{1}{5}\tan\left(\dfrac{\pi}{12} - 5x\right) + c$ h) $\cos\left(\dfrac{1}{2}x\right) + c$

 i) $-3\sin\left(4 - \dfrac{1}{3}x\right) + c$

2. a) $\dfrac{1}{2}$ b) $\dfrac{1}{\sqrt{2}}$ c) 0.488

 d) 0 e) -1.5 f) $\dfrac{\sqrt{2}}{3}$

 g) 0.770 h) -0.510 i) 1.04

3. $y = \tan x + 1$ **4.** 2

5. 1 **6.** $\dfrac{1}{2\sqrt{3}} = 0.289$

Exercise 5.4 page 106

1. a) $\dfrac{1}{8}\sin 4x + \dfrac{1}{2}x + c$ b) $\dfrac{1}{2}x - \dfrac{1}{12}\sin 6x + c$

 c) $\dfrac{1}{2}\tan 2x + c$ d) $-\dfrac{1}{4}\cos 2x + c$

 e) $\dfrac{1}{8}x - \dfrac{1}{32}\sin 4x + c$ f) $x - \dfrac{1}{2}\cos 2x + c$

2. a) $\dfrac{\pi}{8}$ **b)** $\dfrac{\pi}{4}$ **c)** $\dfrac{1}{2\sqrt{3}}$ **d)** $\dfrac{3\pi}{4}$

e) $3\pi + 8 = 17.4$ **f)** $\dfrac{\pi}{2} - 1 = 0.571$

3. $y = \dfrac{1}{2}x + \dfrac{1}{4}\sin 2x$ **4.** $\dfrac{2}{\sqrt{3}}$

5. $\dfrac{1}{2}$

6. a) Proof **b)** $-\dfrac{1}{32}$

7. a) Proof **b)** $\dfrac{13\pi}{2} + 5$

8. a) Proof **b)** $\dfrac{\pi^2 - 4}{8}$

9. a) Proof **b)** $\dfrac{\pi}{12} - \dfrac{\sqrt{3}}{16}$

10. a) Proof **b)** $\dfrac{\pi}{12} + \dfrac{\sqrt{3}}{16}$

c) $\dfrac{\sqrt{3}}{2}$

Exercise 5.5 page 112

1. a) 0.335 **b)** 1.42 **c)** 0.695
 d) 1.46 **e)** 1.13 **f)** 0.429

2. a) 0.333 **b)** 1.53 **c)** 0.663
 d) 1.50 **e)** 1.11 **f)** 0.416

3. a) 2.97

b) The approximation in **a)** is an overestimate as the gradient of the curve is increasing over the interval so the chords are always above the curve and give a (slightly) greater area in each strip than the curve itself would.

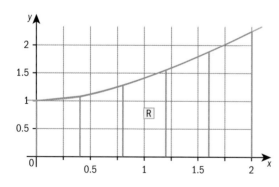

4. a) 0.590

b) The approximation in **a)** is an underestimate as the chords are always below the curve and give a (slightly) smaller area in each strip than the curve itself would.

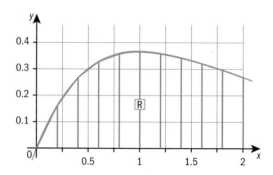

5. a) i) 1.17 (3 s.f.) **ii)** 1.10 (3 s.f.)

b) 1.10 (3 s.f.)

Summary exercise 5 page 113

1. a) $\dfrac{1}{2}e^{2x-3} + c$ **b)** $-\dfrac{1}{2}\ln|6x+2| + c$

c) $-\dfrac{1}{4}\cos 4x + c$ **d)** $\dfrac{x}{2} - \dfrac{\sin 12x}{24} + c$

e) $-2e^{\left(1 - \frac{1}{2}x\right)} + c$ **f)** $4\ln|3x - 4| + c$

g) $\dfrac{5}{2}x + \cos 2x + \dfrac{3}{4}\sin 2x + c$

h) $-2\ln|3x+5| + c$ **i)** $-\dfrac{1}{3}\sin\left(3x + \dfrac{\pi}{3}\right) + c$

2. a) $\dfrac{1}{5}e^{10} - \dfrac{1}{5}$ **b)** $\ln 3$ **c)** $\dfrac{1}{\sqrt{3}}$

d) $\dfrac{11\pi}{8} - 3$ **e)** $2\cos\left(\dfrac{\pi}{8}\right) - \sqrt{2}$ **f)** $\dfrac{5}{2}$

g) $-\dfrac{\sqrt{3}}{4}$ **h)** $-\dfrac{1}{2}(e^4 - e^{-2})$ **i)** $\dfrac{5}{3}\ln\dfrac{1}{4}$

3. $\dfrac{1}{2}e - \dfrac{1}{2}e^{-3}$

4. $\ln 1.5$

5. $y = \frac{1}{2}x - \frac{1}{4}\sin 2x + \frac{1}{4} - \frac{\pi}{8}$

6. $\ln\left(\frac{5\sqrt{7}}{2}\right) = 1.89$

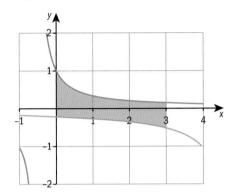

7. 5.29

8. **a)** $\frac{1}{2}$ **b)** $\frac{2}{\sqrt{e}}$ **c)** $\frac{2}{\sqrt{e}}$

9. $4\pi(e^{-4} - e^{-8}) = 0.226$

10. **a)** Proof   **b)** $\frac{3\sqrt{3}}{14} = 0.371$

11. **a)** Proof   **b)** 2.43

12. **a)** Proof   **b)** $\sqrt{\frac{3}{2}}$

13. **a)** 3.61

b) Underestimates the true value as the curve always lies above the trapezium.

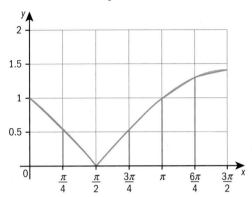

14. **a)** An integral gives the *signed* area, so integrating between 0 and 2 would subtract the area below the axis – between 0 and 1 – from the area above – between 1 and 2.

    b) 0.824

> **Note:** The value of the function at 0 is taken as zero in using the trapezium rule to estimate R. To do the integral analytically would require it to be treated as an improper integral.

15. 9.5

16. **a)** $\frac{1}{6}\sin 6\theta + \theta + c$ **b)** $\ln(\sqrt{11})$

6 Numerical solutions of equations

Skills check page 117

1. a) 1.80   b) 2.24   c) 1
2. a) 2.31   b) 2.62   c) Not defined
3. $x = \frac{1}{2}\left(\ln(y+3)-1\right)$

Exercise 6.1 page 122

1. Root is 1.2 to 1 d.p.   2. Root is 3.9 to 1 d.p.
3. Root is 1.0 to 1 d.p.   4. Root is 0.9 to 1 d.p.
5. a) Root is 0.3 to 1 d.p.
   b) Roots are −1.9 and 1.5 to 1 d.p.
6. Root is 0.77 to 2 d.p.   7. Root is 0.87 to 2 d.p.
8. Root is 3.96 to 2 d.p.

Exercise 6.2 page 127

1. a) $x^2 + 3x - 1 = 0$ b) $x^4 + 5x^3 + 9x^2 - 2 = 0$
   c) $e^{-2x} - x = 0$ d) $2x - e^{3-x} = 0$
   e) $\cos x + \frac{\pi}{4} - x = 0$ f) $x^{10} - x^6 - 3 = 0$

2. The lists below should not be taken as exhaustive of all possibilities:

a) $x = \sqrt[4]{3x-2}$, $x = \frac{x^4+2}{3}$, $x = \frac{3x-2}{x^3}$,

   $x = \sqrt{\frac{3x-2}{x^2}}$, $x = \sqrt[3]{3 - \frac{2}{x}}$

b) $x = \sqrt[7]{4x^3-1}$, $x = \frac{x^7+1}{4x^2}$, $x = \sqrt[3]{\frac{x^7+1}{4}}$, $x = \sqrt[3]{\frac{1}{4-x^4}}$

c) $x = \frac{x^2+1}{x^4}$, $x = \sqrt[5]{x^2+1}$, $x = \frac{x^5-1}{x} = x^4 - \frac{1}{x}$,

   $x = \sqrt{x^5 - 1}$, $x = \frac{1}{x(x^3-1)}$, $x = \sqrt{\frac{1}{(x^3-1)}}$

d) $x = \frac{x^2-e^{-x}}{3}$, $x = \frac{3x+e^{-x}}{x}$, $x = \sqrt{3x+e^{-x}}$,

   $x = -\ln\left(x^2-3x\right)$, $x = \frac{e^{-x}}{x-3}$

e) $x = \frac{1}{3}e^{x^2}$, $x = \sqrt{\ln 3x}$, $x = \frac{\ln 3x}{x}$

f) $x = \tan^{-1}\left(\frac{1}{x^2-3}\right)$, $x = \frac{\cot x + 3}{x}$, $x = \sqrt{\cot x + 3}$

3. a) $x_1 = 0.5$, $x_2 = 0.28571$, $x_3 = 0.30434$,
   $x_4 = 0.30263$, $x_5 = 0.30278$
  b) $x_1 = 0.5$, $x_2 = 0.41649$, $x_3 = 0.42134$
   $x_4 = 0.42100$, $x_5 = 0.42103$
  c) $x_1 = 0.5$, $x_2 = 0.36787$, $x_3 = 0.47914$,
   $x_4 = 0.38355$, $x_5 = 0.46435$
  d) $x_1 = 1.75$, $x_2 = 1.74723$, $x_3 = 1.74881$,
   $x_4 = 1.74791$, $x_5 = 1.74843$
  e) $x_1 = 1.1$, $x_2 = 1.23899$, $x_3 = 1.11114$,
   $x_4 = 1.22903$, $x_5 = 1.12054$
  f) $x_1 = 1.2$, $x_2 = 1.19595$, $x_3 = 1.19474$,
   $x_4 = 1.19439$, $x_5 = 1.19428$

4. a) Proof   b) Proof
  c) Root is 1.47 to 2 d.p.

5. a) Proof   b) Proof
  c) $x_2 = 2.42037$, $x_3 = 2.38492$, $x_4 = 2.36911$,
   $x_5 = 2.36205$

6. a) Proof
  b) $\theta_2 = 1.70670 ...$, $\theta_3 = 1.74946 ...$,
   $\theta_4 = 1.70724 ...$, $\theta_5 = 1.74893 ...$
  c) The odd-numbered values are decreasing and the even-numbered values are increasing – moving towards one another, so eventually will converge to the root.

7. a) Proof
  b) 1.37 to 2 d.p.

8. a) Proof
  b) $x_{n+1} = \cos x_n$, $x_1 = 0.75$, $x_2 = 0.73168 ...$,
   $x_3 = 0.74404 ...$, $x_4 = 0.73573 ...$, $x_5 = 0.74133 ...$
   So this iterative relationship converges to 0.74 (2 d.p.).
   $x_{n+1} = \cos^{-1} x_n$, $x_1 = 0.75$, $x_2 = 0.72273 ...$,
   $x_3 = 0.76304 ...$, $x_4 = 0.70278 ...$, $x_5 = 0.79149 ...$
   So this iterative relationship does not converge.

9. a) Proof
  b) i) $x_{n+1} = \sqrt{\frac{5}{x_n^2-1}}$, $x_1 = 1.7$, $x_2 = 1.6265 ...$,

$x_3 = 1.743\,15\,...,\ x_4 = 1.566\,10\,...,\ x_5 = 1.855\,23\,...$

So this iterative relationship does not converge.

ii) $x_{n+1} = \dfrac{x_n^{\,2}+5}{x_n^{\,3}},\ x_1 = 1.7,\ x_2 = 1.605\,94\,...,$

$x_3 = 1.829\,88\,...,\ x_4 = 1.362\,49\,...,\ x_5 = 2.710\,75\,...$

So this iterative relationship does not converge.

iii) $x_{n+1} = \sqrt[4]{x_n^{\,2}+5},\ x_1 = 1.7,\ x_2 = 1.675\,98\,...,\ x_3 = 1.671\,65\,...,$

$x_4 = 1.670\,88\,...,\ x_5 = 1.670\,74\,...$

So the root is 1.67 (2 d.p.).

Summary exercise 6 page 131

1. 0.3 (1 d.p.)

2. −0.7, 2.6 (1 d.p.)

3. i) 1.3 (1 d.p.) ii) −4.1, −1.3 (1 d.p.)

4. 1.3 (1 d.p.)

5. i) $2x^2 + 4x = 1$ ii) $5x^2 = x^3 - 1$

6. These are not exhaustive lists of rearrangements.

i) $x = \sqrt[3]{4x^2 - 3}$

$x = \dfrac{4x^2 - 3}{x^3}$

$x = \sqrt{\dfrac{4x^2 - 3}{x^2}}$

ii) $x = \dfrac{x^5 + 2}{3x}$

$x = \sqrt{\dfrac{x^5 + 2}{3}}$

$x = \dfrac{3x^2 - 2}{x^4}$

$x = \sqrt[5]{3x^2 - 2}$

7. i) $x_1 = 0.5$
   $x_2 = 0.222\,22\,...$
   $x_3 = 0.209\,30\,...$
   $x_4 = 0.208\,73\,...$
   $x_5 = 0.208\,71\,...$

   ii) $x_1 = 0.5$
   $x_2 = 0.327\,78\,...$
   $x_3 = 0.328\,89\,...$
   $x_4 = 0.328\,86\,...$
   $x_5 = 0.328\,86\,...$

8. i) Proof ii) Proof iii) 2.84

9. i) Proof ii) 1.13

10. i) Proof ii) 1.36 radians

11. i) Proof ii) Proof iii) Proof iv) 0.36

12. i) 3.10 ii) $\alpha = \sqrt[3]{30}$

13. i) $\dfrac{2x(1+e^{2x}) - 2x^2 e^{2x}}{(1+e^{2x})^2}$; Proof ii) Proof

   iii) $m = 1.109$

14. i) Proof ii) Proof iii) Proof iv) 2.28

7 Further algebra

Skills check page 136

1. $A = 1$ $B = -4$ $C = -9$
2. $A = 2$ $B = -4$ $C = 0$
3. $A = -7$ $B = \dfrac{2}{3}$ $C = -2$
4. $A = 3$ $B = \dfrac{1}{2}$ $C = 0$
5. $A = 3$ $B = 0$ $C = -1$

Exercise 7.1A page 138

1. $\dfrac{3}{x-1} - \dfrac{2}{x-2}$

2. $\dfrac{1}{5(x+1)} + \dfrac{4}{5(x-4)}$

3. $\dfrac{1}{x+3} + \dfrac{2}{x-1}$

4. $\dfrac{-1}{x} + \dfrac{4}{2x+1}$

5. $\dfrac{1}{x+2} - \dfrac{2}{x+3} + \dfrac{3}{x-3}$

Exercise 7.1B page 140

1. $\dfrac{1}{x-2} + \dfrac{2}{(x-2)^2}$

2. $\dfrac{3}{x} - \dfrac{9}{3x-1} + \dfrac{9}{(3x-1)^2}$

3. $\dfrac{1}{x-1} - \dfrac{1}{x-2} + \dfrac{2}{(x-2)^2}$

4. $\dfrac{1}{4(x+1)} - \dfrac{1}{4(x-1)} + \dfrac{1}{2(x-1)^2}$

5. $\dfrac{1}{x} - \dfrac{1}{x-1} + \dfrac{1}{(x-1)^2}$

Exercise 7.1C page 141

1. $\dfrac{5}{x^2+2} - \dfrac{2}{1-x}$

2. $\dfrac{1}{x} - \dfrac{3x}{2+x^2}$

3. $\dfrac{5}{x+2} + \dfrac{3x-4}{2x^2-1}$

4. $\dfrac{1-x}{x^2+4} + \dfrac{1}{x+1}$

5. $\dfrac{3}{5-x} - \dfrac{4x-7}{4x^2-3}$

6. $\dfrac{2}{1-2x} + \dfrac{x+1}{1+x^2}$

7. $\dfrac{3-2x}{3-x^2} + \dfrac{2}{x-2}$

8. $\dfrac{3-5x}{2+3x^2} - \dfrac{4}{2x+1}$

Exercise 7.1D page 143

1. $1 - \dfrac{9}{x-2} + \dfrac{14}{x-3}$

2. $1 - \dfrac{1}{2(x+1)} + \dfrac{1}{2(x-1)}$

3. $3 + \dfrac{5x}{x^2+4} - \dfrac{1}{x-3}$

4. $-1 - \dfrac{2}{x+3} + \dfrac{4}{(x+3)^2}$

5. $A = -2, B = 3, C = -1$

Exercise 7.1E page 144

1. $\dfrac{2}{x+3} - \dfrac{1}{x+1}$

2. $1 + \dfrac{8}{x+1} - \dfrac{15}{x+2}$

3. $\dfrac{2}{x-5} + \dfrac{3}{(x-5)^2}$

4. $\dfrac{5}{1-5x} + \dfrac{x-2}{1+x^2}$

5. $\dfrac{3}{x+1} - \dfrac{2}{x-2}$

6. $3 - \dfrac{1}{x} + \dfrac{15}{x-3}$

7. $\dfrac{1}{x} + \dfrac{2}{x-1} + \dfrac{4}{(x-1)^2}$

8. $\dfrac{7x}{2(x^2+2)} - \dfrac{3}{2x}$

9. $\dfrac{-1}{x} + \dfrac{2}{x+1}$

10. $1 + \dfrac{5}{2(x-3)} - \dfrac{1}{2(x+3)}$

11. $\dfrac{5}{x-2} + \dfrac{3}{x+1} - \dfrac{2}{(x+1)^2}$

12. $-\dfrac{1}{x-1} + \dfrac{x+2}{x^2+1}$

13. $\dfrac{1}{x-1} - \dfrac{1}{x-2} + \dfrac{2}{(x-2)^2}$

14. $1 + \dfrac{1}{x-1} - \dfrac{1}{x+1}$

15. $\dfrac{1}{x+2} + \dfrac{x}{x^2+3}$

Exercise 7.2 page 146

1. a) $1 + 4x + 10x^2 + 20x^3 + \dots$ for $-1 < x < 1$

   b) $1 - 15x + 135x^2 - 945x^3 + \dots$ for $-\dfrac{1}{3} < x < \dfrac{1}{3}$

   c) $1 + \dfrac{1}{3}x - \dfrac{1}{9}x^2 + \dfrac{5}{81}x^3 + \dots$ for $-1 < x < 1$

   d) $1 - \dfrac{3}{2}x + \dfrac{3}{2}x^2 - \dfrac{5}{4}x^3 + \dots$ for $-2 < x < 2$

   e) $x + x^2 + x^3 + \dots$ for $-1 < x < 1$

   f) $1 + 10x + 75x^2 + 500x^3 + \dots$ for $-\dfrac{1}{5} < x < \dfrac{1}{5}$

2. $1 - \dfrac{1}{2}x^2 + \dfrac{3}{8}x^4 + \dots$ for $-1 < x < 1$

3. a) $1 + 2x + 3x^2 + 4x^3$ for $-1 < x < 1$ b) $50x^{49}$

4. a) $1 - x - \dfrac{1}{2}x^2 - \dfrac{1}{2}x^3 - \dfrac{5}{8}x^4 + \dots$ for $-\dfrac{1}{2} < x < \dfrac{1}{2}$

   b)   0.8944

5. $1 - \dfrac{3}{2}x + \dfrac{7}{8}x^2 + \dots$ for $-1 < x < 1$

6. $1 - 8x^2 + 48x^4 - 256x^6 + \dots$ for $-\dfrac{1}{2} < x < \dfrac{1}{2}$

7. $1 + 4x + 7x^2 + 10x^3 + \dots$ for $-1 < x < 1$

8. a) $1 + \dfrac{1}{2}x - \dfrac{1}{8}x^2 + \dfrac{1}{16}x^3 + \dots$ for $-1 < x < 1$

   b)   1.0392

9. a) $a = -1, n = -4$ b)   20

10. $1 - 3x - x^2 + 3x^3 + x^4 + \dots$ for $-1 < x < 1$

11. a) $k = -\dfrac{3}{2}, n = -8$ b) $405\,x^3$

12. $1 - 2x - 2x^2 + 4x^3 + 4x^4 + \dots$

Exercise 7.3 page 148

1. a) $\dfrac{1}{5}(1-2x)^{-\frac{1}{2}}$ b) $\left(1 - \dfrac{x}{3}\right)^{-2}$ c) $3\left(1 + \dfrac{4x}{27}\right)^{\frac{1}{3}}$

2. $\dfrac{1}{32} + \dfrac{5}{64}x + \dfrac{15}{128}x^2 + \dfrac{35}{256}x^3 + \dots$ for $-2 < x < 2$

3. $2 - \dfrac{1}{4}x - \dfrac{1}{64}x^2 + \dots$ for $-4 < x < 4$

4. $\dfrac{1}{27}$

5. $\dfrac{1}{2} - \dfrac{x^2}{4} + \dfrac{x^4}{8} - \dfrac{x^6}{16} + \ldots$ for $-\sqrt{2} < x < \sqrt{2}$

6. **a)** $\dfrac{1}{16} + \dfrac{1}{32}x + \dfrac{3}{256}x^2 + \dfrac{1}{256}x^3 + \ldots$ for $-4 < x < 4$

    b) $-\dfrac{1}{256}$

7. $3 + \dfrac{53}{6}x - \dfrac{109}{216}x^2 + \ldots$

8. $\dfrac{1}{2} - \dfrac{1}{8}x^2 + \dfrac{3}{64}x^4 - \dfrac{5x^6}{256} + \ldots$ for $-\sqrt{2} < x < \sqrt{2}$

9. **a)** $p = -24,\ q = 108$ **b)** $1728x^3$

10. $k = 2$

Exercise 7.4 page 150

1. **a)** $\dfrac{2}{(1-x)} + \dfrac{3}{(1+3x)}$

    b) $5 - 7x + 29x^2 + \ldots$ for $-\dfrac{1}{3} < x < \dfrac{1}{3}$

2. **a)** $\dfrac{1}{(1+2x)} - \dfrac{1}{(1+x)^2}$ **b)** -4

3. **a)** $\dfrac{3}{2(x-1)} - \dfrac{1}{2(x+1)}$

    b) $-2 - x - 2x^2 - x^3 + \ldots$ for $-1 < x < 1$

4. **a)** $\dfrac{2}{(1-x)} + \dfrac{1}{(1-x)^2} + \dfrac{2}{(2-x)}$

    b) $4 + \dfrac{9}{2}x + \dfrac{21}{4}x^2 + \ldots$ for $-1 < x < 1$

5. **a)** $-\dfrac{1}{(1-x)} - \dfrac{1}{(x-2)}$ **b)** $-\dfrac{7}{8}$

6. **a)** $\dfrac{1}{2(1-x)} + \dfrac{x+1}{2(1+x^2)}$

    b) $1 + x + x^4 + \ldots$ for $-1 < x < 1$

7. **a)** $\dfrac{2}{(1-2x)} + \dfrac{1+x}{(1+x^2)}$

    b) $3 + 5x + 7x^2 + \ldots$

Summary exercise 7 page 151

1. $\dfrac{4}{x-3} - \dfrac{1}{x+2}$ **2.** $\dfrac{3x+1}{2(x^2+1)} - \dfrac{3}{2(x+1)}$

3. $1 + \dfrac{3}{x-2} + \dfrac{1}{x+1}$ **4.** $\dfrac{5}{9(x-1)} + \dfrac{4}{3(x-1)^2} - \dfrac{5}{9(x+2)}$

5. $\dfrac{3}{5(x-2)} - \dfrac{1}{2(x+3)} - \dfrac{1}{2x-1}$

6. $\dfrac{1}{x-4} - \dfrac{6}{x+4} + \dfrac{4}{(x+4)^2}$ **7.** $\dfrac{3x-1}{x^2-2} + \dfrac{4}{x+3}$

8. $A = 5,\ B = -3,\ C = 2,\ D = -2$

9. $\dfrac{1}{4} + \dfrac{5}{4}x + \dfrac{75}{16}x^2 + \dfrac{125}{8}x^3 + \ldots$ for $-\dfrac{2}{5} < x < \dfrac{2}{5}$

10. **a)** $1 - \dfrac{1}{2}x - \dfrac{1}{8}x^2 - \dfrac{1}{16}x^3 + \ldots$ for $-1 < x < 1$

    b) 0.499

11. $\dfrac{1}{64} - \dfrac{9}{256}x + \dfrac{27}{512}x^2 + \ldots$ for $-\dfrac{4}{3} < x < \dfrac{4}{3}$

12. **a)** $\dfrac{1}{2} - \dfrac{1}{8}x + \dfrac{3}{64}x^2 - \dfrac{5}{256}x^3 + \ldots$ for $-2 < x < 2$

    b) 0.49875467

13. $2 - 3x - \dfrac{1}{2}x^3 + \ldots$ for $-\dfrac{1}{2} < x < \dfrac{1}{2}$

14. **a)** $\dfrac{3}{1+x} - \dfrac{1}{2+x}$

    b) $\dfrac{5}{2} - \dfrac{11}{4}x + \dfrac{23}{8}x^2 - \dfrac{47}{16}x^3 + \ldots$ for $-1 < x < 1$

15. **a)** $\dfrac{2}{(1+x)} - \dfrac{1}{(1-x)} + \dfrac{3}{(1-x)^2}$

    b) $4 + 3x + 10x^2 + \ldots$ for $-1 < x < 1$

8 Further integration

Skills check page 154

1. a) $1+\dfrac{1}{x+1}+\dfrac{3}{x-2}$

b) $\dfrac{-5}{9(x+2)}+\dfrac{5}{9(x-1)}+\dfrac{4}{3(x-1)^2}$

c) $\dfrac{3x-1}{x^2-2}+\dfrac{4}{x+3}$

2. $\dfrac{1}{2x+1}$

Exercise 8.1 page 159

1. a) $\ln\left|\dfrac{2x-1}{x+3}\right|+c$ **b)** $\ln\left|\dfrac{x+5}{x+7}\right|+c$

c) $\ln\left|\dfrac{x-2}{x+5}\right|+c$ **d)** $\ln\left|\dfrac{\sqrt{2x+3}}{x+5}\right|+c$

e) $\dfrac{1}{3}\ln|2+3x|-\dfrac{1}{4}\ln|1-4x|+c$

f) $\ln\left|(x-2)\sqrt{3-2x}\right|+c$ **g)** $2\ln\left(\dfrac{15}{14}\right)$

h) $\ln 6$ **i)** $\ln\left(\dfrac{98}{9}\right)$

2. a) $\ln\left|\dfrac{x}{x-1}\right|-\dfrac{1}{(x-1)}+c$ **b)** $\ln\left|\dfrac{x}{x+1}\right|+\dfrac{1}{(x+1)}+c$

c) $\ln\left|\dfrac{x+1}{x-3}\right|-\dfrac{12}{(x-3)}+c$ **d)** $\dfrac{1}{5}-\ln 2$

e) $\ln\left(\dfrac{6}{7}\right)+\dfrac{2}{7}$ **f)** $\ln\left(\dfrac{2}{3}\right)+1$

3. a) $x+\ln\left|\dfrac{(x-2)^2}{(x+1)^3}\right|+c$ **b)** $2x+\ln\left|\dfrac{(x-3)^4}{(x+1)}\right|+c$

c) $1+2\ln\left(\dfrac{72}{49}\right)$ **d)** $5+\ln\left(\dfrac{4^2\times\sqrt{3}}{3^2}\right)=6.13$

4. Proof **5.** Proof

6. $\ln\left(\dfrac{4}{3}\right)$ **7.** $0.5+\ln\left(\dfrac{25}{8}\right)$

8. Proof

Exercise 8.2 page 163

1. a) $\dfrac{1}{2}\tan^{-1}\left(\dfrac{x}{2}\right)+c$ **b)** $\tan^{-1}\left(\dfrac{x}{5}\right)+c$

c) $\tan^{-1}\left(\dfrac{x}{9}\right)+c$

2. a) $\dfrac{1}{3}\tan^{-1}(3x)+c$ **b)** $\dfrac{1}{4}\tan^{-1}(4x)+c$

c) $\dfrac{1}{7}\tan^{-1}(7x)+c$

3. a) $\dfrac{1}{\sqrt{3}}\tan^{-1}\left(\sqrt{3}x\right)+c$ **b)** $\dfrac{1}{2\sqrt{2}}\tan^{-1}\left(2\sqrt{2}x\right)+c$

c) $\dfrac{1}{\sqrt{6}}\tan^{-1}\left(\sqrt{6}x\right)+c$

4. a) $\dfrac{1}{6}\tan^{-1}\left(\dfrac{3x}{2}\right)+c$ **b)** $\tan^{-1}\left(\dfrac{4x}{3}\right)+c$

c) $\dfrac{1}{63}\tan^{-1}\left(\dfrac{7x}{9}\right)+c$

5. a) $\dfrac{\pi}{6}$ **b)** $\dfrac{\pi}{12}$ **c)** π

6. a) $\dfrac{\pi}{6}$ **b)** $\dfrac{\pi}{10}$ **c)** $\dfrac{\pi}{3}$

Exercise 8.3 page 165

1. a) $2\ln|x^3+3|+c$ **b)** $\dfrac{1}{2}\ln|1+\sin 2x|+c$

c) $\dfrac{1}{2}\ln|x^2-4x+3|+c$ **d)** $\dfrac{1}{6}\ln|4+2e^{3x}|+c$

e) $\dfrac{1}{2}\ln|x^2+4|+c$ **f)** $\ln|\sin x|+c$

g) $\dfrac{1}{2}\ln\left(1+\dfrac{\sqrt{3}}{2}\right)$ **h)** $\dfrac{1}{2}\ln\left(\dfrac{1+e^2}{2}\right)$

i) $\ln 2$

2. 0

3. a) Proof **b)** $\ln|x+\sin^2 x|+c$

4. a) Proof **b)** $\dfrac{44\sqrt{(3)}}{27}$

5. a) Proof **b)** 1.38

6. $\dfrac{1}{2}\ln\left|e^{2x}+e^{-2x}\right|+c$

7. a) Proof **b)** $\dfrac{1}{2}\ln 2$

8. a) Proof **b)** 0.186

Exercise 8.4 page 172

1. a) $2x\,e^{2x}-e^{2x}+c$ **b)** $-\dfrac{3}{2}x\cos 2x+\dfrac{3}{4}\sin 2x+c$

c) $(2x-1)e^x+c$

d) $\dfrac{1}{3}(x+3)\sin 3x+\dfrac{1}{9}\cos 3x+c$

e) $\frac{1}{3}xe^{3x-1} - \frac{1}{9}e^{3x-1} + c$ **f)** $\frac{1}{2}x^2\ln 5x - \frac{1}{4}x^2 + c$

g) $\frac{x^2}{2}\tan^{-1}x - \frac{1}{2}x + \frac{1}{2}\tan^{-1}x + c$

2. a) $1 + e^2$ **b)** $\frac{3\pi}{4}$ **c)** $3e^2 + 1$

d) 1.05 **e)** 23.8 **f)** 1

3. a) $6x^2e^{3x} - 4x\,e^{3x} + \frac{4}{3}e^{3x} + c$

b) $\frac{3}{2}x^2\sin 2x + \frac{3}{2}x\cos 2x - \frac{3}{4}\sin 2x + c$

c) 31.9 **d)** 0.718 **e)** 1.14

f) 0.319 **g)** 0.0741 **h)** 1.39

4. 1

5. 0.227

6. a) 0.102 **b)** 0.0745

Exercise 8.5 page 178

1. a) $\left(\sqrt{2x+1}\right)^3 - 3\sqrt{2x+1} + c$

b) $\frac{2}{9}(1 + x^3)^{\frac{3}{2}} + c$ **c)** $\sin^{-1}\left(\frac{x}{3}\right) + c$

d) $\frac{2}{3}(x+2)^{\frac{3}{2}} - 4(x+2)^{\frac{1}{2}} + c$

e) $\frac{1}{10}\tan^{-1}\left(\frac{2x}{5}\right) + c$ **f)** $\ln\left|e^{2x} + x\right| + c$

2. a) 2 **b)** $5\frac{7}{9}$ **c)** $\frac{\pi}{6}$

d) $\sqrt{2} - 1$ **e)** $\sqrt{3}$ **f)** 2.13

g) 4.5 **h)** 1

3. 1.14 **4.** $\frac{1}{5}$

5. 1.08 **6.** 0.366

Summary exercise 8 page 180

1. a) $\ln\left|A\left(\frac{x-2}{x+3}\right)\right|$ **b)** $\ln\left|A\sqrt{2x-1}(x+4)^2\right|$

c) $\ln\left|A\left(\frac{x+1}{x}\right)^2\right| - \frac{3}{x+1}$ **d)** $\ln\left|A\frac{x+1}{\sqrt{2x-1}}\right| + \frac{1}{(x+1)}$

e) $3x - \ln\left|A(x-1)(x+1)^4\right|$

2. $\ln\left|\frac{Ax}{\sqrt{x^2+4}}\right|$

3. a) $\ln\left|x^4 + 6\right| + c$ **b)** $-\frac{1}{6}\ln\left|3 - 2\sin 3x\right| + c$

c) $\frac{1}{2}\ln\left|3x^2 - 4x + 7\right| + c$

d) $-\frac{1}{25}\ln\left|6 - 5e^{5x}\right| + c$ **e)** $\ln\left|x^2 + 9\right| + c$

f) $\frac{1}{3}\ln\left|x^3 + 10\right| + c$ **g)** $\frac{1}{3}\ln\frac{3}{2}$

h) $\frac{1}{3}\ln\left(\frac{2 + e^3}{3}\right)$ **i)** $\ln 3$

4. a) $\frac{1}{3}xe^{3x} - \frac{1}{9}e^{3x} + c$ **b)** $2x\sin 2x + \cos 2x + c$

c) $(2x - 5)e^x + c$

d) $\frac{1}{3}(x^2 - x + 3)\sin 3x + \frac{1}{9}(2x - 1)\cos 3x - \frac{2}{27}\sin 3x + c$

e) 0.264 **f)** 13.4

5. a) $\frac{4}{9}\left(\sqrt{3x-1}\right)^3 + \frac{4}{3}\left(\sqrt{3x-1}\right) + c$

b) $\frac{1}{6}\left(4 + x^4\right)^{\frac{3}{2}} + c$ **c)** $\frac{1}{4}\sin^{-1}\left(\frac{4}{5}x\right) + c$

d) $\frac{1}{6}(2x + 7)^{\frac{3}{2}} - \frac{7}{2}(2x + 7)^{\frac{1}{2}} + c$

e) $\frac{16}{3}$ **f)** 1.06 **g)** $\frac{1}{7}$ **h)** $\sqrt{3} - 1$

6. $1.5 + \ln 4$ **7.** $\frac{\pi}{36}$ **8.** $\frac{3\pi}{8}$ **9.** $\frac{13}{6}$

9 Vectors

Skills check page 182

1. **a)** $\overrightarrow{AB} = \begin{pmatrix} 3 \\ 4 \end{pmatrix}$, $\overrightarrow{BA} = \begin{pmatrix} -3 \\ -4 \end{pmatrix}$

b) $\overrightarrow{AB} = \begin{pmatrix} 3 \\ -1 \end{pmatrix}$, $\overrightarrow{BA} = \begin{pmatrix} -3 \\ 1 \end{pmatrix}$

c) $\overrightarrow{AB} = \begin{pmatrix} -1 \\ -2 \end{pmatrix}$, $\overrightarrow{BA} = \begin{pmatrix} 1 \\ 2 \end{pmatrix}$

2. **a)** $\begin{pmatrix} -1 \\ 2 \end{pmatrix}$ **b)** $\begin{pmatrix} -3 \\ -3 \end{pmatrix}$ **c)** $\begin{pmatrix} 6 \\ 11 \end{pmatrix}$

Exercise 9.1 page 186

1. **a)** $\begin{pmatrix} 2 \\ -1 \end{pmatrix}$ **b)** $\begin{pmatrix} 5 \\ -5 \end{pmatrix}$ **c)** $\begin{pmatrix} 4 \\ 0 \\ 5 \end{pmatrix}$ **d)** $\begin{pmatrix} -5 \\ 4 \\ -3 \end{pmatrix}$

2. **a)** $3i - j$ **b)** $-3i + j$
c) $i + 3j + k$ **d)** $-3i - 3j - 5k$

3. $\overrightarrow{OE} = 6i + 6k$, $\overrightarrow{OF} = 6i + 6j + 6k$,
$\overrightarrow{EG} = -6i + 6j$, $\overrightarrow{CE} = 6i - 6j + 6k$

4. **a)** $(7, 8, 2)$ **b)** $\left(6, \frac{17}{2}, \frac{3}{2}\right)$ **c)** $\begin{pmatrix} -1 \\ -5 \\ 3 \end{pmatrix}$

5. **a)** $\overrightarrow{AB} = \begin{pmatrix} 1 \\ 2 \\ -6 \end{pmatrix}$, $\overrightarrow{BA} = \begin{pmatrix} -1 \\ -2 \\ 6 \end{pmatrix}$

b) $\overrightarrow{AB} = \begin{pmatrix} 3 \\ 1 \\ -3 \end{pmatrix}$, $\overrightarrow{BA} = \begin{pmatrix} -3 \\ -1 \\ 3 \end{pmatrix}$

c) $\overrightarrow{AB} = \begin{pmatrix} 1 \\ 2 \\ -0.5 \end{pmatrix}$, $\overrightarrow{BA} = \begin{pmatrix} -1 \\ -2 \\ 0.5 \end{pmatrix}$

d) $\overrightarrow{AB} = \begin{pmatrix} \frac{3}{2} \\ -\frac{1}{2} \\ 2 \end{pmatrix}$, $\overrightarrow{BA} = \begin{pmatrix} -\frac{3}{2} \\ \frac{1}{2} \\ -2 \end{pmatrix}$

6. **a)** $i + j + 5k, -i - j - 5k$
b) $8i - 4j - 7k, -8i + 4j + 7k$

c) $4i + 4j + 4k, -4i - 4j - 4k$
d) $-2i - tj + 2tk, 2i + tj - 2tk$

7. **a)** $8i + j - 3k$ **b)** $-8i - j + 3k$

8. $x = 2, y = -2, z = 2$

Exercise 9.2 page 189

1. **a)** $\sqrt{34} = 5.83$ **b)** $\sqrt{45} = 3\sqrt{5} = 6.71$
c) $\sqrt{50} = 5\sqrt{2} = 7.07$ **d)** $\sqrt{90} = 3\sqrt{10} = 9.49$
e) 15

2. **a)** $\frac{1}{\sqrt{14}}(2i - 3j + k)$ **b)** $\frac{1}{\sqrt{3}}\begin{pmatrix} 1 \\ -1 \\ 1 \end{pmatrix}$

c) $\frac{1}{3}(-i + 2j - 2k)$ **d)** $\frac{1}{\sqrt{14}}\begin{pmatrix} -1 \\ 2 \\ -3 \end{pmatrix}$

3. **(b)** and **(c)**

4. **a)** 1 **b)** 28 **c)** $\frac{15}{2}$

5. **a)** $a = 3, b = -4, c = 1$ **b)** $a = 2, b = -1, c = 0$

6. **a)** AB and CD **b)** CD

7. **a)** $(0, 2, 4)$ **b)** $\frac{1}{\sqrt{21}}\begin{pmatrix} -1 \\ 4 \\ 2 \end{pmatrix}$ or $\frac{1}{\sqrt{21}}(-i + 4j + 2k)$

c) $\frac{5}{3}\begin{pmatrix} 2 \\ -2 \\ -1 \end{pmatrix}$ or $\frac{5}{3}(2i - 2j - k)$

Exercise 9.3 page 192

1. **a)** $\begin{pmatrix} 4 \\ -1 \\ 5 \end{pmatrix}$ **b)** $\begin{pmatrix} 12 \\ 2 \\ 13 \end{pmatrix}$ **c)** $\begin{pmatrix} 3 \\ 2 \\ 3 \end{pmatrix}$

d) $\begin{pmatrix} 8 \\ 3 \\ 8 \end{pmatrix}$ **e)** $\begin{pmatrix} -21 \\ -12 \\ -12 \end{pmatrix}$ **f)** $\begin{pmatrix} 3 \\ \frac{1}{2} \\ \frac{3}{2} \end{pmatrix}$

2. $2(-i - 2j + 2k)$

3. **a)** **i)** $4i$ **ii)** $i + 3j + 5k$ **iii)** $-i - 3j - 5k$
b) $3i - 3j - 5k$
c) **i)** $4i + 4j$ **ii)** $3i + j - 5k$ **iii)** $i - j + 5k$
d) $\sqrt{27} = 3\sqrt{3}$

4. Proof

5. a) $\dfrac{7}{2}i-\dfrac{1}{2}j-\dfrac{1}{2}k=\dfrac{1}{2}\left(7i-j-k\right)$

b) $\dfrac{15}{4}i+\dfrac{1}{4}j-\dfrac{5}{4}k=\dfrac{1}{4}(15i+j-5k)$

6. a) i) $\begin{pmatrix}6\\0\\0\end{pmatrix}$ **ii)** $\begin{pmatrix}0\\0\\2\end{pmatrix}$ **iii)** $\begin{pmatrix}0\\4\\0\end{pmatrix}$

b) $\begin{pmatrix}6\\-4\\-2\end{pmatrix}$

c) i) $\begin{pmatrix}-3\\4\\2\end{pmatrix}$ **ii)** $\begin{pmatrix}3\\0\\-2\end{pmatrix}$ **iii)** $\begin{pmatrix}3\\4\\2\end{pmatrix}$ **iv)** $\begin{pmatrix}-3\\-4\\2\end{pmatrix}$

d) $\dfrac{1}{\sqrt{14}}\begin{pmatrix}3\\-2\\1\end{pmatrix}$

7. $4i-j$

Exercise 9.4 page 196

1. a) i) $(2, 3, -1)$ **ii)** $(6, -2, 9)$
iii) $(-10, 18, -31)$

b) i) $(-4, 6, -3)$ **ii)** $\left(\frac{1}{2},\frac{3}{2},0\right)$
iii) $(14, -12, 9)$

2. a) $r=\begin{pmatrix}4\\-1\end{pmatrix}+t\begin{pmatrix}-3\\5\end{pmatrix}$

b) $r=\begin{pmatrix}2\\5\end{pmatrix}+t\begin{pmatrix}4\\-6\end{pmatrix}$ **c)** $r=t\begin{pmatrix}1\\0\end{pmatrix}$

d) $r=\begin{pmatrix}2\\-5\end{pmatrix}+t\begin{pmatrix}2\\8\end{pmatrix}$ **e)** $r=t\begin{pmatrix}1\\3\end{pmatrix}$

f) $r=\begin{pmatrix}0\\1\end{pmatrix}+t\begin{pmatrix}1\\-1\end{pmatrix}$

3. a) $r=(2-t)i+(1-t)j+(5+t)k$
b) $\mathbf{r} = (3-t)\mathbf{i} + (2+4t)\mathbf{j} + (4-4t)\mathbf{k}$
c) $r=tj$
d) $\mathbf{r} = (-3-2t)\mathbf{i} + (7+3t)\mathbf{j} + 2t\mathbf{k}$
e) $\mathbf{r} = t\mathbf{i} + t\mathbf{j} + t\mathbf{k}$
f) $\mathbf{r} = -3\mathbf{i} + 2\mathbf{j} + (6+t)\mathbf{k}$

4. a) $r=\begin{pmatrix}1\\2\\3\end{pmatrix}+t\begin{pmatrix}-2\\0\\-4\end{pmatrix}$ **b)** $r=\begin{pmatrix}1\\2\\3\end{pmatrix}+t\begin{pmatrix}5\\0\\6\end{pmatrix}$

c) $r=\begin{pmatrix}0\\2\\1\end{pmatrix}+t\begin{pmatrix}2.5\\0\\3\end{pmatrix}$

5. P and Q

6. a) $r=\begin{pmatrix}-2\\1\\0\end{pmatrix}+t\begin{pmatrix}6\\-1\\4\end{pmatrix}$ **b)** Proof

7. a) $r=\begin{pmatrix}2\\-1\\3\end{pmatrix}+t\begin{pmatrix}-1\\2\\-4\end{pmatrix}$ **b)** $\alpha=-3,\beta=7$

8. a) $r=\begin{pmatrix}2\\3\\4\end{pmatrix}+t\begin{pmatrix}-1\\1\\-1\end{pmatrix}$ **b)** $x=-2, y=7$

Exercise 9.5 page 200

1. a) Parallel **b)** Parallel **c)** Not parallel
d) Not parallel **e)** Not parallel **f)** Not parallel
g) Parallel **h)** Parallel

2. a) Intersect at $(3, 2, -2)$ **b)** No intersection

c) Intersect at $\begin{pmatrix}10\\1\\11\end{pmatrix}$ **d)** $\begin{pmatrix}-9\\\frac{7}{2}\\-\frac{3}{2}\end{pmatrix}$

3. a) $\begin{pmatrix}3\\0\\0\end{pmatrix}$ **b)** $2i-j+k$

4. $\begin{pmatrix}-1\\6\end{pmatrix}$

5. Proof

6. $c=-2;\ (1, 5, -3)$

7. Proof

Exercise 9.6 page 204

1. **a)** 22.2° **b)** 50.0° **c)** 90°
    d) 50.8° **e)** 99.6° **f)** 180°

2. **a)** 38.3° **b)** 86.6° **c)** 94.8° **d)** 0°

3. **a** and **d** 4. $p = 5$

5. **a)** -3 **b)** $\dfrac{4}{3}$

6. $A = 58.7°$ $B = 90.0°$ $C = 31.3°$

7. Proof

Exercise 9.7 page 205

1. **a)** 83.8° **b)** 84.1° **c)** 45.6°
    d) 72.0° **e)** 22.2°

2. Proof

3. **a)** 35.1° **b)** 50.5° **c)** 64.8°

Exercise 9.8 page 208

1. **a)** $r = \begin{pmatrix} 3 \\ -1 \\ 2 \end{pmatrix} + \lambda \begin{pmatrix} -1 \\ 1 \\ 0 \end{pmatrix}$

    b) $(5, -3, 2)$ **c)** $4\sqrt{6}$

2. **a)** $\sqrt{2}$ **b)** 3

3. $10\sqrt{\frac{2}{3}}$

4. $\sqrt{\dfrac{377}{9}}$

5. **a)** $r = \begin{pmatrix} 4 \\ 0 \\ 0 \end{pmatrix} + \lambda \begin{pmatrix} -4 \\ 5 \\ 2 \end{pmatrix}$

    b) $\dfrac{10}{3}$

6. **a)** $\begin{pmatrix} 2 \\ -1 \\ 6 \end{pmatrix} + \lambda \begin{pmatrix} 1 \\ -1 \\ 1 \end{pmatrix}$

    b) Proof **c)** $3\sqrt{3}$

Summary exercise 9 page 209

1. $\sqrt{14} = 3.74 \, (3 \text{ s.f.})$ 2. $-\dfrac{4}{3}$

3. $p = 1, q = 5$ 4. $a = 2, b = -1$

5. $\dfrac{1}{3}(2\mathbf{i} - \mathbf{j} + 2\mathbf{k})$ 6. $\begin{pmatrix} 6 \\ 0 \\ -8 \end{pmatrix}$

7. $\begin{pmatrix} 2 \\ 1 \\ 5 \end{pmatrix}$ or any non-zero multiple of this vector

8. **a)** **i)** $5\mathbf{i} + 5\mathbf{j} + 5\mathbf{k}$ **ii)** $-5\mathbf{i} + 5\mathbf{j} + 5\mathbf{k}$
    b) 70.5°

9. 45°

10. $\dfrac{1}{\sqrt{69}} \begin{pmatrix} -7 \\ 2 \\ -4 \end{pmatrix}$

11. $A = 54.8°$, $B = 82.7°$, $C = 42.5°$

12. **a)** $OA = \sqrt{6} \,(= 2.45), \, AB = \sqrt{14} \,(= 3.74),$
    $BC = \sqrt{6} \,(= 2.45), \, OC = \sqrt{14} \,(= 3.74)$
    b) Parallelogram
    c) $\angle O = \angle B = 40.2°, \, \angle A = \angle C = 139.8°$
    d) $\sqrt{\dfrac{35}{6}}$

13. $10 + \sqrt{66} = 18.1 \,(3 \text{ s.f.})$

14. **a)** $\dfrac{1}{5}$ **b)** 28.6° **c)** $\sqrt{\dfrac{6}{5}}$

15. $3\mathbf{j} + 5\mathbf{k}$

16. **a)** 65.9° **b)** 144.7° so acute angle is 35.3°
    c) 65.9°

17. **a)** $c = 2$ **b)** $\dfrac{1}{\sqrt{38}} \begin{pmatrix} 2 \\ 5 \\ 3 \end{pmatrix}$

18. **a)** $k = 5$ **b)** 12.6°

19. **a)** $AC = ED = 6$
    $BF = CD = AE = 8$
    $AB = BC = EF = DF = 5$
    b) 29.6°

20. **a)** $\dfrac{1}{3}(2\mathbf{i} + \mathbf{j} - 2\mathbf{k}), \dfrac{1}{3}(-2\mathbf{i} + \mathbf{j} + 2\mathbf{k})$ **b)** $\mathbf{j}$

21. **a)** $r = \begin{pmatrix} 1 \\ 2 \\ -1 \end{pmatrix} + \lambda \begin{pmatrix} 1 \\ 1 \\ 0 \end{pmatrix}$ **b)** $\left(-\dfrac{1}{2}, \dfrac{1}{2}, -1\right)$

22. $\dfrac{1}{10}$

23. **a)** $(1, 1, 1)$ **b)** 54.7°

24. **a)** $(1, 1, 2)$ **b)** 70.5°

25. $r = \dfrac{1}{11}(27\mathbf{i} + 19\mathbf{j} + 14\mathbf{k}) + \lambda(\mathbf{i} - 3\mathbf{j} - \mathbf{k})$

26. i) $(1, 3, 2)$ **ii)** $\theta = 123.1°$

iii) 2.87

27. i) $73.4°$ **ii)** $\begin{pmatrix} 1 \\ 1 \\ 3 \end{pmatrix} + \lambda \begin{pmatrix} 2 \\ -1 \\ -2 \end{pmatrix}$ **iii)** Proof

iv) $\begin{pmatrix} \dfrac{-5}{9} \\ \dfrac{16}{9} \\ \dfrac{13}{9} \end{pmatrix}$

28. i) $\overrightarrow{BC} = -5\mathbf{i} - 4\mathbf{j} + 6\mathbf{k}$

$\overrightarrow{BA} = 3\mathbf{i} - 4\mathbf{j}$

ii) $88.7°$

iii) 8.77

29. i) $\begin{pmatrix} 5p \\ -p \\ -p \end{pmatrix} + \lambda \begin{pmatrix} -4p \\ -4p \\ 8p \end{pmatrix}$ **ii)** Proof

iii) Proof   **iv)** $\begin{pmatrix} 3p \\ -3p \\ 3p \end{pmatrix}$

30. i) $(0, 3, -1)$ **ii)** 2

10 Differential equations

Skills check page 215

1. **a)** $c = 4$ **b)** $c = -103$
   c) $c = -3$ **d)** $c = 2$

2. $1 - 2\sin^2\theta$

3. $\dfrac{1}{x-1} - \dfrac{1}{x+1}$

Exercise 10.1 page 220

1. $\dfrac{\mathrm{d}T}{\mathrm{d}t} = -k(T - T_0)$

2. $\dfrac{\mathrm{d}x}{\mathrm{d}t} = k(30 - x)$

3. $\dfrac{\mathrm{d}N}{\mathrm{d}t} = kN(P - N)$

4. $\dfrac{\mathrm{d}N}{\mathrm{d}t} = kN; \quad k = 0.07$

5.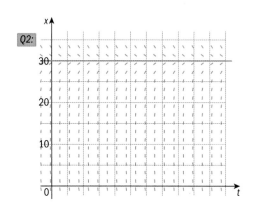

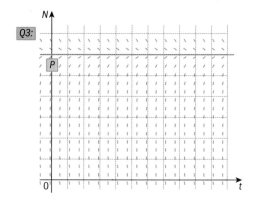

Exercise 10.2 page 224

1. $y^2 = x^2 + c$
2. $2y^2 = x^4 + c2y^2 = x^2 + c$
3. $2y^3 = 2x^3 + 3x^2 + 6x + c$
4. $y = Ae^{\frac{1}{2}x^2 + 3x}$
5. $\sin y = 2x + c$
6. $\dfrac{1}{2}y^2 = \ln x + c$
7. $y^2 = x^4 + c$
8. $\dfrac{1}{4y^2} = \dfrac{1}{x} + c$
9. $\ln|y| = \sin x + c$
10. $e^{-y} = -x^2 + c$
11. $\sin y = x^2 + c$
12. $-\dfrac{1}{y} = \dfrac{1}{2}\ln\left|\dfrac{x}{x+2}\right| + c$
13. $\sin y = \sin x + c$
14. $\tan y = \ln|x| + c$
15. $\ln\left|\dfrac{y-1}{y+1}\right| = x^4 + c$
16. $\tan y = \tan x + c$
17. $\sin y = xe^x - e^x + c$
18. $\dfrac{y-1}{y+1} = A\dfrac{x}{x+1}$
19. $-\dfrac{1}{y} = \tan x + c$
20. $\dfrac{1}{4}\ln(1 + y^4) = \dfrac{1}{2}x^2 + c$
21. $\dfrac{1}{y} = \ln\left|A\dfrac{(x+1)^2}{x^3}\right|$
22. $\dfrac{1}{y^2} = \tan^{-1}(e^x) + c$
23. $\dfrac{1}{2}y^2 = -\dfrac{1}{3}\cos^3 x + \dfrac{1}{5}\cos^5 x + c$
24. $\dfrac{1}{2}y^2 = 9\tan^{-1}\left(\dfrac{x}{3}\right) + c$

Exercise 10.3 page 227

1. $y^2 = x^2 + 3$
2. $2y^2 = x^4 + 7$
3. $\dfrac{1}{3}y^3 = \dfrac{1}{3}x^3 + \dfrac{1}{2}x^2 + x + 9$
4. $y = 5e^{\left(\frac{1}{2}x^2 + 3x\right)}$
5. $\sin y = 2x + 1$
6. $\dfrac{1}{2}y^2 = \ln x + 2$
7. $y^2 = x^4 - 7$
8. $\dfrac{1}{4y^2} = \dfrac{1}{x}$
9. $y = e^{(\sin x - 1)}$
10. $y^2 = x^2 - 4$

11. $y^2 = \ln\left(\dfrac{x^2}{e}\right)$ **12.** $y^2 = x^2 + x + 2$

13. $\sqrt{y} = \sqrt{x+1} + 1$ **14.** $\sin y = -\dfrac{1}{2x^2} + 1$

15. $e^y = 2\tan x - 1$ **16.** $\dfrac{2}{3}y^{\frac{3}{2}} = 2x^{\frac{1}{2}} + 14$

17. $e^y = \ln\left|x^2 + 1\right| + 1$ **18.** $y = \left|\dfrac{6x}{x+1}\right|$

19. $e^{-y} = \dfrac{1}{3}e^{3x} + \dfrac{2}{3}$ **20.** $\tan y = -\dfrac{1}{2}x^{-2} + \dfrac{3}{2}$

21. $y^2 = e^{x^2} + 8$

Exercise 10.4 page 235

1. a) $\dfrac{dT}{dx} = -0.16x$ **b)** $T = 400 - 0.08x^2$

    c) $200°C$ **d)** $68.9\,\text{cm}$

2. a) $\dfrac{1}{y}\dfrac{dy}{dt} = -3e^{-3t}$ **b)** $\ln|y| = e^{-3t} + \ln(20e^{-1})$

    c) $t \to \infty$, $\ln|y| \to \ln(20e^{-1})$, $y \to 20e^{-1}$

    Initially there was 20 grams of Y, so the proportion which will remain is 37%.

3. a) $\dfrac{1}{m}\dfrac{dm}{dt} = -km = Ae{-}kt$

    b) Proof

    c) $k = 1.21 \times 10^{-4}$ **d)** $40\,000$ years

4. a) $\left(\dfrac{1}{d-150}\right)\dfrac{dd}{dt} = -10^{-5}$ **b)** $|d-150| = 142e^{-0.00001t}$

    c) $d = 8.21$ metres

5. a) $\dfrac{1}{N}\dfrac{dN}{dt} = k$ **b)** $N = 2e^{0.0372t}$

    c) 43.3 hours

6. a) $\dfrac{1}{v}\dfrac{dv}{dt} = -k$ **b)** $v = 25e^{-0.0223t}$

    c) 31.1 seconds

7. a) $\dfrac{dn}{dt} = kn(P - n)$ **b)** $\dfrac{n}{P-n} = \dfrac{1}{19}e^{0.249t}$

    c) 11.8 days

    d) As $t \to \infty$, $n \to P$ so in the long run the whole population will be infected.

Summary exercise 10 page 237

1. a) $\dfrac{1}{4}y^4 = \dfrac{1}{3}x^3 + c$ **b)** $\tan y = x^2 + c$

    c) $\sin y = \dfrac{1}{3}e^{3x} + c$ **d)** $\cos y = \dfrac{A}{x}$

    e) $e^y = \dfrac{1}{3}e^{3x} + c$ **f)** $y = B\sqrt{x^2 + 1}$

g) $y = B\left(x^2 + 6\right)^4$

h) $\dfrac{-1}{3}\ln|\cos 3\theta| = \dfrac{1}{2}x^2 + 2x + c$

2. a) $y^4 = x^3 + \dfrac{1}{2}x^2 - 9$ **b)** $y = \dfrac{1}{3 + 2\cos x}$

    c) $e^y = \dfrac{1}{3}e^{3x} + \dfrac{2}{3}$ **d)** $2v^{\frac{1}{2}} = -e^{-t} + 7$

    e) $e^y = xe^x - e^x + 2$

    f) $y\ln y - y = 2\ln(x + 1) - \ln x - 1 - \ln 4$

3. $\ln y = 2(x + 1)^{\frac{1}{2}} - 2\sqrt{2}$

4. $\sin y = xe^x - e^x + 1.5$

5. $ye^y - e^y = \tan x - 2$

6. $\dfrac{2}{3}y^{\frac{3}{2}} = \ln\left|\dfrac{3(x-1)}{x+1}\right|$

7. $2y^2 = 2x + \sin 2x - \dfrac{\pi}{2} - 1$

8. $\ln(y^2 + 1) + \tan^{-1}y = \ln x$

9. a) $\dfrac{dT}{dt} = -k(T - 20)$ **b)** $T = 20 + 70\,e^{-0.14t}$

    c) $37.3°\,C$ **d)** 18.9 minutes

10. a) $\dfrac{dv}{dt} = -kv^2$ **b)** $v = \dfrac{300}{t+6}$

    c) $18.75\,\text{m s}^{-1}$ **d)** 24 seconds

11. a) $\dfrac{dN}{dt} = kN(2000 - N)$ **b)** $N = \dfrac{2000\left(\sqrt{\frac{117}{17}}\right)^t}{39 + \left(\sqrt{\frac{117}{17}}\right)^t}$

    c) The majority have heard the rumour when $N = 1001$. When $N = 1001$, $t \approx 3.8$ hours.

12. a) Proof   **b)** Proof

    c) $v^2 = 100 - 4x^2$ **d)** $v = 8\,\text{m s}^{-1}$

13. a) $\dfrac{1}{m}\dfrac{dm}{dt} = -0.05$ **b)** $m = 12e^{-0.05t}$

    c) 8.1 minutes

14. $y = (5 + e^{2x})^2$

15. $\ln(x + 4) = \dfrac{1}{2}\theta - \dfrac{1}{12}\sin 6\theta + \ln 4; x = 2.44$

16. a) $\ln\left(\dfrac{5x}{3-x}\right) = \dfrac{1}{2}\ln(10)t$; proof

    b) $t = 2.80$ hours

17. a) Proof   **b)** Proof

11 Complex numbers

Skills check page 242

1. **a)** $2 + 2\sin\theta$ **b)** $\sin^4\theta$ **c)** 1
2. **a)** Proof   **b)** Proof   **c)** Proof
3. **a)** 17 **b)** 81 **c)** -71

Exercise 11.1A page 244

1. c) and d)
2. **a)** $\pm 2i$ **b)** $\pm 10i$ **c)** $\pm\sqrt{13}\,i$
   d) $\pm\sqrt{75}\,i$ or $\pm 5\sqrt{3}\,i$
3. **a)** i **b)** $-i$ **c)** $-i$
   d) i **e)** 1
4. **a)** -1 **b)** $-128i$ **c)** $-27i$
   d) -1024
5. **a)** ± 8 **b)** $\pm 8i$ **c)** $\pm\sqrt{19}\,i$
   d) $\pm\sqrt{32}\,i$ or $\pm 4\sqrt{2}\,i$
6. **a)** $\pm 7i$ **b)** $\pm\sqrt{18}\,i$ or $\pm 3\sqrt{2}\,i$
   c) $\pm\sqrt{2}\,i$ **d)** $\pm\sqrt{5}\,i$
7. **a)** -2 **b)** 49
   c) $-\dfrac{1}{27}i$ **d)** $\dfrac{1}{256}$
8. **a)** $x + y = 10,\ xy = 40$ **b)** Proof
   c) Proof

Exercise 11.1B page 246

1. **a)** $5 + i$ **b)** $4 - \sqrt{6}\,i$
   c) $-2 + 2\sqrt{10}\,i$ **d)** $-1 - 9i$
2. **a)** $4 + 2i$ **b)** $-5 - i$
   c) $-9 + 7i$ **d)** $-6 - 3i$
3. **a)** $z^* = 10 - 12i$ **b)** $z = 3 + 3i$
   c) $z^* = -5i$ **d)** $z = 8$
4. **a)** 2 real roots   **b)** 2 complex roots
   c) 2 real roots   **d)** 2 real roots
   e) 2 complex roots   **f)** 2 complex roots
5. **a)** $2 \pm i$ **b)** $-\dfrac{1}{4} \pm \dfrac{\sqrt{3}}{4}i$
   c) $\dfrac{9 \pm \sqrt{105}}{4}$ **d)** $\dfrac{2}{3}, -1$
   e) $\dfrac{1}{6} \pm \dfrac{\sqrt{11}}{6}i$ **f)** $-\dfrac{1}{4} \pm \dfrac{\sqrt{7}}{4}i$
   g) $\pm\dfrac{\sqrt{3}}{2} - \dfrac{1}{2}i$

6. **a)** $1 \pm \sqrt{2}\,i$ **b)** $-1 \pm \sqrt{\dfrac{7}{3}}\,i$
   c) $2 \pm \dfrac{1}{\sqrt{2}}$ **d)** $\dfrac{3}{2} \pm \sqrt{\dfrac{23}{12}}$
   e) $-1 \pm i$ **f)** $-3 \pm i$
   g) $\pm 2 + i$
7. $\dfrac{1}{i}$ or $\dfrac{-2}{i}$

Exercise 11.2 page 249

1. **a)** $5 + 6i$ **b)** $11 + 4i$
   c) -3 **d)** $3 + 3i$
   e) $1 + 5i$ **f)** $-1 - 8i$
2. **a)** $2 + 2i$ **b)** $2 + 6i$
   c) $9 - 6i$ **d)** $11 - 8i$
   e) $-6 + 2i$ **f)** $2 + 6i$
3. **a)** $21 + 11i$ **b)** $-2 - 10i$
   c) $-5 - 15i$ **d)** $24 + 41i$
   e) $3 - 33i$ **f)** $-2 - 5i$
4. **a)** $9 + 7i$ **b)** $23 - 2i$
   c) $1 + 7i$ **d)** $5 + 10i$
   e) $-14 + 10i$ **f)** $15 + 45i$
5. **a)** $\dfrac{14}{5} + \dfrac{3}{5}i$ **b)** $\dfrac{16}{5} - \dfrac{3}{5}i$
   c) $-\dfrac{11}{61} + \dfrac{60}{61}i$ **d)** $\dfrac{1}{4}$
   e) $\dfrac{4}{5} - \dfrac{3}{5}i$ **f)** $\dfrac{8}{25} + \dfrac{6}{25}i$
   g) $\dfrac{1}{2} - \dfrac{3}{2}i$ **h)** $\dfrac{1}{7} - \dfrac{4\sqrt{3}}{7}i$
6. **a)** $1 - 2\sqrt{2}\,i$
   b) i) $2\sqrt{2}$ **ii)** $-4i$
        iii) 15 **iv)** $-\dfrac{2\sqrt{2}}{3} - \dfrac{1}{3}i$
7. **a)** $8 - 16i$ **b)** $-2 + 8i$
   c) $-33 - 56i$ **d)** $\dfrac{63}{169} + \dfrac{16}{169}i$
8. **a)** $3 - 4i$ **b)** $2 - 11i$
   c) $-7 - 24i$ **d)** $-278 + 29i$
9. **a)** $\dfrac{3}{10} + \dfrac{1}{10}i$ **b)** $\dfrac{23}{65} - \dfrac{11}{65}i$

10. a) i **b)** $\dfrac{3}{17} - \dfrac{5}{17}$ i **c)** $\dfrac{3}{5} + \dfrac{4}{5}$ i

  d) $\dfrac{a}{a^2+b^2} - \dfrac{b}{a^2+b^2}$ i **e)** $\dfrac{52}{25} + \dfrac{33}{25}$ i vw

11. a) $\text{Re}(u^2) = a^2 - b^2,\ \text{Im}(u^2) = 2ab$
  b) $\text{Re}(u - u^*) = 0,\ \text{Im}(u - u^*) = 2b$
  c) $\text{Re}(u^3 - (u^*)^3) = 0,\ \text{Im}(u^3 - (u^*)^3) = 6a^2b - 2b^3$

12. a) $\cos 2\theta + i\sin 2\theta$
  b) $\cos\theta - i\sin\theta$
  c) $\cos 2\theta + i\sin 2\theta$

Exercise 11.3A page 253

1. a) ± 20i **b)** $\pm 2\sqrt{15}$ i
  c) $1 + 2i, -1 - 2i$ **d)** $2 + i, -2 - i$
  e) $2 + 3i, -2 - 3i$ **f)** $5 + 2i, -5 - 2i$
  g) $2 - 5i, -2 + 5i$ **h)** $6 - i, -6 + i$
  i) $-3\sqrt{2} - 3\sqrt{2}\,i,\ 3\sqrt{2} + 3\sqrt{2}\,i$

2. a) $2 - i, -2 + i$ **b)** $-i, -2$

3. $2 + 3i$

4. Square roots of i are $\dfrac{\sqrt{2}}{2} + \dfrac{\sqrt{2}}{2}i$ and $-\dfrac{\sqrt{2}}{2} - \dfrac{\sqrt{2}}{2}i$.

Square roots of $-i$ are $\dfrac{\sqrt{2}}{2} - \dfrac{\sqrt{2}}{2}i$ and $-\dfrac{\sqrt{2}}{2} + \dfrac{\sqrt{2}}{2}i$.

5. a) $\dfrac{5\sqrt{2}}{2} + \dfrac{5\sqrt{2}}{2}i,\ \dfrac{-5\sqrt{2}}{2} + \dfrac{-5\sqrt{2}}{2}i$

  b) $-2\sqrt{2} + 2\sqrt{2}\,i,\ 2\sqrt{2} - 2\sqrt{2}\,i$ **c)** $\pm 2i$

6. $a = 16, b = -2$

7. $1 + 4i$

8. a) $\dfrac{1}{2} + \dfrac{1}{2}i$ **b)** $\pm\dfrac{\sqrt{3}}{2} + \dfrac{1}{2}i$

9. $3 - i$

10. $a = 5, b = 30$ or $a = -5, b = -30$

Exercise 11.3B page 254

1. a) $1 - \sqrt{3}\,i$ **b)** $z^3 - 4z^2 + 8z - 8 = 0$

2. a) $\sqrt{2} \pm \sqrt{2}\,i$ **b)** Proof

3. a) $i, -5i$ **b)** $-\dfrac{1}{2}i$ **c)** $\pm\dfrac{\sqrt{3}}{6} - \dfrac{1}{2}i$

4. $z = 3 + i, w = 2 - 3i$

5. $z = 4i, w = -3i$

6. a) $z = 1$ **b)** $z - 1$ **c)** $1, -\dfrac{1}{2} \pm \dfrac{\sqrt{3}}{2}i$

7. $3, -\dfrac{3}{2} \pm \dfrac{3\sqrt{3}}{2}i$

8. a) -4 **b)** $1 \pm i, -1 \pm i$

9. $-\dfrac{5}{2}, 5 - 2i$

10. a) Proof **b)** $2, -1 - 2i$

11. a) Proof **b)** $\pm 1, 2 - 3i$

12. a) -2 **b)** Proof **c)** $-2, \pm i, 1 \pm 2i$

Exercise 11.4 page 257

1.

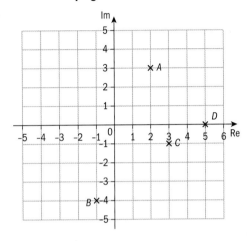

2. $u = 5 + 3i, u^* = 5 - 3i, 2u = 10 + 6i, -u = -5 - 3i$

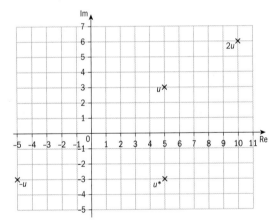

3. $z_1 = 4 + 5i, z_2 = 3i, z_3 = 2 - 6i, z_4 = -2i, z_5 = -4,$
  $z_6 = -5 - 4i, z_7 = -3 + 4i$

4. a) $3 + 4i, -4 + 3i, 4 - 3i, 4 + 3i$

5. a) $-2 + 5i$

b) $iz_1z_2 = -13 - 11i$

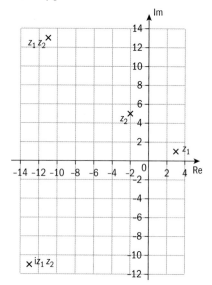

c) $z + w = -3, z - w = -7 - 6i$

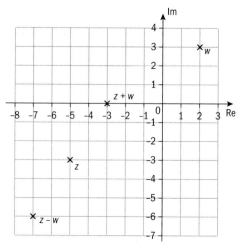

6. a) $z + w = 5 + 5i, z - w = 3 - i$

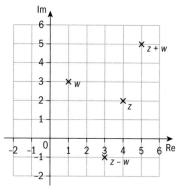

d) $z + w = 4 - 3i, z - w = 2 - 5i$

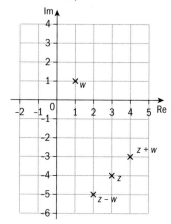

b) $z + w = -2 + 6i, z - w = 6 + 4i$

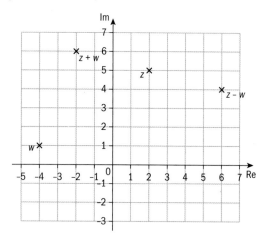

7. $\dfrac{1}{z} = \dfrac{1}{4} + \dfrac{1}{4}i, \ z + \dfrac{1}{z} = \dfrac{9}{4} - \dfrac{7}{4}i$

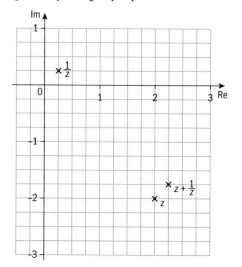

Exercise 11.5 page 263

1. **a)** $1, 0$ **b)** $1, \pi$
   c) $\sqrt{2}, \frac{\pi}{4}$ **d)** $\sqrt{2}, -\frac{\pi}{4}$
   e) $3, \frac{\pi}{2}$ **f)** $13, 1.18$ rad
   g) $7, -1.03$ rad   **h)** $1, \frac{\pi}{4}$

2. **a)** $1, \frac{\pi}{3}$ **b)** $1, -\frac{\pi}{3}$
   c) $1, \frac{2\pi}{3}$ **d)** $1, -\frac{2\pi}{3}$
   e) $1, -\frac{2\pi}{3}$ **f)** $1, \pi$
   g) $1, 0$ **h)** 0, not defined
   i) $1, -\frac{\pi}{3}$ **j)** $2, -\frac{2\pi}{3}$

3. **a) i)** 5 **ii)** 25
       iii) 125
   b) i) 0.644 rad   **ii)** 1.29 rad
       iii) 1.93 rad

4. **a)** $2i$ **b)** $-5 + 0i$
   c) $2 + 2\sqrt{3}i$ **d)** $\frac{3\sqrt{3}}{2} + \frac{3}{2}i$
   e) $5\sqrt{2} + 5\sqrt{2}i$ **f)** $-\frac{1}{2}i$
   g) $-\frac{\sqrt{2}}{2} - \frac{\sqrt{2}}{2}i$ **h)** $-3 + 3\sqrt{3}i$
   i) $7 + 0i$

5.
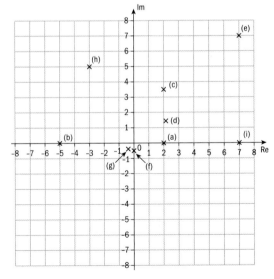

6. **a)** $|wz| = 2\sqrt{2}, \arg(wz) = -\frac{\pi}{12}$
   b) $\left|\frac{w}{z}\right| = \sqrt{2}, \arg\left(\frac{w}{z}\right) = -\frac{7\pi}{12}$
   c) $|iz| = \sqrt{2}, \arg(iz) = \frac{3\pi}{4}$
   d) $\left|\frac{1}{w*}\right| = \frac{1}{2}, \arg\left(\frac{1}{w*}\right) = -\frac{\pi}{3}$

7. **a)** -1 **b)** $-i$ **c)** $2\sqrt{2} + 2\sqrt{2}i$

8. **a)** $\sqrt{2}e^{\frac{i\pi}{4}}$ **b)** $3e^{i\pi}$ **c)** $\sqrt{5}\,e^{2.03i}$
   d) $7e^{\frac{i\pi}{2}}$ **e)** $10e^{0.927i}$ **f)** $10e^{-0.927i}$
   g) $2e^{\frac{-3\pi i}{4}}$ **h)** $2\sqrt{13}\,e^{-0.197i}$ **i)** $e^{\frac{\pi i}{3}}$

9. **a)** $\left(6, \frac{\pi}{2}\right)$ **b)** $\left(12, \frac{3\pi}{4}\right)$
   c) $\left(\frac{2}{3}, -\frac{\pi}{6}\right)$ **d)** $\left(4, \frac{\pi}{2}\right)$

10. **a)** $9, \frac{\pi}{2}$ **b)** $27, \frac{3\pi}{4}$ **c)** $81, \pi$
    d) $6561, 0$ **e)** $6, \frac{5\pi}{12}$ **f)** $12, \frac{7\pi}{12}$
    g) $216, -\frac{3\pi}{4}$ **h)** $\frac{3}{2}, \frac{\pi}{12}$ **i)** $\frac{1}{9}, -\frac{\pi}{2}$
    j) $2\sqrt{3}, 0$ **k)** $2, \frac{\pi}{2}$ **l)** $\frac{1}{6}, \frac{5\pi}{12}$

11. Proof

12. **a)**
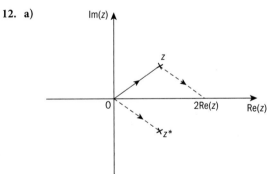

b)

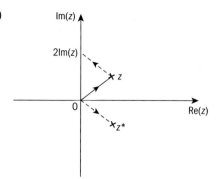

c) *z* lies on a circle, centre (−5, 0), radius 2.

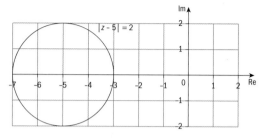

d) *z* lies on or inside a circle, centre (2, −2), radius 2.

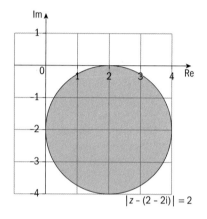

c)

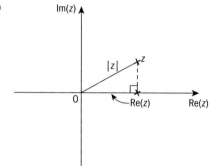

Exercise 11.6 page 268

1. **a)** *z* lies on a circle, centre (0, 0), radius 1.

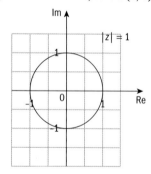

b) *z* lies on a circle, centre (3, 0), radius 1.

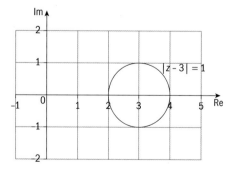

e) *z* lies on or outside a circle, centre (−1, −3), radius 1.

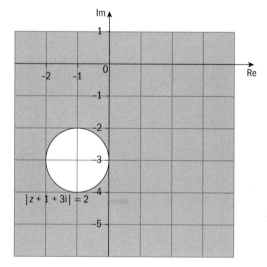

f) *z* lies on a circle, centre (*a*, 0), radius *a*.

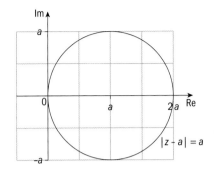

2. a) i) $|z| = 3$ **ii)** $|z - (2+2i)| = 2$ or $|z - 2 - 2i| = 2$

   iii) $|z - (-4 - 3i)| = 5$ or $|z + 4 + 3i| = 5$

iii)

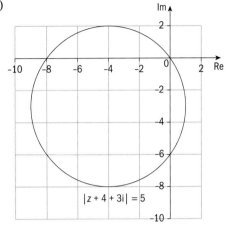

b) i)

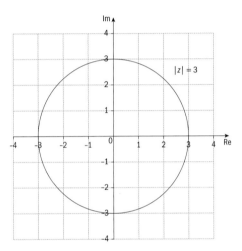

ii)

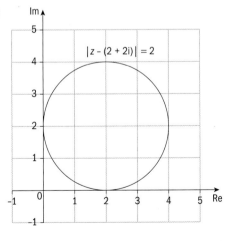

3. a) *z* lies on the half line from (1, 0) at an angle of $\frac{\pi}{2}$ with the positive real axis.

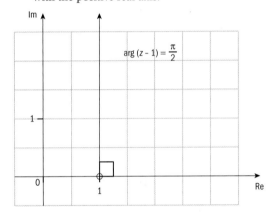

b) *z* lies on the half line from (−1, 0) at an angle of $\frac{\pi}{3}$ with the positive real axis.

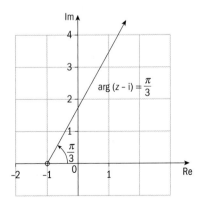

c) *z* lies on the half line from (0, −1) at an angle of $\frac{-2\pi}{3}$ with the positive real axis.

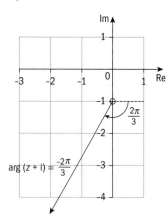

$\text{arg}(z + i) = \frac{-2\pi}{3}$

d) *z* lies on the half line from (2, 3) at an angle of $\frac{3\pi}{4}$ with the positive real axis.

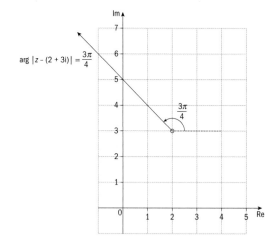

$\text{arg}|z - (2 + 3i)| = \frac{3\pi}{4}$

4. a) *z* lies on the line $\text{Re}(z) = \frac{5}{2}$ (or $x = \frac{5}{2}$).

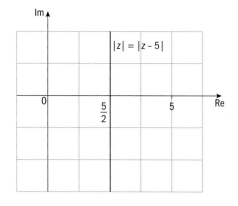

$|z| = |z - 5|$

b) *z* lies on the line Im(*z*) = 3 (or *y* = 3).

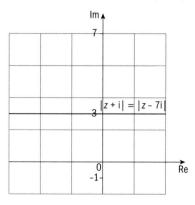

$|z + i| = |z - 7i|$

c) *z* lies on the perpendicular bisector (or mediator) of the line joining (−4, 0) and (0, 2).

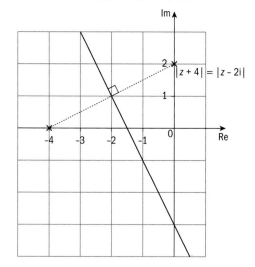

$|z + 4| = |z - 2i|$

d) *z* lies on the perpendicular bisector (or mediator) of the line joining (−2, −2) and (2, 2), i.e. on the line Re(*z*) = −Im(*z*) or *y* = −*x*.

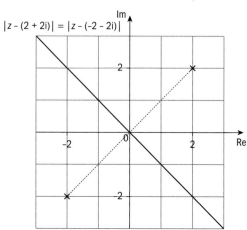

$|z - (2 + 2i)| = |z - (-2 - 2i)|$

e) z lies on the perpendicular bisector (or mediator) of the line joining $(1, 2)$ and $(5, 2)$ (or $x = 3$).

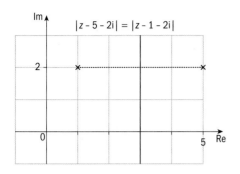

5. a)

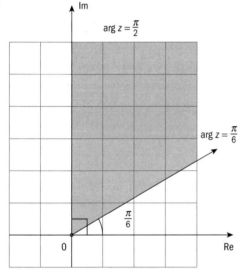

b)

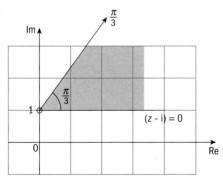

c)

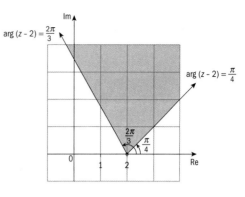

d)

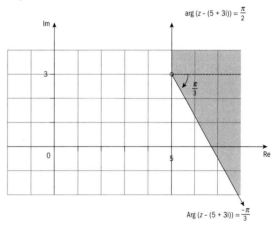

6. a) z lies on or inside a circle, centre $(6, 0)$, radius 3.

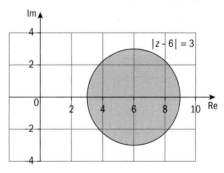

b) z lies on or outside a circle, centre $(1, 3)$, radius 1.

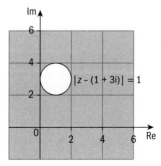

c) z lies on the line $x = 2$ or within the shaded region, i.e. z lies on the perpendicular bisector (mediator) of the line joining $(1, 0)$ and $(3, 0)$ or to the left of the line.

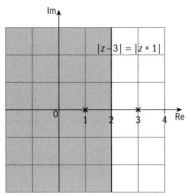

d) z lies on the line $y = -x$ or within the shaded region, i.e. z lies on the perpendicular bisector (mediator) of the line joining $(-1, 0)$ and $(0, 1)$ or below the line.

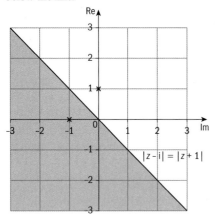

e) z lies on the perpendicular bisector (or mediator) of the line joining $(-1, -1)$ and $(3, 3)$ or above the line, i.e. on the line $x + y = 2$ or above the line.

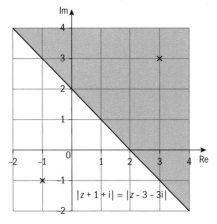

f) z lies on the perpendicular bisector (or mediator) of the line joining $(2, -3)$ and $(0, 2)$ or below the line.

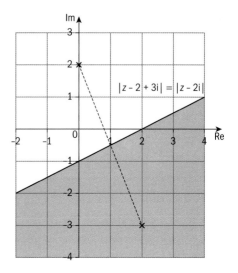

7. z lies within the shaded region.

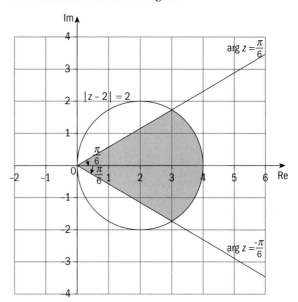

8. *z* lies within the shaded region.

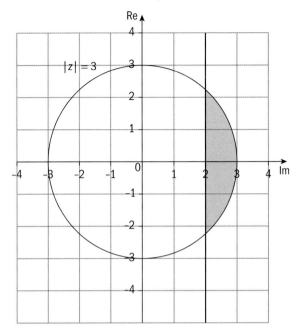

9. a) *z* lies within the shaded region.

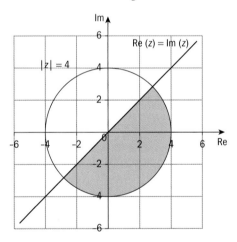

b) Greatest value of $|z|$ is 4, least value of $|z|$ is 0.

10. a) *z* lies within the shaded region.

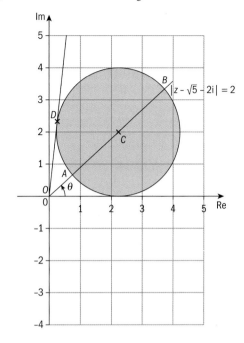

b) Greatest value of $|z|$ is 5.

c) Greatest value of arg *z* is 1.46 rad.

Summary exercise 11 page 270

1. $x = \frac{9}{41}, y = -\frac{42}{41}$

2. **a)** Both roots have modulus $\sqrt{10}$. **b)** $-1.89, 1.89$

3. **a)** Proof **b)** $-\frac{1}{2} + \frac{\sqrt{3}}{2}$ i, $-\frac{1}{2} - \frac{\sqrt{3}}{2}$ i

4. **a)** $-4 - 2\sqrt{5}$ i **b)** $p = 2$

5. **a)** $|z - 3 - 3i| \leq 2$ and $|z + 1 + 0| \leq 1$

   i.e. $|z - (3 + 3i)| \leq 2$ and $|z - (-1 - i)| \leq 1$
   $|z - 3 - 3i| = 2$
   $|w - (-1 - i)| = 1$

Loci of z and w are within or on the shaded circles.

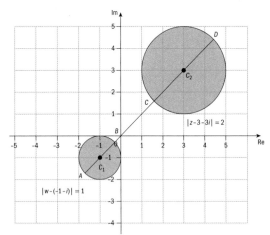

b) Least value is $4\sqrt{2} - 3$, greatest value is $4\sqrt{2} + 3$.

6. Proof

7. a) $-4, -1$ **b)** $\pm i, \pm 2i$

8. a)

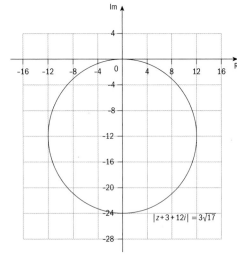

$|z + 3 + 12i| = 3\sqrt{17}$

z lies on the circle.

b) Greatest value of $|z|$ is $6\sqrt{17}$, least value of $|z|$ is 0.

9. a)

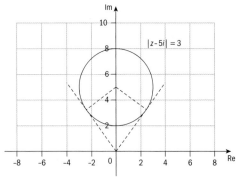

$|z - 5i| = 3$

b) Greatest value of $|z|$ is 8, least value of $|z|$ is 2.
Greatest value of arg z is 2.21 rad,
least value of arg z is 0.927 rad.

10. Im

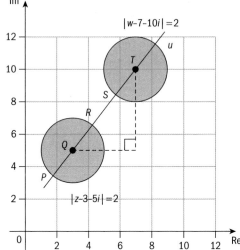

Greatest value of $|z - w|$ is $\sqrt{41} + 4$.

Least value of $|z - w|$ is $\sqrt{41} - 4$.

11. a) Proof   **b)** $1 - i$ **c)** $1 \pm i, 2 \pm 3i$

12. $2 + \sqrt{5}i, -4 \pm \sqrt{13}$

13. a) Proof   **b)** $1 \pm 2i, 1$

14. $-\dfrac{3\sqrt{3}}{2} + \dfrac{3}{2}i, \dfrac{3\sqrt{3}}{2} + \dfrac{3}{2}i, -3i$

15. a) $-\dfrac{1}{2} - \dfrac{\sqrt{3}}{2}i$ **b)** Proof

    c) i) $\left(1, \dfrac{2\pi}{3}\right), \left(1, -\dfrac{2\pi}{3}\right)$

       ii) An Argand diagram shows that the addition
of 1, w and w^2 forms an equilateral triangle,
starting and ending at the origin.

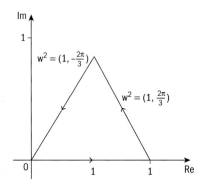

16. a) Proof   **b)** Proof

17. a) $-5 - i$

b)

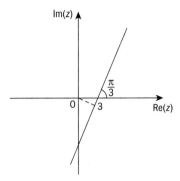

c) $\dfrac{\sqrt[3]{3}}{2}$

18. i) Proof   **ii)** $-\dfrac{1}{2}; 1 - i$

iii)

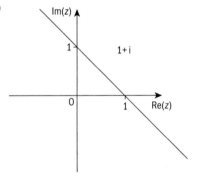

19. i) 6   **ii)** $-3 - 4i$ **iii)** $(5, 2.21)$

iv)

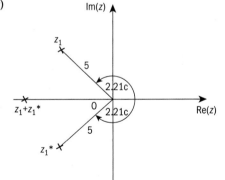

Exam-style paper 2A page 278

1. a)

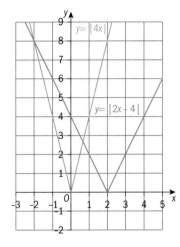

b) $x < -2$ or $x > \dfrac{2}{3}$

2. $x = 2.00$

3. a) $2x^2 - x - 1$ **b)** $x = 4$ or -3 or $-\dfrac{1}{2}$ or 1

4. a) $\left(2, \dfrac{3e^2}{4}\right)$ **b)** Minimum

5. a) Proof   **b)** $\theta = 60°, 131.8°, 228.2°, 300°$

6. a) Proof   **b)** $\dfrac{5\pi}{2} - 3$

7. a) $\dfrac{9}{10} - 6\ln 10$ **b)** $-\dfrac{5}{9}$

8. a) $p + \dfrac{1}{2}e^{-2p} - \dfrac{1}{2}$ **b)** Proof

     c) $p_2 = 1.4751, p_3 = 1.4738, p_4 = 1.4738, p = 1.47$ (2 d.p.)

Exam-style paper 2B page 280

1. $x > \dfrac{1}{2}$

2. 0.35

3. $A = 25.8, b = 1.7$

4. a) $2x - 5$ **b)** $(2x - 5)(2x - 1)(x + 5)$

5. a) $12\cos\theta - 5\sin\theta = 13\cos(\theta + 22.62°)$

     b) $\theta = 17.1°, 297.7°$ **c)** -13

6. **a)** Proof **b)** Proof

7. **a)**

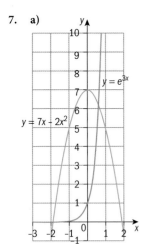

Two points of intersection means two roots of the equation

b) Proof **c)** Proof

d) $x_1 = 0.6125$, $x_2 = 0.6108$, $x_3 = 0.6111$, $x_4 = 0.6110$, $x = 0.61$ (2 d.p.)

8. **a)** Proof
b) $y = 2x - 6$

Exam-style paper 3A page 282

1. $x = -0.0937$ (3 s.f.)

2. **a)** $1 + x + \dfrac{3}{2}x^2 + \dfrac{5}{2}x^3$

b) $\dfrac{1}{2}$

3. **a)** $x_2 = 3.6905$, $x_3 = 3.6866$, $x_4 = 3.6851$, $x_5 = 3.6844$, $x_6 = 3.6842$, $\alpha = 3.68$ (2 d.p.)

b) $\alpha^3 = 50$, $\alpha = \sqrt[3]{50}$

4. $\dfrac{17}{4}$, $-\dfrac{1}{4}$

5. **a)** $a = 5$, $b = -2$

b) $(x-1)$, $(x-1)$

6. **a)** $8\sin x - 6\cos x = 10\sin(x - 36.87°)$

b) $\theta = 33.4°$, $93.4°$ **c)** 30

7. **a)** $|w| = 5$, $\arg w = 0.644$ rad

b)

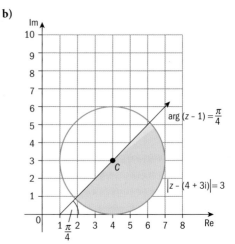

c) 8

8. **a)** $109.1°$ **b)** -1 **c)** -3 and 9

9. **a)** $\dfrac{25}{y^2(5-y)} = \dfrac{1}{y} + \dfrac{5}{y^2} + \dfrac{1}{(5-y)}$ **b)** $x = \ln\left|\dfrac{4y}{5-y}\right| - \dfrac{5}{y} + 5$

10. **a)** 1 **b)** $\left(e^{\frac{1}{3}}, \dfrac{1}{3e}\right)$ **c)** $\dfrac{1}{4} - \dfrac{3}{4e^2}$

Exam-style paper 3B page 284

1. **a)**

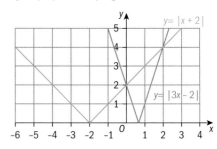

b) $0 \le x \le 2$

2. **a)** $\cos^2 3x = \dfrac{\cos 6x + 1}{2}$ **b)** $\dfrac{\sqrt{3}}{24} + \dfrac{\pi}{36}$

3. Proof **4.** $\theta = 0°$, $63.4°$, $180°$, $243.4°$, $360°$

5. **a)** $\dfrac{dy}{dx} = \dfrac{2(2-t)}{(t+1)^2}$ **b)** $\dfrac{4}{25}$

6. **a)** $2 - i$ **b)** $p = -1$, $q = -7$ **c)** -3

7. **a)** Proof **b)** $\theta = 1.97$ (2 d.p.), $AB = 20.0$ cm (1 d.p.)

8. **a)** $\dfrac{2}{1-x} - \dfrac{4x}{2+x^2}$ **b)** $2 + 2x^2 + 3x^3$

9. **a)** Proof

b) $-\ln|30 - x| = 0.1t - \ln 30$, or other form

c) $x = 19.0$ (1 d.p.)

d) x gets close to 30 as t becomes very large.

10. **a)** $\overrightarrow{OP} = 3\mathbf{i} + 10\mathbf{j}$, $\overrightarrow{PD} = -10\mathbf{j} + 4\mathbf{k}$

b) $27.2°$ **c)** Proof

Glossary

Command words

calculate   Work out from given facts, figures and information.

describe   Give the characteristics and main features.

determine   Establish with certainty.

evaluate   Judge or calculate the quality/ importance/amount/value of something.

explain   Set out purposes/reasons/mechanisms, or make the relationship between things clear, with supporting evidence.

identify   Name/select/recognise.

justify   Support a case with evidence/argument.

show that   Provide structured evidence that leads to a given result.

sketch   Make a simple freehand drawing showing the key features.

state   Express in clear terms.

verify   Confirm that a given statement/result is true.

Mathematical terms

Argand diagram   Geometric representation of complex numbers on coordinate axes, where the real part is plotted as the x-coordinate and the imaginary part is plotted as the y-coordinate.

argument   Angle θ between the real axis ($\theta = 0$) and the line joining the origin to the point (r, θ) representing a complex number.

binomial expansion   The result of multiplying out a two-term expression raised to a power.

boundary condition   An extra piece of information which can be used to determine the solution of a differential equation, e.g. when $x = 3$, $y = 5.6$.

Cartesian form   Way of expressing the position of a point or equation of a curve (or straight line) using x and y coordinates.

complex conjugates   A pair of complex numbers of the form $a + bi$ and $a - bi$.

complex number   Number of the form $a + bi$, where a and b are real numbers and $i = \sqrt{-1}$.

converge   Approach a definite value.

cosecant (cosec)   Trigonometric function that is the reciprocal of sine (sin).

cotangent (cot)   Trigonometric function that is the reciprocal of tangent (tan).

decimal search   A process using the sign-change rule to identify the location of the root of an equation as lying between neighbouring integers, then between neighbouring values correct to 1, 2, … decimal places until the required accuracy is achieved.

definite integral   The difference between the values of the integral at the upper and lower limits of the variable.

differential equation   An equation in which at least one derivative appears, e.g. $\frac{dy}{dx}$.

discriminant   The value of $b^2 - 4ac$ in the quadratic formula.

displacement vector   Vector that gives the movement (translation) from one position to another position.

diverge   Does not converge, and may approach $\pm\infty$.

dividend   The amount you want to divide up.

divisor   The number by which another number, a dividend, is divided.

explicit function   A function in which the dependent variable is expressed explicitly in terms of the independent variable.

exponential decay   Occurs for a geometric progression where the common ratio, r, between successive terms satisfies $0 < r < 1$, and for an exponential function b^x where $0 < b < 1$.

exponential form   Representation of a complex number in the form $re^{i\theta}$, where r is its modulus and θ is its argument.

exponential function   A function in the form $f(x) = b^x$ for all real numbers x, where $b > 0, b \neq 1$. The special case $f(x) = e^x$ has the property that the rate of change of the function (its derivative) is equal to the function itself.

exponential growth   Occurs for a geometric progression where the common ratio, r, between successive terms satisfies $r > 1$, and for an exponential function b^x where $b > 1$.

factors   Numbers that can be multiplied together to get another number.

factor theorem   A special case of the remainder theorem. For a polynomial $g(x)$, if $g(c) = 0$ then $(x - c)$ is a factor of $g(x)$.

first-order differential equation   A differential equation that contains only a first derivative.

geometric progression   A sequence of numbers that has a common ratio between successive terms.

imaginary number   Number of the form bi, where b is a real number and $i = \sqrt{-1}$.

implicit differentiation   A method of finding $\frac{dy}{dx}$ when the dependent variable y is not expressed explicitly in terms of the independent variable x.

implicit function   A function in which the dependent variable is not expressed explicitly in terms of the independent variable, e.g. $x^2 + 3xy^2 - 3y = 0$.

improper algebraic fraction   A rational algebraic fraction where the degree of the numerator is greater than or equal to the degree of the denominator.

improper integral   An integral where either at least one of the limits is infinite, or the function to be integrated is not defined at a point in the interval of integration.

indefinite integral   An integral with no upper or lower values of the variable.

initial condition   An extra piece of information relating to a start time (or position) which can be used to determine the solution of a differential equation, e.g. when $t = 0$, $P = 250$.

integration by parts   A technique that allows some products of functions to be integrated.

inverse function   The function that maps the range back onto the domain. If $y = f(x)$ then f^{-1} is the inverse function of f if $x = f^{-1}(y)$. There are conditions on f which need to be satisfied (such as being a one-to-one function) for the inverse to be well-defined.

iteration   One step in the process of using an iterative relation to find an approximate value to the solution of an equation.

iterative relation   A relation in the form $x_{n+1} = F(x_n)$, together with an initial value x_1. The output of each iteration is used as the input of the next.

logarithmic function   The inverse of the exponential function.

magnitude   Size or length (e.g. of a vector).

modulus (of a complex number)   The length of OP where P is the point representing the complex number on an Argand diagram.

modulus (of a real number)   The magnitude of the number.

modulus function   The modulus function $f(x) = x$ is defined as $f(x) = x$ for $x \geq 0$, and $f(x) = -x$ for $x < 0$.

modulus–argument form   Polar coordinate form for a complex number.

monotonic   Moving in one direction only – either increasing or decreasing. $f(x)$ is monotonically increasing means that $f(x_2) \geq f(x_1)$ if $x_2 > x_1$.

natural logarithm   Logarithm in base e.

parameter   If variables x and y are related via a third variable, e.g. t, then t is called a parameter.

parametric differentiation   A method of finding $\dfrac{dy}{dx}$ when x and y are related via a third variable.

parametric equation   Any equation expressed in terms of parameters.

partial fraction   Each of two or more fractions into which a more complex fraction can be decomposed as a sum.

polar coordinate form   Coordinates of the form (r, θ), where r is the length of the line joining the origin to the point and θ $(-\pi < \theta \le \pi)$ is the angle between this line and the line $\theta = 0$ (positive x-axis).

position vector   Vector that gives the movement (translation) from the origin to the position of a point.

product rule   If y is a product of two algebraic expressions (u and v) we use the product rule to differentiate: if $y = uv$, then $\dfrac{dy}{dx} = u\dfrac{dv}{dx} + v\dfrac{du}{dx}$.

quotient   The result of dividing one number by another.

quotient rule   If y is a division of two algebraic expressions (u and v) we use the quotient rule to differentiate: if $y = \dfrac{u}{v}$, then $\dfrac{dy}{dx} = \dfrac{v\dfrac{du}{dx} - u\dfrac{dv}{dx}}{v^2}$.

remainder theorem   If $g(x)$ is divided by $(x - c)$, the remainder is given by $g(c)$.

scalar   Quantity that has magnitude (size) only.

scalar product   For two vectors of magnitude a and b that have an angle θ between their directions, the value of $ab \cos \theta$ is the scalar product.

secant (sec)   Trigonometric function that is the reciprocal of cosine (cos).

separable variables   A differential equation which can be rearranged to the form $g(y)\dfrac{dy}{dx} = f(x)$ is said to have separable variables.

sign-change rule   If $f(x)$ is continuous in an interval $\alpha \le x \le \beta$ and $f(\alpha)$ and $f(\beta)$ have different signs, then $f(x) = 0$ has at least one root between α and β.

skew   Geometric description of two lines in three dimensions that do not meet and are not parallel.

trapezium rule   Gives an approximation to a definite integral by using a number of trapezia to estimate the area under the curve.

trigonometric function   A function of an angle.

unit vector   Vector that has a magnitude of one unit.

vector   Quantity that has magnitude (size) and direction.

vector equation   Way to express the equation of a straight line using vectors.

Index

A

accuracy
  and approximate roots 120–1,
    125
  and numerical integration 107–11
addition
  of complex numbers 248, 249, 257
  of vectors 183, 190–1
addition formulae 50–3
air resistance 231, 276
algebra 2
  binomial expansions 144–6, 147–8,
    149–50
  factor theorem 12–14
  modulus function 3–5, 6–7
  partial fractions 137–8, 139,
    140–1
  as improper fractions 142–3
  polynomial division 8–9, 10–11
  real-world applications 136
  remainder theorem 10–11
angles
  between lines 205
  between vectors 201–4, 205
approximate roots 118–22
  iterative relationships 123–7,
    129–30
area on graphs 93, 100–1, 162
  trapezium rule for 107–11
Argand diagrams 255–7, 259
  loci in 264–8
argument 259–60, 261, 262
asymptotes
  on reciprocal functions 95, 96, 97
  on secant graphs 42

B

base 21, 22, 25, 26
binomial expansions 144–6, 147–8
  and partial fractions 149–50
boundary conditions 224

C

Cartesian coordinates 259
Cartesian form 193, 259–60, 262
chain rule 68, 82, 84, 173
circles as loci 264–5, 266, 267
coefficients, equating 136
  and partial fractions 137–9, 140–1,
    142–3
coincident lines 197, 198, 199
column vectors 183–4, 185, 190–1
complementary angle identities 49
complex conjugates 245, 252
  on Argand diagrams 256
  as roots of polynomial equations
    254
complex numbers 241, 243–4, 245–6
  Argand diagrams 255–7, 259,
    264–8
  calculating with 248–9, 261–3
  on Argand diagrams 257
  equations with 251–3, 254
  as loci 264, 265, 266, 267
  forms of 259–61
  in real-life 241, 276–7
  square roots of 252–3
composite functions 45, 68
compound angle formula 52
constant of integration 92, 95
  and differential equations 217, 221,
    223
  particular solutions 224–6, 230
constant of proportionality 230
convergence
  of binomial expansions 144
  of iterative functions 123–6,
    129–30
cosecant 41–2, 44–5
  geometric interpretation of 49
  graphs of 43
  in trigonometric identities 47, 49
cosine 40–1, 58–60
  addition formulae 50–3
  differentiating 79–80

double angle formula 55–6
  geometric interpretation of 49
  graphs of 42
  in identities 47, 49
  and integration 102, 103–5
  integration 99–101
  in differential equations 222, 226
  and identities 102, 103–5
  by parts 168
  by substitution 177–8
cotangent 41–2, 43–5
  geometric interpretation of 49
  graphs of 43
  in trigonometric identities 47, 49
  and integration 102, 105
critical values 5, 29
curves
  gradient of tangents 83, 86
  parametric equations of 84

D

dam construction 91
damped oscillations 277
decay, exponential 19–20, 33, 221
decimal search 120–1
definite integrals 92–3
  and integration by parts 170
  reciprocal functions 95–6, 97–8
  and substitution 173, 174–5
  trigonometric functions 100–1,
    103–5
differential equations 215, 216–20
  modelling with 229–35, 277
  particular solutions to 224–7
  with separable variables 221–3
differentiation 68, 91
  of exponential functions 69–70
  implicit differentiation 81–4, 160,
    161
  and integration 92
  of logarithmic functions 71–2
  parametric 84–6
  product rule 73–5, 167